T0181138

Euclidean Distance Matrices and Their Applications in Rigidity Theory

Abdo Y. Alfakih

Euclidean Distance Matrices and Their Applications in Rigidity Theory

 Springer

Abdo Y. Alfakih
Department of Mathematics and Statistics
University of Windsor
Windsor, ON, Canada

ISBN 978-3-030-07417-3   ISBN 978-3-319-97846-8   (eBook)
https://doi.org/10.1007/978-3-319-97846-8

This Springer imprint is published by the registered company Springer Nature Switzerland AG
The registered company address is: Gewerbestrasse 11, 6330 Cham, Switzerland

In loving memory of my mother.

Preface

This monograph is devoted to a unified, up-to-date, and accessible exposition of Euclidean distance matrices (EDMs) and rigidity theory of bar-and-joint frameworks. EDMs, which comprise the first part of the monograph, are those matrices whose entries can be realized as the squared Euclidean interpoint distances of a point configuration. Such matrices arise in many areas of science and engineering including statistics, computational biochemistry, and computer science, to name a few. The second part of the monograph focuses on rigidity theory of bar-and-joint frameworks. Given a subset E of the interpoint distances of a point configuration, rigidity theory is concerned with the existence of a second point configuration having the same interpoint distances as those of E. Various rigidity notions correspond to various conditions on the second point configuration. Rigidity theory has a long and rich history going at least as far back as Cauchy (1813) and is of great interest to structural engineers and mathematicians. EDMs and rigidity theory fall in the general area of distance geometry. Distance geometry has important applications in statistics (multidimensional scaling [48]), computational biochemistry (molecular conformations [66]), and computer science (sensor networks).

The last four decades have seen a growing body of literature on EDMs and rigidity theory. Much of this literature, unfortunately, is available mainly in scattered form in journals of various disciplines. This, coupled with the lack of a unified notation, makes it difficult to get a firm grasp of the published literature and acts as a barrier to new researchers entering the field. This monograph is an attempt to rectify this situation by presenting a unified account of EDMs and rigidity theory based on the one-to-one correspondence between EDMs and projected Gram matrices. Accordingly, the machinery of semidefinite programming is a common thread that runs throughout the monograph. As a result, two parallel approaches to rigidity theory are presented. The first one is traditional and more intuitive and is based on a vector representation of a point configuration. The second one is novel and less intuitive and is based on a Gram matrix representation of a point configuration. Each of these two approaches, obviously, has its advantages and disadvantages.

The monograph is self-contained and should be accessible to a wide audience including students and researchers in statistics, computational biochemistry, engineering, computer science, operations research, and mathematics. The notation used here is standard in the semidefinite programming literature. Chapters 1 and 2 provide the necessary background for the rest of the chapters. The focus of Chap. 1 is on pertinent results from matrix theory, graph theory, and convexity theory, while Chap. 2 is devoted entirely to positive semidefinite (PSD) matrices due to the key role these matrices play in our approach. Chapters 3–7 are devoted to a detailed study of EDMs, and in particular their various characterizations, classes, eigenvalues, entries, and geometry. Chapters 9 and 10 are devoted to local and universal rigidities of bar-and-joint frameworks. The literature on rigidity theory is vast. We chose to include only those two notions of rigidity because they lend themselves easily to semidefinite programming machinery used throughout the monograph. Moreover, due to space limitation, we discuss only the most significant results and results directly relevant to EDMs. Finally, Chap. 8 is a transitional chapter that links rigidity theory to EDMs by viewing various rigidity problems as EDM completion uniqueness problems.

Finally, I would like to express my sincere thanks and gratitude to Katta G. Murty (thesis advisor) and Henry Wolkowicz (postdoctoral advisor). Murty introduced me to Henry and Henry introduced me to EDMs.

Windsor, ON, Canada   Abdo Y. Alfakih

Contents

List of Notation

$\mathbf{0}$	The zero vector or matrix of appropriate dimensions
$\|S\|$	The cardinality of set S
$\langle .,. \rangle$	Inner product
T^*	The adjoint of linear transformation T
$T^{-1}(S)$	The preimage of set S under linear transformation T
$\mathscr{L}^\perp$	The orthogonal complement of subspace $\mathscr{L}$
$\|\|x\|\|$	The norm of x
x^i	The ith vector in a set of vectors $x^1, \ldots, x^k$
x_i	The ith component of vector x
e_n^i	The standard ith unit vector in $\mathbb{R}^n$
e_n	The vector of all 1's in $\mathbb{R}^n$
V	The $n \times (n-1)$ matrix whose columns form an orthonormal basis of $e_n^\perp$
I_n	The identity matrix of order n
E_n	$= e_n e_n^T$
$E_{n,m}$	$= e_n e_m^T$
J	$= I_n - E_n/n$
$E^{ij}, (i \neq j)$	$= e_n^i (e_n^j)^T + e_n^j (e_n^i)^T$
$M^{ij}, (i \neq j)$	$= -V^T E^{ij} V / 2$
$F^{ij}, (i \neq j)$	$= (e^i - e^j)(e^i - e^j)^T$
$\mathscr{S}^n$	The space of $n \times n$ real symmetric matrices
$\mathscr{S}_+^n$	The set of $n \times n$ real symmetric positive semidefinite matrices
$\mathscr{S}_{++}^n$	The set of $n \times n$ real symmetric positive definite matrices
$\mathscr{D}^n$	The set of $n \times n$ Euclidean distance matrices
$A_{.j}$	The jth column of matrix A
$A_{i.}$	The ith row of matrix A
$\mathrm{Diag}(x)$	The diagonal matrix formed by vector x
$\mathrm{diag}(A)$	The vector consisting of the diagonal entries of matrix A
$\chi_A(\lambda)$	The characteristic polynomial of matrix A
$m_A(\lambda)$	The minimal polynomial of matrix A
$\|\|A\|\|_F$	The Frobenius norm of matrix A

$\rho(A)$	The spectral radius of a square matrix A
$\text{null}(A)$	The null space of matrix A
$\text{col}(A)$	The column space of matrix A
$\text{gal}(D)$	The Gale space of EDM D
$\bar{r} = n - 1 - r$	The dimension of $\text{gal}(D)$
$A^{\dagger}$	The Moore–Penrose inverse of matrix A
$A \circ B$	The Hadamard product of matrices A and B
$A \otimes B$	The Kronecker product of matrices A and B
$A \otimes_s A$	The symmetric Kronecker product of matrix A and itself
$\mathscr{L}_1 \oplus \mathscr{L}_2$	The direct sum of subspaces $\mathscr{L}_1$ and $\mathscr{L}_2$
$V(G)$	The vertex set of graph G
$E(G)$	The edge set of simple graph G of cardinality m
$\bar{E}(\bar{G})$	The edge set of the complement graph $\bar{G}$, or the set of missing edges of G, of cardinality $\bar{m}$
$\mathscr{E}(y)$	$= \sum_{\{i,j\} \in \bar{E}(\bar{G})} y_{ij} E^{ij}$
$\mathscr{M}(y)$	$= \sum_{\{i,j\} \in \bar{E}(\bar{G})} y_{ij} M^{ij}$
$\deg(i)$	The degree of node i of a graph
$\deg$	The vector consisting of the degrees of all nodes of a graph
K_n	The complete graph on n nodes
$\text{conv}(S)$	The convex hull of set S
$\text{aff}(S)$	The affine hull of set S
$\text{int}(S)$	The interior of set S
$\text{relint}(S)$	The relative interior of set S
$\text{cl}(S)$	The closure of set S
∂S	The boundary of set S
$\text{rbd}(S)$	The relative boundary of set S
K°	The polar of cone K
$\backslash$	The set theoretic difference
$N_S(\hat{x})$	The normal cone of set S at point $\hat{x}$
$T_S(\hat{x})$	The tangent cone of set S at point $\hat{x}$
$\text{face}(x, S)$	The minimal face of set S containing x
$A \succ \mathbf{0}$	Real matrix A is symmetric positive definite
$A \succeq \mathbf{0}$	Real matrix A is symmetric positive semidefinite

Chapter 1
Mathematical Preliminaries

In this chapter, we briefly review some of the mathematical preliminaries that will be needed throughout the monograph. These include a brief review of the most pertinent concepts and results in the theories of vector spaces, matrices, convexity, and graphs. Proofs of several of these results are included to make this chapter as self-contained as possible.

1.1 Vector Spaces

The notion of a vector space plays an important role in Euclidean geometry. In this monograph we are interested only in finite-dimensional real vector spaces.

Let $\mathcal{V}$ be a nonempty set equipped with the operations of addition and scalar multiplication. Then $\mathcal{V}$ is a real vector space (or a real linear space) if the following conditions are satisfied:

1. $x + y = y + x$ for all x, y in $\mathcal{V}$.
2. $x + (y + z) = (x + y) + z$ for all x, y, z in $\mathcal{V}$.
3. There exists a unique $\mathbf{0} \in \mathcal{V}$ such that $x + \mathbf{0} = x$ for all $x \in \mathcal{V}$.
4. For each $x \in \mathcal{V}$, there exists a unique $(-x) \in \mathcal{V}$ such that $x + (-x) = \mathbf{0}$.
5. $(\alpha + \beta)x = \alpha x + \beta x$ for all α, β in $\mathbb{R}$ and all $x \in \mathcal{V}$.
6. $\alpha(x + y) = \alpha x + \alpha y$ for all α in $\mathbb{R}$ and all x, y in $\mathcal{V}$.
7. $(\alpha\beta)x = \alpha(\beta x)$ for all α, β in $\mathbb{R}$ and all x in $\mathcal{V}$.
8. $1x = x$ for all $x \in \mathcal{V}$.

The vector spaces of interest to us are the ones where $\mathcal{V} = \mathbb{R}^n$ and $\mathcal{V} = \mathscr{S}^n$, the set of $n \times n$ real symmetric matrices. The elements of real vector space $\mathcal{V}$ are called *vectors*. If the origin $\mathbf{0}$ is of no particular interest to us, then the elements of $\mathcal{V}$ are called *points*. Let $\mathcal{V}$ be a real vector space and let $\mathcal{V}' \subset \mathcal{V}$. If $\mathcal{V}'$ is a real vector space in its own right, then $\mathcal{V}'$ is called a *linear subspace*, or a subspace for short, of $\mathcal{V}$. It is easy to see that a nonempty subset of $\mathcal{V}$ is a subspace of $\mathcal{V}$ iff it is closed under linear combinations.

© Springer Nature Switzerland AG 2018
A.Y. Alfakih, *Euclidean Distance Matrices and Their Applications in Rigidity Theory*, https://doi.org/10.1007/978-3-319-97846-8_1

An *inner product* on a real vector space $\mathscr{V}$, denoted by $\langle .,. \rangle$, is a real-valued function on $\mathscr{V} \times \mathscr{V}$ that satisfies the following properties:

1. $\langle x, x \rangle \geq 0$ for all $x \in \mathscr{V}$ and $\langle x, x \rangle = 0$ iff $x = \mathbf{0}$.
2. $\langle \alpha x + \beta y, z \rangle = \alpha \langle x, z \rangle + \beta \langle y, z \rangle$ for all x, y, z in $\mathscr{V}$ and all α, β in $\mathbb{R}$.
3. $\langle x, y \rangle = \langle y, x \rangle$ for all x, y in $\mathscr{V}$.

A vector space on which an inner product is defined is called an *inner product space*.

A *norm* on a real vector space $\mathscr{V}$, denoted by $||x||$, is a function $\mathscr{V} \to \mathbb{R}$ that satisfies the following properties:

1. $||x|| \geq 0$ for all $x \in \mathscr{V}$, and $||x|| = 0$ iff $x = \mathbf{0}$.
2. $||\alpha x|| = |\alpha| \, ||x||$ for all x in $\mathscr{V}$ and all α in $\mathbb{R}$.
3. $||x + y|| \leq ||x|| + ||y||$ for all x, y in $\mathscr{V}$.

A vector space equipped with a norm is called a *normed vector space*. Every inner product naturally induces a norm of the form $||x|| = \langle x, x \rangle^{1/2}$. Our interest in this monograph is in the norm in $\mathbb{R}^n$ induced by $\langle x, y \rangle = x^T y$ and the norm in $\mathscr{S}^n$ induced by $\langle X, Y \rangle = \mathrm{trace}(XY)$.

Theorem 1.1 (Cauchy–Schwarz inequality) *Let x and y be two vectors in a real vector space $\mathscr{V}$ equipped with inner product $\langle .,. \rangle$. Then*

$$|\langle x, y \rangle| \leq \langle x, x \rangle^{1/2} \langle y, y \rangle^{1/2},$$

where equality holds if and only if $x - \alpha y = \mathbf{0}$ for some scalar α.

Proof. Let $||x||^2 = \langle x, x \rangle$. Then, it follows from the definition of inner product that $\langle x - ty, x - ty \rangle = t^2 ||y||^2 - 2t \langle x, y \rangle + ||x||^2 \geq 0$ for all $t \in \mathbb{R}$. Now if $y = \mathbf{0}$, then the result follows trivially. Thus, assume that $||y||^2 \neq 0$ and let

$$\hat{t} = \frac{\langle x, y \rangle}{||y||^2}.$$

Then

$$\langle x - \hat{t}y, x - \hat{t}y \rangle = ||x||^2 - \frac{\langle x, y \rangle^2}{||y||^2} \geq 0.$$

Consequently, $\langle x, y \rangle^2 \leq ||y||^2 ||x||^2$, with equality iff $x - \hat{t}y = \mathbf{0}$.

$\square$

Cauchy–Schwarz inequality is used to establish the continuity of the inner product.

Lemma 1.1 *Let $\{x^k\}_{k \in \mathbb{N}}$, $\{y^k\}_{k \in \mathbb{N}}$ be two sequences in $\mathscr{V}$ that converge to x and y, respectively. Then*

$$\lim_{k \to \infty} \langle x^k, y^k \rangle = \langle x, y \rangle.$$

That is, the inner product $\langle x, y \rangle$ is a continuous function.

Proof.

$$\begin{aligned}
|\langle x^k, y^k \rangle - \langle x, y \rangle| &= |\langle x^k, y^k \rangle - \langle x^k, y \rangle + \langle x^k, y \rangle - \langle x, y \rangle| \\
&= |\langle x^k, (y^k - y) \rangle + \langle (x^k - x), y \rangle| \\
&\leq |\langle x^k, (y^k - y) \rangle| + |\langle (x^k - x), y \rangle| \\
&\leq \|x^k\| \, \|y^k - y\| + \|x^k - x\| \, \|y\|,
\end{aligned}$$

where the last inequality follows from Cauchy–Schwarz inequality. Now $\|x^k\|$ is bounded by the convergence of $\{x^k\}$. Thus, $\langle x^k, y^k \rangle \to \langle x, y \rangle$ as $x^k \to x$ and $y^k \to y$. $\qquad\square$

Let $\mathcal{V}_1$ and $\mathcal{V}_2$ be two vector spaces and let $T : \mathcal{V}_1 \to \mathcal{V}_2$ be a linear transformation. The *adjoint* of T, denoted by T^*, is the unique transformation $T^* : \mathcal{V}_2 \to \mathcal{V}_1$ that satisfies

$$\langle y, T(x) \rangle = \langle T^*(y), x \rangle \text{ for all } x \in \mathcal{V}_1 \text{ and for all } y \in \mathcal{V}_2.$$

For example, let $\mathrm{Diag}(x)$ denote the diagonal matrix formed by vector x and let $\mathrm{diag}(A)$ denote the vector consisting of the diagonal entries of matrix A. Let $\mathbb{R}^n$ be endowed with inner product $\langle x, y \rangle = x^T y$ and let $\mathscr{S}^n$ be endowed with the trace inner product $\langle A, B \rangle = \mathrm{trace}(AB)$. Further, let $T : \mathbb{R}^n \to \mathscr{S}^n$, where $T(x) = \mathrm{Diag}(x)$. Then for any $A \in \mathscr{S}^n$, we have

$$\mathrm{trace}(A\mathrm{Diag}(x)) = \sum_{i=1}^{n} a_{ii} x_i = x^T \mathrm{diag}(A).$$

Therefore, the adjoint of T is $T^* : \mathscr{S}^n \to \mathbb{R}^n$, where $T^*(A) = \mathrm{diag}(A)$.

1.2 Matrix Theory

In this monograph we deal only with real matrices. Let A be an $n \times n$ matrix. The matrix obtained from A by deleting $n - k$ rows and $n - k'$ columns, where $1 \leq k, k' \leq n$, is a $k \times k'$ *submatrix* of A. A *principal submatrix* of A is the square submatrix obtained from A by deleting similarly indexed rows and columns; i.e., if the ith row of A is deleted, then so is the ith column. The determinant of a principal submatrix of A is called a *principal minor* of A. The kth *leading principal submatrix* of A is the square submatrix obtained by deleting the last $n - k$ columns and rows of A. Note that the nth leading principal submatrix of A is A itself. The determinant of the kth leading principal submatrix of A is called the kth *leading principal minor* of A. It easily follows that an $n \times n$ matrix has $2^n - 1$ principal minors and n leading principal minors.

1.2.1 The Characteristic and the Minimal Polynomials

Let A be an $n \times n$ matrix and let x be a nonzero vector in $\mathbb{R}^n$. Then x is said to be an *eigenvector* of A if

$$Ax = \lambda x,$$

for some scalar λ, in which case, λ is said to be the *eigenvalue* of A corresponding to x. The pair (λ, x) is called an *eigenpair* of A. An immediate consequence of this definition is that the eigenvalues of A are the roots of the polynomial

$$\chi_A(\lambda) = \det(A - \lambda I). \tag{1.1}$$

$\chi_A(\lambda)$ is called the *characteristic polynomial* of A. An important fact to bear in mind is that A and A^T have the same characteristic polynomial and hence the same eigenvalues. Since $\chi_A(\lambda)$ is of degree n, it follows that A has n eigenvalues some of which may be complex even if A is real. On the other hand, A may or may not have n linearly independent eigenvectors. A is said to be *diagonalizable* if there exists a nonsingular matrix S such that

$$A = S\Lambda S^{-1},$$

where Λ is the diagonal matrix consisting of the eigenvalues of A, in which case, the columns of S are the eigenvectors of A, and the rows of S^{-1} are the eigenvectors of A^T. It is easy to see that A is diagonalizable if and only if it has n linearly independent eigenvectors. As will be shown next, real symmetric matrices are always diagonalizable, and more importantly, they are diagonalizable by orthogonal matrices.

Theorem 1.2 *Let A be an $n \times n$ real symmetric matrix. Then A has n real eigenvalues and n orthonormal eigenvectors.*

Proof. Let $Ax = \lambda x$. Then $A\bar{x} = \bar{\lambda}\bar{x}$, where $\bar{x}$ is the complex conjugate of x. Therefore, $\bar{x}^T Ax - x^T A\bar{x} = (\lambda - \bar{\lambda})\bar{x}^T x = 0$. Thus, $\bar{\lambda} = \lambda$ since $\bar{x}^T x \neq 0$. Hence, λ is real and thus x can be chosen real.

Now let $Ax^1 = \lambda_1 x^1$ and $Ax^2 = \lambda_2 x^2$, where $\lambda_1 \neq \lambda_2$. Then $x^{2^T} Ax^1 - x^{1^T} Ax^2 = (\lambda_1 - \lambda_2) x^{2^T} x^1 = 0$. Thus $x^{2^T} x^1 = 0$. Hence, the eigenvectors of A corresponding to distinct eigenvalues are orthogonal. Now if an eigenvalue λ of A is repeated, then the eigenvectors corresponding to λ can be chosen to be orthogonal.

□

Theorem 1.3 (The Spectral Theorem) *Let A be a real $n \times n$ matrix. Then A is symmetric if and only if*

$$A = Q\Lambda Q^T, \tag{1.2}$$

where Λ is the diagonal matrix consisting of the eigenvalues of A and Q is an orthogonal matrix whose columns are the corresponding eigenvectors.

Equation (1.2) is called the *spectral decomposition* of A. Let $(\lambda_1, q^1), \ldots, (\lambda_n, q^n)$ be the eigenpairs of A. Then Eq. (1.2) can be written as $A = \sum_{i=1}^n \lambda_i q^i (q^i)^T$.

Rayleigh–Ritz Theorem gives a variational characterization of the largest and the smallest eigenvalues of a real symmetric matrix.

Theorem 1.4 (Rayleigh–Ritz) *Let A be an $n \times n$ real symmetric matrix and let $\lambda_1 \geq \cdots \geq \lambda_n$ be the eigenvalues of A. Further, let x^1 and x^n be eigenvectors of A corresponding to λ_1 and λ_n, respectively. Then*

$$\lambda_1 = \max_{x \neq 0} \frac{x^T A x}{x^T x} \text{ and } \lambda_n = \min_{x \neq 0} \frac{x^T A x}{x^T x}.$$

Moreover, the maximum is attained at x^1 and the minimum is attained at x^n.

Proof. We only present a proof of the maximum case. The proof of the minimum case is similar. Let $A = Q \Lambda Q^T$ be the spectral decomposition of A. Then for all $x \neq 0$ we have

$$\frac{x^T A x}{x^T x} = \frac{y^T \Lambda y}{y^T y} = \frac{\sum_{i=1}^n \lambda_i y_i^2}{\sum_{i=1}^n y_i^2} \leq \lambda_1$$

since $\sum_{i=1}^n \lambda_i y_i^2 \leq \sum_{i=1}^n \lambda_1 y_i^2$. The result follows since $(x^1)^T A x^1 = \lambda_1 (x^1)^T x^1$. $\qquad \square$

We will find the following corollary of Rayleigh–Ritz Theorem useful in later chapters.

Corollary 1.1 *Let A be an $n \times n$ real symmetric matrix. Let $\lambda_1, \ldots, \lambda_n$ be the eigenvalues of A with corresponding orthonormal eigenvectors $q^1, \ldots, q^n$. Assume that $\lambda_1 > \lambda_i$ for all $i = 2, \ldots, n$ and let x be a unit vector such that $x^T A x = \lambda_1$. Then $x = \pm q^1$.*

Proof. Let $x = \sum_{i=1}^n \alpha_i q^i$. Then $\sum_{i=1}^n \alpha_i^2 = 1$ and $\lambda_1 = \sum_{i=1}^n \alpha_i^2 \lambda_i$. Hence,

$$\sum_{i=2}^n \alpha_i^2 \lambda_i = \lambda_1 (1 - \alpha_1^2) = \lambda_1 \sum_{i=2}^n \alpha_i^2.$$

Therefore, $\sum_{i=2}^n \alpha_i^2 (\lambda_1 - \lambda_i) = 0$. But $(\lambda_1 - \lambda_i) > 0$ for all $i = 2, \ldots, n$. Hence, $\alpha_2 = \cdots = \alpha_n = 0$ and thus $x = \pm q^1$. $\qquad \square$

Theorem 1.5 *Let A be a real symmetric $n \times n$ matrix and let $\mathscr{L}$ be a k-dimensional subspace of $\mathbb{R}^n$ such that $x^T A x \leq 0$ for all $x \in \mathscr{L}$. Then A has at least k nonpositive eigenvalues.*

Proof. Let $q^1, \ldots, q^n$ be orthonormal eigenvectors of A with corresponding eigenvalues $\lambda_1 \geq \cdots \geq \lambda_n$. Let $S = \text{span} \{q^1, \ldots, q^{n-k+1}\}$. Then $\dim(S) = n - k + 1$. Thus $\mathscr{L} \cap S \neq \emptyset$ and hence let x be a unit vector in $\mathscr{L} \cap S$. Hence, $\lambda_{n-k+1} \leq x^T A x \leq 0$. Therefore, $\lambda_n \leq \cdots \leq \lambda_{n-k+1} \leq 0$. $\qquad \square$

The inertia of a real symmetric matrix A is the ordered triple (n_+, n_-, n_0), where n_+, n_-, and n_0 are, respectively, the numbers of positive, negative, and zero eigenvalues of A. Thus, $\text{rank}(A) = n_+ + n_-$.

Theorem 1.6 (Sylvester Law of Inertia) *Let A be an $n \times n$ real symmetric matrix and let S be a nonsingular $n \times n$ matrix. Then A and SAS^T have the same inertia.*

Theorem 1.7 (Cauchy Interlacing Theorem) *Let $\mu_1 \geq \cdots \geq \mu_n$ be the eigenvalues of an $n \times n$ real symmetric matrix A. Let B be any $(n-1) \times (n-1)$ principal submatrix of A and let $\lambda_1 \geq \cdots \geq \lambda_{n-1}$ be the eigenvalues of B. Then the eigenvalues of A are interlaced by those of B; i.e.,*

$$\mu_k \geq \lambda_k \geq \mu_{k+1} \text{ for } k = 1, \ldots, n-1.$$

The coefficients of the characteristic polynomial can be expressed in terms of the principal minors.

Theorem 1.8 *Let A be an $n \times n$ matrix and let c_k be the coefficient of λ^k in $\chi_A(\lambda)$, the characteristic polynomial of A. Then for $k \leq n-1$, we have*

$$c_k = (-1)^k \times \text{the sum of all principal minors of A of order } n-k.$$

Proof. Let e^i be the ith standard unit vector in $\mathbb{R}^n$ and let $A_{.j}$ be the jth column of A. Then

$$\chi_A(\lambda) = \det(A - \lambda I) = \det([(A_{.1} - \lambda e^1) \ (A_{.2} - \lambda e^2) \ \cdots \ (A_{.n} - \lambda e^n)]). \quad (1.3)$$

Since the determinant is linear in each column separately, the coefficient of λ^k in (1.3) is

$$c_k = (-1)^k \sum \det([x(1) \ x(2) \ldots x(k) \ x(k+1) \ldots x(n)]), \quad (1.4)$$

where the sum is taken over all possible ways to replace $x(j)$ with $A_{.j}$ or e^j such that the total number of e^j's is k. For instance, $\det([e^1 \ e^2 \ldots e^k \ A_{.k+1} \ldots A_{.n}])$ is one of the terms in the sum in (1.4). But this term is precisely the principal minor of A of order $n - k$, obtained by deleting the first k rows and columns. Hence, the sum in (1.4) is the sum of all principal minors of A of order $n - k$.

$\square$

The coefficient c_n is called the *leading coefficient* of $\chi_A(\lambda)$ and it is equal to $(-1)^n$. Also, Theorem 1.8 immediately implies that $c_0 = \det(A)$, $c_{n-1} = (-1)^{n-1}$ trace(A). Note that the determinant and the trace are equal, respectively, to the product and the sum of the eigenvalues counting multiplicities.

Theorem 1.9 (Cayley–Hamilton) *Every square matrix satisfies its characteristic polynomial.*

The *geometric multiplicity* of an eigenvalue λ is equal to the maximum number of linearly independent eigenvectors corresponding to λ. On the other hand, the *algebraic multiplicity* of λ is equal to the number of times λ is repeated as a root of the characteristic polynomial. Note that the geometric multiplicity of an eigenvalue is always less than or equal to its algebraic multiplicity. Moreover, matrix A is diagonalizable if and only if, for each eigenvalue of A, the geometric multiplicity is equal to the algebraic multiplicity.

Let $\lambda_1, \ldots, \lambda_k$ be the distinct eigenvalues of A with respective algebraic multiplicities $m_1, \ldots, m_k$. Then

$$\chi_A(\lambda) = (\lambda_1 - \lambda)^{m_1} \ldots (\lambda_k - \lambda)^{m_k}.$$

Hence, Cayley–Hamilton Theorem implies that

$$\chi_A(A) = (\lambda_1 I - A)^{m_1} \ldots (\lambda_k I - A)^{m_k} = \mathbf{0}.$$

A polynomial is called *monic* if its leading coefficient is 1. The *minimal polynomial* of A, denoted by $m_A(\lambda)$, is the smallest degree monic polynomial that annihilates A, i.e., $m_A(A) = \mathbf{0}$. Consequently,

$$m_A(\lambda) = (\lambda_1 - \lambda)^{r_1} \ldots (\lambda_k - \lambda)^{r_k},$$

where $r_i \le m_i$ for all $i = 1, \ldots, k$.

Theorem 1.10 *Let A be an $n \times n$ matrix. Then A is diagonalizable if and only if its minimal polynomial, $m_A(\lambda)$, is the product of linear terms, i.e., iff $r_1 = \cdots = r_k = 1$.*

The norm of matrix A, denoted by $||A||$, is a real-valued function that satisfies the following three properties: (i) $||A|| \ge 0$ for all A and $||A|| = 0$ iff $A = \mathbf{0}$, (ii) $||\alpha A|| = |\alpha| \, ||A||$ for all A and for all scalars α, and (iii) $||A + B|| \le ||A|| + ||B||$ for all A and B. In addition, if a matrix norm satisfies the property that $||AB|| \le ||A|| \, ||B||$ for all A and B, then this norm is said to be *consistent* or *submultiplicative*.

The *Frobenius norm* of an $m \times n$ real matrix A, denoted by $||A||_F$, is defined by $||A||_F = \sqrt{\text{trace}(A^T A)}$. It is not hard to show that the Frobenius norm is submultiplicative. Every vector norm $||x||$ *induces* a matrix norm as follows:

$$||A|| = \max_{x \ne 0} \frac{||Ax||}{||x||}.$$

Thus, $||A|| \, ||x|| \ge ||Ax||$ for any x. Accordingly, every induced matrix norm is submultiplicative since

$$||ABx|| \le ||A|| \, ||B|| \, ||x|| \text{ for any } x \in \mathbb{R}^n.$$

Furthermore, it follows from Rayleigh–Ritz Theorem that $||A||_2$, the matrix norm induced by the Euclidean vector norm, is given by $||A||_2 = \sqrt{\lambda_{\max}(A^T A)}$. Consequently, $||A||_2 \le ||A||_F$ for any matrix A.

1.2.2 The Perron Theorem

A vector x in $\mathbb{R}^n$ is said to be *positive*, denoted by $x > \mathbf{0}$, if $x_i > 0$ for all $i = 1, \ldots, n$. Similarly, an $n \times n$ real matrix A is said to be *positive (nonnegative)*, denoted by $A > \mathbf{0} \, (\ge \mathbf{0})$, if $a_{ij} > 0 \, (\ge 0)$ for all $i, j = 1, \ldots, n$. A nonnegative matrix A is said to be *primitive* if $A^k > 0$ for some positive integer k. Clearly, positive matrices form a subset of primitive matrices. The *spectral radius* of A is $\rho(A) = \max\{|\lambda_i| : \lambda_i$ is an eigenvalue of $A\}$.

Theorem 1.11 (Perron) *Let A be an $n \times n$ primitive matrix and let $\rho(A)$ be its spectral radius. Then*

1. *There exists an eigenvalue $\lambda_1 = \rho(A)$ with a corresponding eigenvector $x^1 > \mathbf{0}$.*
2. *λ_1 has algebraic multiplicity 1.*
3. *x^1 is the only positive eigenvector of A.*
4. *$\lambda_1 > |\lambda|$ for all eigenvalues $\lambda \neq \lambda_1$ of A.*

The eigenpair (λ_1, x^1) is called the Perron eigenpair.

To keep the proof simple, we assume that matrix A is symmetric. Also, we assume that A is positive. We comment later on the proof when A is primitive.

Proof.   Assume that A is symmetric and positive. Thus $\rho(A) > 0$. Let $\lambda_1 \geq \cdots \geq \lambda_n$ be the eigenvalues of A. Notice that $\lambda_1 > 0$ since trace$(A) > 0$. Therefore, $\rho(A)$ is either equal to λ_1 or $|\lambda_n|$. Let y^1 and y^n be the normalized eigenvectors corresponding to λ_1 and λ_n, respectively, and let $x^1 = |y^1|$, i.e., $x_i^1 = |y_i^1|$ for $i = 1, \ldots, n$. Then

$$|\lambda_n| = |(y^n)^T A y^n| = |\sum_j a_{ij} y_i^n y_j^n| \leq \sum_j a_{ij} |y_i^n| \, |y_j^n| \leq \lambda_1,$$

where the last inequality follows from Rayleigh–Ritz Theorem. Thus $\lambda_1 = \rho(A)$. Moreover,

$$\lambda_1 = |\lambda_1| = |(y^1)^T A y^1| \leq (x^1)^T A x^1 \leq \lambda_1.$$

Therefore, $(x^1)^T A x^1 = \lambda_1$ and thus $A x^1 = \lambda_1 x^1$. Furthermore, since $x^1 \geq \mathbf{0}$ and since $x^1 = A x^1 / \lambda_1$, it follows that $x^1 > \mathbf{0}$. This proves Statement 1.

To prove Statement 2, assume that there exists y such that $Ay = \lambda_1 y$. Let

$$\alpha = \min\{\frac{x_i^1}{y_i} : y_i > 0\} = \frac{x_{i_0}^1}{y_{i_0}} > 0.$$

Let $z = x^1 - \alpha y$. Then $z \geq \mathbf{0}$ and $z_{i_0} = 0$. Assume that $z \neq \mathbf{0}$. Then $Az = \lambda_1 z$ and hence, $z = Az/\lambda_1$ must be $> \mathbf{0}$ since $z \geq \mathbf{0}$, a contradiction. Therefore, $z = \mathbf{0}$ and y is a multiple of x^1 and hence the geometric multiplicity of λ_1 is 1. Statement 2, thus, follows since A is diagonalizable.

Statement 3 follows from the fact that eigenvectors corresponding to distinct eigenvalues of a real symmetric matrix are orthogonal. Indeed, assume, to the contrary, that there exists an eigenpair (λ, y), where $\lambda \neq \lambda_1$ and $y > \mathbf{0}$. Thus, on one hand $y^T x^1 > 0$, and on the other hand, y and x^1 are orthogonal, a contradiction.

To prove Statement 4, it suffices to prove that $\lambda_n \neq -\lambda_1$. To this end, assume to the contrary that $Ay^n = -\lambda_1 y^n$. Then $(y^n)^T x^1 = 0$. Moreover, $A^2 y = \lambda_1^2 y$ and $A^2 x^1 = \lambda_1^2 x^1$. Thus, for the symmetric positive matrix A^2, the algebraic multiplicity of the Perron eigenvalue λ_1^2 is 2, a contradiction.

□

The proof of the case where A is primitive uses the above proof applied to the symmetric positive matrix A^k. It also uses the following facts. First, if $\lambda_1, \ldots, \lambda_n$ are the eigenvalues of A, then $\lambda_1^k, \ldots, \lambda_n^k$ are the eigenvalues of A^k. Consequently, if $|\lambda_1|^k \geq \cdots \geq |\lambda_n|^k$, then $|\lambda_1| \geq \cdots \geq |\lambda_n|$. Second, since the algebraic multiplicity of λ_1^k is 1, and thus its geometric multiplicity is also 1, it follows that the Perron eigenvector x^1 of A^k is also a Perron eigenvector of A. Furthermore, $\lambda_1 > 0$ since $Ax^1 = \lambda_1 x^1$.

1.2.3 The Null Space, the Column Space, and the Rank

Let A be an $m \times n$ real matrix. The *null space* of A is $\text{null}(A) = \{x \in \mathbb{R}^n : Ax = \mathbf{0}\}$ and the *column space* of A is $\text{col}(A) = \{y \in \mathbb{R}^m : y = Az \text{ for some } z \text{ in } \mathbb{R}^n\}$. Both $\text{null}(A)$ and $\text{col}(A)$ are, respectively, subspaces of $\mathbb{R}^n$ and $\mathbb{R}^m$ of dimensions $n - \text{rank}(A)$ and $\text{rank}(A)$. The subspace $\text{null}(A^T)$ is often called the *left null space* of A.

Every k-dimensional subspace $\mathscr{L}$ of $\mathbb{R}^n$ can be represented either as the column space of an $n \times k$ matrix A, or as the null space of an $(n - k) \times n$ matrix B. The columns of A form a basis of $\mathscr{L}$, while the rows of B form a basis of $\mathscr{L}^\perp$, the orthogonal complement of $\mathscr{L}$ in $\mathbb{R}^n$.

The *Moore–Penrose inverse* of A, denoted by $A^\dagger$, is the unique matrix that satisfies: (i) $AA^\dagger A = A$, (ii) $A^\dagger AA^\dagger = A^\dagger$, (iii) $(A^\dagger A)^T = A^\dagger A$, and (iv) $(AA^\dagger)^T = AA^\dagger$. Obviously, if A is nonsingular, then $A^\dagger = A^{-1}$. The following two facts are easy to verify. First, if A has full column rank, then $A^\dagger = (A^T A)^{-1} A^T$. Second, if $A = Q\Lambda Q^T$ where Q is orthogonal, then $A^\dagger = Q\Lambda^\dagger Q^T$.

A matrix P satisfying $P^2 = P$ is called a *projection matrix*. If such P is symmetric, then it is called an *orthogonal projection matrix*. Otherwise, it is called an *oblique projection matrix*. It easily follows that $AA^\dagger$ is the orthogonal projection matrix onto $\text{col}(A)$. Notice that $AA^\dagger$ is symmetric. Thus, if the system of equations $Ax = b$ is consistent, i.e., if $AA^\dagger b = b$, then $x = A^\dagger b + (I - A^\dagger A)z$, where z is an arbitrary vector, is a solution of this system since $Ax = AA^\dagger b = b$.

The following technical result will be used repeatedly in this monograph.

Proposition 1.1 *Let A and B be two real symmetric square matrices such that $AB = 0$. Further, assume that A is singular and let U be the matrix whose columns form an orthonormal basis of $\text{null}(A)$. Then $B = U\Phi U^T$, where $\Phi = U^T BU$.*

Proof. Let $A = W\Lambda W^T$ be the spectral decomposition of A, where Λ is the diagonal matrix consisting of the nonzero eigenvalues of A. Therefore, $W^T B = \mathbf{0}$ since $W\Lambda$ has full column rank. Let $Q = [W \ U]$. Then Q is orthogonal and $Q^T BQ = \begin{bmatrix} \mathbf{0} & \mathbf{0} \\ \mathbf{0} & \Phi \end{bmatrix}$, where $\Phi = U^T BU$. Consequently, $B = U\Phi U^T$.

□

Let A and B be two $m \times n$ matrices. Then $col([A\ B]) = col(A) + col(B)$. Hence, $\dim(col([A\ B])) = \dim(col(A)) + \dim(col(B)) - \dim(col(A) \cap col(B))$. Therefore,

$$\text{rank}([A\ B]) = \text{rank}(A) + \text{rank}(B) - \dim(col(A) \cap col(B)). \tag{1.5}$$

On the other hand, $col(A + B) = \{y : y = (A + B)z \text{ for some } z \in \mathbb{R}^n\} = \{y : y = [A\ B]\begin{bmatrix} z \\ z \end{bmatrix} \text{ for some } z \in \mathbb{R}^n\} \subseteq col([A\ B])$. Therefore,

$$\text{rank}(A + B) \leq \text{rank}([A\ B]). \tag{1.6}$$

As a result, it follows from Eqs. (1.5) and 1.6 that

$$\text{rank}(A + B) \leq \text{rank}(A) + \text{rank}(B). \tag{1.7}$$

The following theorem establishes a necessary and sufficient condition for equality to hold in Eq. (1.7).

Theorem 1.12 (Marsaglia and Styan [140]) *Let A and B be two $m \times n$ matrices. Further, let $\alpha = dim(col(A) \cap col(B))$ and $\beta = dim(col(A^T) \cap col(B^T))$. Then*

$$rank(A + B) = rank(A) + rank(B)$$

if and only if $\alpha = \beta = 0$.

We present the proof for the case where A and B are symmetric.

Proof.   It follows from Eqs. (1.5) and 1.6 that

$$\text{rank}(A + B) \leq \text{rank}([A\ B]) \leq \text{rank}(A) + \text{rank}(B) - \alpha \leq \text{rank}(A) + \text{rank}(B).$$

Thus, if $\text{rank}(A + B) = \text{rank}(A) + \text{rank}(B)$, then $\alpha = 0$.

To prove the reverse direction, let $A = W_1 \Lambda_1 W_1^T$ and $B = W_2 \Lambda_2 W_2^T$ be the spectral decompositions of A and B, where Λ_1 and Λ_2 are the diagonal matrices consisting of the nonzero eigenvalues of A and B. Thus, W_1 and W_2 are orthonormal bases of $col(A)$ and $col(B)$. Let $\Lambda = \begin{bmatrix} \Lambda_1 & 0 \\ 0 & \Lambda_2 \end{bmatrix}$. Then $\text{rank}(\Lambda) = \text{rank}(A) + \text{rank}(B)$.

If $\alpha = 0$, then $W = [W_1\ W_2]$ has full column rank and hence, has a left inverse $W^\dagger = (W^T W)^{-1} W^T$; i.e., $W^\dagger W = I$. Moreover,

$$A + B = [W_1\ W_2] \begin{bmatrix} \Lambda_1 & 0 \\ 0 & \Lambda_2 \end{bmatrix} \begin{bmatrix} W_1^T \\ W_2^T \end{bmatrix}.$$

Thus,

$$\text{rank}(A + B) = \text{rank}(W \Lambda W^T) \geq \text{rank}(W^\dagger W \Lambda W^T (W^\dagger)^T) = \text{rank}(\Lambda)$$

and the proof is complete.

$\square$

As an illustration of Theorem 1.12, let A and B be two real symmetric matrices such that $AB = \mathbf{0}$. Then clearly $\text{col}(B)$ is perpendicular to $\text{col}(A)$. Thus, by Theorem 1.12, $\text{rank}(A + B) = \text{rank}(A) + \text{rank}(B)$.

A real-valued function $f(x)$ is said to be *lower semicontinuous* if the set $\{x : f(x) > a\}$ is open for every $a \in \mathbb{R}$.

Lemma 1.2 *Let $\mathscr{S}^{mn}$ denote the set of $m \times n$ real matrices and let $S \subseteq \mathbb{R}^n$. Let $A(x) : S \to \mathscr{S}^{mn}$ be continuous. Then $\text{rank}(A(x))$ is lower semicontinuous.*

Proof. Let $a \in \mathbb{R}$ and let $S' = \{x \in S : \text{rank}(A(x)) > a\}$. If $S' = \emptyset$, then S' is open. Therefore, assume that $S' \neq \emptyset$ and let $x^0 \in S'$. Assume that $\text{rank}(A(x^0)) = k$, then there exists a $k \times k$ submatrix of $A(x^0)$, say $A_{\mathscr{I} \mathscr{J}}(x^0)$, such that $\det(A_{\mathscr{I} \mathscr{J}}(x^0)) \neq 0$. Hence, by the continuity of the determinant function, there exists a neighborhood U of x^0 such that $\det(A_{\mathscr{I} \mathscr{J}}(x)) \neq 0$ for all $x \in U$. Consequently, $\text{rank}(A(x)) \geq k > a$ for all $x \in U$ and thus $U \subset S'$. As a result, S' is open and the result follows.

□

Hence, for a sufficiently small perturbation of A, $\text{rank}(A)$ either stays the same or increases. That is, for a sufficiently small neighborhood U of x^0, $\text{rank}(A(x)) \geq \text{rank}(A(x^0))$ for all $x \in U$.

We end this section with a useful property of rank-2 symmetric matrices. Vectors u and v in $\mathbb{R}^n$ are *parallel* if $u = cv$ for some nonzero scalar c. Thus, u and v are parallel if $u = v = \mathbf{0}$.

Proposition 1.2 *Let a and b be two nonzero, nonparallel vectors in $\mathbb{R}^n$, $n \geq 2$, and let $C = ab^T + ba^T$. Then C has exactly one positive eigenvalue λ_1 and one negative eigenvalue λ_n, where*

$$\lambda_1 = a^T b + ||a|| \, ||b|| \text{ and } \lambda_n = a^T b - ||a|| \, ||b||.$$

Here, $||.||$ is the Euclidean norm.

Proof. Assume that $n = 2$ and let the eigenvectors of C be of the form $xa + yb$, where x and y are scalars. Then $C(xa + yb) = \lambda(xa + yb)$ leads to the following system of equations:

$$\begin{bmatrix} a^T b & ||b||^2 \\ ||a||^2 & a^T b \end{bmatrix} \begin{bmatrix} x \\ y \end{bmatrix} = \lambda \begin{bmatrix} x \\ y \end{bmatrix}. \tag{1.8}$$

Hence, the eigenvalues of C are precisely the eigenvalues of $\begin{bmatrix} a^T b & ||b||^2 \\ ||a||^2 & a^T b \end{bmatrix}$, which are $\lambda_1 = a^T b + ||a|| \, ||b||$ and $\lambda_r = a^T b - ||a|| \, ||b||$.

Now assume that $n \geq 3$ and let $u^1, \ldots, u^{n-2}$ be an orthonormal basis of the null space of $\begin{bmatrix} a^T \\ b^T \end{bmatrix}$. Then obviously, $u^1, \ldots, u^{n-2}$ are orthonormal eigenvectors of C corresponding to eigenvalue 0. Thus, we have two remaining eigenvectors of C of the form $xa + yb$, where x and y satisfy Eq. (1.8). Therefore, the remaining two eigenvalues of C are λ_1 and λ_n as given above. The fact that $\lambda_1 > 0$ and $\lambda_n < 0$ follows from Cauchy–Schwarz inequality since a and b are nonzero and nonparallel.

□

1.2.4 Hadamard and Kronecker Products

Let A and B be two $m \times n$ matrices. The *Hadamard product* of A and B, denoted by $A \circ B$, is the $m \times n$ matrix C such that $c_{ij} = a_{ij}b_{ij}$ for all $i = 1, \ldots, m$ and $j = 1, \ldots, n$. An $n \times n$ symmetric matrix A is said to be *positive definite (positive semidefinite)* if and only if all of its eigenvalues are positive (nonnegative). Chapter 2 is devoted to a detailed study of these matrices.

Theorem 1.13 (Schur Product Theorem) *Let A and B be two $n \times n$ symmetric positive semidefinite matrices. Then $A \circ B$ is symmetric positive semidefinite.*

It should be pointed out that Schur Product Theorem follows from Theorem 1.15 below. Let A and B be two $m \times n$ and $p \times q$ matrices, respectively. The *Kronecker product* of A and B, denoted by $A \otimes B$, is the $mp \times nq$ matrix

$$A \otimes B = \begin{bmatrix} a_{11}B & a_{12}B & \cdots & a_{1n}B \\ a_{21}B & a_{22}B & \cdots & a_{2n}B \\ \vdots & \vdots & \ddots & \vdots \\ a_{m1}B & a_{m2}B & \cdots & a_{mn}B \end{bmatrix}.$$

The following basic lemma follows immediately from the definition.

Lemma 1.3 *Let A, B, C, and D be matrices of appropriate sizes. Then*

$$(A \otimes B)(C \otimes D) = (AC) \otimes (BD).$$

Let A and B be two $n \times n$ and $m \times m$ matrices and let (λ, x) and (μ, y) be two eigenpairs of A and B, respectively. Then it immediately follows from Lemma 1.3 that $(\lambda\mu, x \otimes y)$ is an eigenpair of $A \otimes B$. Moreover, every eigenvalue of $A \otimes B$ is of the form $\lambda_i\mu_j$ where λ_i and μ_j are eigenvalues of A and B. Hence, we have the following two theorems.

Theorem 1.14 *Let A and B be two $n \times n$ and $m \times m$ matrices. Then*

1. *$det(A \otimes B) = (det(A))^m (det(B))^n$,*
2. *$trace(A \otimes B) = trace(A) \, trace(B)$.*

Theorem 1.15 *Let A and B be two symmetric positive semidefinite matrices. Then $A \otimes B$ is positive semidefinite.*

Schur Product Theorem follows from Theorem 1.15 since $A \circ B$ is a principal submatrix of $A \otimes B$. See Chap. 2 for more details.

For more details concerning the topics discussed in this section, see, e.g., [112, 113, 41].

1.3 Graph Theory

In this monograph, we are interested in connected simple (no loops and no multiple edges) graphs. For a simple graph $G = (V, E)$, we denote by $V(G)$ its node set and by $E(G)$ its edge set. We assume that $V(G) = \{1, \ldots, n\}$ and that the number of edges of G is m. The complement graph of G is $\bar{G} = (V(G), \bar{E}(\bar{G}))$, where $\{i, j\} \in \bar{E}(\bar{G})$ iff $i \neq j$ and $\{i, j\} \notin E(G)$. The cardinality of $\bar{E}(\bar{G})$ is denoted by $\bar{m}$ and hence $\bar{m} = n(n-1)/2 - m$. The edges of $\bar{G}$ are referred to as the *missing edges* of G.

The *adjacency matrix* of G is the $n \times n$ symmetric $(0-1)$ matrix $A = (a_{ij})$ such that $a_{ij} = 1$ iff $\{i, j\} \in E(G)$. The *degree* of node i, denoted by $\deg(i)$, is the number of edges incident with i. The vector consisting of all the degrees is denoted by $\deg$. As a result, $\deg = Ae$, where e is the vector of all 1's in $\mathbb{R}^n$. Nodes of degree one are called *leaves*. It is easy to see that the sum of the degrees in a graph is equal to twice the number of its edges, i.e., $e^T \deg = 2m$.

Let us orient each edge $\{i, j\}$ arbitrarily as (i, j) and refer to i and j as the *tail* and the *head* of (i, j), respectively. The *node-edge incidence matrix* of G is the $n \times m$ matrix $M = (m_{ij})$ such that

$$
m_{ij} = \begin{cases} 1 & \text{if node } i \text{ is the tail of edge } j, \\ -1 & \text{if node } i \text{ is the head of edge } j, \\ 0 & \text{otherwise.} \end{cases}
$$

Obviously, $M^T e = 0$. Moreover, it is easy to prove that $\text{rank}(M) = n - 1$ if and only if G is connected. The matrix

$$
L = \text{Diag}(\deg) - A = \text{Diag}(Ae) - A
$$

is called the *Laplacian* of G. It is a well-known result in algebraic graph theory [42] that $L = MM^T$. Consequently, $Le = 0$ and L is positive semidefinite. Moreover, $\text{rank}(L) = n - 1$ iff G is connected.

Graph G is *complete* if its adjacency matrix is $A = E - I$, where E is the $n \times n$ matrix of all 1's and I is the identity matrix of order n. That is, G is complete if every two of its nodes are adjacent. The complete graph on n nodes is denoted by K_n. A *clique* of G is a complete subgraph of G. Graph G is said to be *k-vertex connected* if either $G = K_{k+1}$, or $|V(G)| \geq k + 2$ and the deletion of any $k - 1$ nodes leaves G connected. Connected graphs with no cycles are called *trees* . It is not hard to see that if T is a tree on n nodes, then T has $n - 1$ edges. Thus for a tree T, $e^T \deg = 2(n-1)$.

A graph is said to be *series-parallel* [47] if it can be obtained from an edge by a sequence of series and parallel extensions. A *series extension* is the subdivision of an edge, while a *parallel extension* is the addition of a new edge joining two adjacent nodes.

Graph G is said to be *chordal* [44, 89] if every cycle of G of length ≥ 4 has a chord, that is, an edge connecting two nonconsecutive nodes on the cycle. Chordal graphs are also known as *triangulated, monotone transitive, rigid circuit,* or *perfect elimination graphs*.

Among the many different characterizations of chordal graphs, the following two are the most useful for our purposes. An ordering $\pi(1),\ldots,\pi(n)$ of the vertices of a graph G is called a *perfect elimination ordering* if for each $i = 1,\ldots,n-1$, the set of vertices $\pi(j)$, $j > i$ that are adjacent to $\pi(i)$, induce a clique in G.

Theorem 1.16 (Fulkerson and Gross [82]) *Graph G is chordal if and only if it has a perfect elimination ordering.*

It should be pointed out that such a perfect elimination ordering can be obtained in polynomial time [161].

Let G_1 and G_2 be two graphs and let $K_1 \subset V(G_1)$ and $K_2 \subset V(G_2)$ be two cliques of the same cardinality. We say that G is a *clique sum* of G_1 and G_2 if it is obtained from G_1 and G_2 by identifying K_1 and K_2, and then deleting duplicate edges in the clique.

Theorem 1.17 (Dirac [72]) *Graph G is a chordal graph if and only if it is a clique sum of complete graphs.*

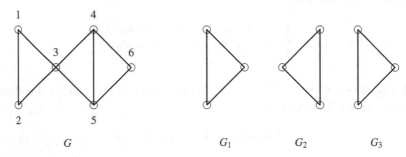

Fig. 1.1 The chordal graph of Example 1.1

Example 1.1 *Consider the chordal graph G depicted in Fig. 1.1. Clearly, the ordering $1,2,\ldots,6$ is a perfect elimination ordering of G. Also, it is clear that G is a clique sum of the three complete graphs $G_1, G_2,$ and G_3.*

A graph is *planar* if it can be drawn in the plane with no two of its edges crossing. Every planar graph admits a planar drawing in which all edges are straight line segments (Fáry's Theorem [78]). A drawing of a connected planar graph divides the plane into regions or faces. The unbounded face is called the *outer face* and all other bounded faces are called *inner faces*.

1.4 Convexity Theory

Convex sets play a prominent role in this monograph. For excellent references on the topics discussed in this section, see, e.g., [160, 109, 166]. Let $\mathcal{V}$ be a finite-dimensional normed real vector space. Our interest here is in the two Euclidean vector spaces: $\mathbb{R}^n$ endowed with the inner product $\langle x, y \rangle = x^T y$, and $\mathscr{S}^n$ endowed with the trace inner product $\langle A, B \rangle = \text{trace}(AB)$. When the origin $\mathbf{0}$ is of no interest to us, we refer to the elements of $\mathscr{V}$ as points. On the other hand, given a point configuration, we can always impose a vector space structure by fixing an origin.

Set $S \subset \mathscr{V}$ is said to be *convex* if the closed line segment joining any two points of S lies entirely in S, i.e., for any two points x^1 and x^2 in S, the point $\lambda x^1 + (1 - \lambda)x^2$ lies in S for all $\lambda : 0 \leq \lambda \leq 1$. The *convex hull* of S, denoted by conv(S), is the smallest convex set containing S. Set S is said to be *affine* if the line passing through any two points of S lies entirely in S, i.e., for any two points x^1 and x^2 in S, the point $\lambda x^1 + (1 - \lambda)x^2$ lies in S for all λ. Every affine set in $\mathscr{V}$ is parallel to a unique vector subspace in $\mathscr{V}$. The dimension of an affine set is equal to the dimension of the vector subspace parallel to it. The *affine hull* of S, denoted by aff(S), is the smallest affine set containing S. The dimension of S is equal to the dimension of aff(S). Finally, set K in $\mathscr{V}$ is said to be a *cone* if for every x in K and for every scalar $\alpha \geq 0$, it follows that αx lies in K. Cone K is *pointed* if $K \cap (-K) = \{\mathbf{0}\}$. The *conic hull* of set S is the set of all conic combinations (i.e., linear combinations with nonnegative coefficients) of vectors in S.

Let S be a convex set and let $x \in S$. We say that x is an *interior point* of S if there exists $r > 0$ such that the set $\{y \in \mathscr{V} : ||y - x|| \leq r\} \subseteq S$. That is, x is an interior point of S if and only if, for every $d \in \mathscr{V}$, there exists $\varepsilon > 0$ such that $x + \varepsilon d$ lies in S. The interior of S, denoted by int(S), is the set of all interior points of S. Set S is said to be *open* if $S = \text{int}(S)$. On the other hand, S is said to *closed* if its complement is open, i.e., if the set $\{x \in \mathscr{V} : x \notin S\}$ is open. The closure of S, denoted by cl(S), is the smallest closed set containing S.

Theorem 1.18 *Let $S \subset \mathscr{V}$. Then S is closed if and only if for every sequence $\{x^k\}_{k \in \mathbb{N}}$ in S that converges to $x \in \mathscr{V}$, it follows that $x \in S$.*

Let S_1 and S_2 be two sets in $\mathscr{V}$. Then $S = S_1 + S_2$ is the set defined as $S = \{x : x = x^1 + x^2, \text{ where } x^1 \in S_1 \text{ and } x^2 \in S_2\}$. S is called the *Minkowski sum* of S_1 and S_2. Set S is said to *bounded* if for all $x \in S$, $||x|| \leq M$ for some finite scalar M. Set S is said to be *compact* if it is both closed and bounded. As an immediate consequence of the definition of convex sets, we have that the intersection of two convex sets is convex, and the Minkowski sum of two convex sets is convex.

Lemma 1.4 *Let S_1 and S_2 be two closed sets in $\mathscr{V}$ and assume that S_2 is bounded. Let $S = S_1 + S_2$. Then S is closed.*

Proof. Let $x \in \text{cl}(S)$ and let the sequence $\{x^k\}$ in S converge to x. Thus it suffices to show that $x \in S$. To this end, there exist sequences y^k in S_1 and z^k in S_2 such that $\{y^k + z^k\}$ converges to x. Since S_2 is compact, there exists a subsequence $\{z^{k_i}\}$ that converges to $z \in S_2$. But since every convergent subsequence converges to the same limit of the sequence, it follows that the subsequence $\{y^{k_i} + z^{k_i}\}$ converges to x and hence $\{y^{k_i}\}$ converges to $x - z$. But since S_1 is closed, it follows that $x - z$ lies in S_1. Therefore, $x = (x - z) + z$ lies in S. $\qquad \square$

Theorem 1.19 *Let S be a compact set in $\mathscr{V}$. Then the convex hull of S is compact.*

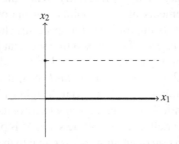

Fig. 1.2 An example of a closed set whose convex hull is not closed

We remark here that the proof of Theorem 1.19 uses Carathéodory's Theorem. Also, note that the condition that S is bounded cannot be dropped. Let $S = \{x \in \mathbb{R}^2 : x_1 \geq 0, x_2 = 0\} \cup \{(0,1)\}$ (see Fig. 1.2). Then S is closed while the convex hull of S is not closed.

The *boundary* of set S, denoted by ∂S, is defined as $\partial S = \text{cl}(S) \backslash \text{int}(S)$. Note that ∂S is closed since it is the intersection of two closed sets, namely $\text{cl}(S)$ and the complement of $\text{int}(S)$.

The notion of relative interior of S is of special interest to us since most of the sets we deal with in this monograph have empty interior. A point $\hat{x} \in S$ is a *relative interior point* of S if there exists $r > 0$ such that $\{x \in \text{aff}(S) : ||x - \hat{x}|| \leq r\} \subseteq S$. The relative interior of S, denoted by $\text{relint}(S)$, is the set of all relative interior points of S. Observe that if $\text{aff}(S) = \mathscr{V}$, then $\text{relint}(S) = \text{int}(S)$. Consequently, either $\text{int}(S)$ is empty or $\text{int}(S) = \text{relint}(S)$. Evidently,

$$\text{relint}(S) \subseteq S \subseteq \text{cl}(S). \tag{1.9}$$

It is a well-known fact that every nonempty convex set has a nonempty relative interior. The relative interior of convex sets can be easily characterized.

Theorem 1.20 *Let S be a convex set in $\mathscr{V}$. Then*

$$relint(S) = \{x \in S : \forall\, y \in S, \exists\, \mu > 1 \text{ such that } \mu x + (1 - \mu)y \in S\}. \tag{1.10}$$

Proof.   Let $x \in S$. Then $x \in \text{relint}(S)$ if and only if for every $y \in S$, there exists $y' \in S$ such that $x = \lambda y' + (1 - \lambda)y$ for some $\lambda : 0 < \lambda < 1$; i.e., for every $y \in S$, there exists $\lambda : 0 < \lambda < 1$ such that $y' = (x/\lambda + (1 - 1/\lambda)y) \in S$. The result follows by setting $\mu = 1/\lambda$.
                                                                                               □

Therefore, x is a relative interior point of S if and only if for each y in S, the line segment $[y,x]$ can be extended slightly beyond x without leaving S. That is, by setting $\mu = 1 + \gamma$, we have that $x \in \text{relint}(S)$ iff for every y in S, there exists $\gamma > 0$ such that $x + \gamma(x - y)$ lies in S.

Lemma 1.5 *Let S_1 and S_2 be two nonempty convex sets in $\mathcal{V}$ and let $S = S_1 + S_2$. Then $relint(S) = relint(S_1) + relint(S_2)$.*

Note that the above lemma is false if the relative interior is replaced by the interior. For example, let $S_1 = \{x \in \mathbb{R}^2 : 0 \le x_1 \le 1, x_2 = 0\}$ and $S_2 = \{x \in \mathbb{R}^2 : x_1 = 0, 0 \le x_2 \le 1\}$. Then $int(S_1) = int(S_2) = \emptyset$, while $int(S_1 + S_2) = \{x \in \mathbb{R}^2 : 0 < x_1 < 1, 0 < x_2 < 1\}$.

Set S is said to be *relatively open* if $relint(S) = S$. Moreover, the *relative boundary* of S, denoted by $rbd(S)$, is defined as $rbd(S) = cl(S) \backslash relint(S)$.

Affine sets are closed. Also, the relative interior of an affine set is the set itself. Hence, affine sets are also relatively open. In particular, a singleton set $\{\hat{x}\}$ is affine and thus is relatively open. In other words, $aff(\{\hat{x}\}) = \{\hat{x}\} = relint(\{\hat{x}\})$.

Theorem 1.21 *Let S be a convex set in $\mathcal{V}_1$ and let $T : \mathcal{V}_1 \to \mathcal{V}_2$ be a linear transformation. Then $T(S)$ is a convex set in $\mathcal{V}_2$. Moreover, $T(relint(S)) = relint(T(S))$.*

In other words, if x lies in $relint(S)$, then $T(x)$ lies in $relint(T(S))$. On the other hand, let x' be in $relint(T(S))$ and let $T^{-1}(x') = \{x \in \mathcal{V}_1 : T(x) = x'\}$. Then $T^{-1}(x') \cap relint(S) \ne \emptyset$; i.e., there exists x in $relint(S)$ such that $x' = T(x)$.

Theorem 1.22 *Let S be a closed convex set in $\mathcal{V}_1$ and let $T : \mathcal{V}_1 \to \mathcal{V}_2$ be a linear transformation. If T has a trivial kernel, then $T(S)$ is a closed convex set.*

Theorem 1.23 *Let S be a nonempty convex set in $\mathcal{V}$. Then $relint(S)$ and $cl(S)$ are convex. Furthermore, the three sets S, $relint(S)$, and $cl(S)$ have the same affine hull.*

Theorem 1.24 *Let S be a nonempty convex set in $\mathcal{V}$. Then $cl(relint(S)) = cl(S)$ and $relint(cl(S)) = relint(S)$.*

As a result, the three convex sets S, $relint(S)$, and $cl(S)$ have the same relative interior and the same closure. Note that for any set S, whether convex or not, $relint(relint(S)) = relint(S)$ and $cl(cl(S)) = cl(S)$. Finally, it should be pointed out that the convexity assumption in Theorem 1.24 cannot be dropped. For example, let $\mathcal{V} = \mathbb{R}$ and let S be the set of rational numbers in $[0, 1]$. Then $aff(S) = \mathbb{R}$ and $relint(S) = int(S) = \emptyset$ since every neighborhood of a rational number must contain irrational numbers. Hence, $cl(relint(S)) = \emptyset$. On the other hand, $cl(S) = [0, 1]$ and thus $relint(cl(S)) = (0, 1)$.

1.4.1 Faces of a Convex Set

Definition 1.1 *Let S be a convex set in $\mathcal{V}$ and let F be a convex subset of S. Then F is said to be a face of S if for every x in F such that $x = \lambda y + (1 - \lambda)z$ for some y and z in S and $0 < \lambda < 1$, it follows that y and z are both in F.*

In other words, a convex subset F is a face of convex set S if every line segment in S with a relative interior point in F must have both endpoints in F. Evidently, $\emptyset$

and S are faces of S. These two faces are called *improper faces* of S, while all other faces of S are called *proper*. The dimension of a face is the dimension of its affine hull. Faces of S of dimensions 0, 1, and $\dim(S) - 1$ are called, respectively, *extreme points*, *edges*, and *facets*.

The following theorems are easy consequences of the definition of a face.

Theorem 1.25 *Let F_1 and F_2 be two faces of a convex set S. Then $F_1 \cap F_2$ is a face of S.*

Theorem 1.26 *Let S be a convex set and let F_2 be a face of S. Let F_1 be a face of F_2. Then F_1 is a face of S.*

As will be shown next, faces of S are characterized by their relative interiors. More precisely, every point x in S belongs to a unique face of S containing x in its relative interior. This unique face is called the *minimal face* of S containing x, which we denote by $\text{face}(x, S)$.

Theorem 1.27 *Let S be a convex set in $\mathcal{V}$ and let K be a subset of S. Then there is a minimal face of S containing K which we denote by $\text{face}(K, S)$.*

Proof. Let $\mathcal{F}$ be the family of all faces of S containing K. $\mathcal{F} \neq \emptyset$ since S is in $\mathcal{F}$. Let F be the intersection of all faces in $\mathcal{F}$. Thus $\text{face}(K, S) = F$.

$\square$

As a result, $\text{face}(K, S)$ is a subset of every face of S that contains K. In particular, if $K_1 \subset K_2 \subset S$, then $\text{face}(K_1, S) \subset \text{face}(K_2, S)$ since $\text{face}(K_2, S)$ is a face of S containing K_1.

Theorem 1.28 *Let S be a convex set in $\mathcal{V}$ and let F be a face of S. Let K be a convex subset of S such that $\text{relint}(K) \cap F \neq \emptyset$. Then $K \subseteq F$.*

Proof. Let $x \in K$ and let $y \in \text{relint}(K) \cap F$. Then there exists $z \in K$ such that $y = \lambda x + (1 - \lambda)z$ for some $\lambda : 0 < \lambda < 1$. Since F is a face of S and since $y \in F$, it follows that x, z are in F. Therefore, $K \subseteq F$.

$\square$

This theorem has three immediate consequences. First, let $x \in S$ and let K and F be two faces of S containing x. Assume that $x \in \text{relint}(K)$. Then $\text{relint}(K) \cap F \neq \emptyset$, and hence, $K \subseteq F$. Consequently, K is the minimal face of S containing x; i.e., $\text{face}(x, S)$ is the face of S containing x in its relative interior. As a result, the faces of S are uniquely determined by their relative interiors. The other two consequences are the following two corollaries.

Corollary 1.2 *Let S be a closed convex set in $\mathcal{V}$ and let $K \subset S$. Let F be a face of S containing K. If $K \cap \text{relint}(F) \neq \emptyset$, then $F = \text{face}(K, S)$.*

Proof. Let $x \in K \cap \text{relint}(F)$. Then $F = \text{face}(x, S)$. Now clearly $\text{face}(K, S) \subseteq F$ and $\text{face}(x, S) \subseteq \text{face}(K, S)$. Therefore, $F = \text{face}(K, S)$.

$\square$

Corollary 1.3 *Let F_1 and F_2 be two faces of a convex set S such that relint$(F_1) \cap$ relint$(F_2) \neq \emptyset$. Then $F_1 = F_2$.*

Proof. This follows from the previous theorem since relint$(F_1) \cap F_2 \neq \emptyset$ and relint$(F_2) \cap F_1 \neq \emptyset$.

□

Theorem 1.29 *Let S be a closed convex set in $\mathcal{V}_1$ and let $T : \mathcal{V}_1 \to \mathcal{V}_2$ be a linear transformation. Let F' be a face of $S' = T(S)$. Then $F = T^{-1}(F')|_S$, the preimage of F' restricted to S, is a face of S. In particular, if $\bar{x}'$ is an extreme point of S', then $T^{-1}(\bar{x}')|_S$ is a face of S.*

Proof. Let x and y be in S such that $x/2 + y/2$ lies in $F = T^{-1}(F')|_S$. Let $x' = T(x)$ and $y' = T(y)$. Thus, x' and y' are in S' and $x'/2 + y'/2$ lies in F'. But F' is a face of S'. Therefore, x' and y' are in F'. Hence, x and y are in F and thus F is a face of S. The second result follows since $F' = \{\bar{x}'\}$ is a face of S'.

□

Let $\mathcal{V}_1 = \mathcal{V}_2 = \mathbb{R}^n$ and let $T(x) = Ax$. Then it is easy to see that the preimage of $\bar{x}'$ is an affine set since $T^{-1}(\bar{x}') = \{x \in \mathbb{R}^n : Ax = \bar{x}'\}$.

Example 1.2 *Let S and $S' = T(S)$ be the sets depicted in Fig. 1.3, where T is the projection on the x_1-axis. Then $T^{-1}(\bar{x}')$ is the affine hull of u and v, while $T^{-1}(\bar{x}')|_S$ is the line segment $[u, v]$. Notice that $\bar{x}'$ is a face of S' and $T^{-1}(\bar{x}')|_S$ is a face of S. Also, notice that $T(relint(S)) = relint(T(S))$.*

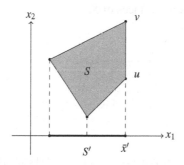

Fig. 1.3 The sets S and S' of Example 1.2

The following theorem is useful in the study of the geometry of Euclidean distance matrices.

Theorem 1.30 *Let S be a closed convex set in $\mathcal{V}_1$ and let $T : \mathcal{V}_1 \to \mathcal{V}_2$ be a linear transformation. Let $x \in S$ and let $S' = T(S)$ be closed. If $y \in face(x, S)$, then $T(y) \in face(T(x), S')$.*

Proof. If face$(x, S) = \{x\}$, then the result follows trivially. Therefore, assume that $y \in$ face(x, S) and $y \neq x$. Since $x \in$ relint(face(x, S)), there exist $\lambda : 0 < \lambda < 1$ and

$z \in \text{face}(x,S)$ such that $x = \lambda y + (1 - \lambda)z$. Hence, $T(x) = \lambda T(y) + (1 - \lambda)T(z)$ and consequently, $T(y) \in \text{face}(T(x), S')$.

$\square$

Let $p \in \mathcal{V}, p \neq \mathbf{0}$, the hyperplane $H = \{x \in \mathcal{V} : \langle p, x \rangle = p_0\}$ is said to be a *supporting hyperplane* of a convex set S if $S \not\subset H$, $H \cap S \neq \emptyset$ and $\langle p, x \rangle \geq p_0$ for all $x \in S$. The set $H^+ = \{x \in \mathcal{V} : \langle p, x \rangle \geq p_0\}$ is called a *closed half space*.

Theorem 1.31 *Let S be a convex set in $\mathcal{V}$ and let $H = \{x \in \mathcal{V} : \langle p, x \rangle = p_0\}$ be a supporting hyperplane of S. Further, let $F = H \cap S$. Then F is a face of S.*

Proof.   The convexity of F is obvious. Now let $x \in F$ and let y and z be in S such that $x = \lambda y + (1 - \lambda)z$ for some $\lambda : 0 < \lambda < 1$. Thus, $\langle p, y \rangle \geq p_0$ and $\langle p, z \rangle \geq p_0$ since H is a supporting hyperplane. Moreover, $\langle p, x \rangle = \lambda \langle p, y \rangle + (1 - \lambda)\langle p, z \rangle = p_0$ or $\lambda(\langle p, y \rangle - p_0) + (1 - \lambda)(\langle p, z \rangle - p_0) = 0$. Thus, $\lambda(\langle p, y \rangle - p_0) = 0$ and $(1 - \lambda)(\langle p, z \rangle - p_0) = 0$. Hence, $\langle p, y \rangle = p_0$ and $\langle p, z \rangle = p_0$, and thus y and z are in F. Therefore, F is a face of S.

$\square$

Definition 1.2 *Let S be a convex set in $\mathcal{V}$ and let F be a proper face of S. Then F is said to be an* exposed face *if $F = S \cap H$ for some supporting hyperplane H of S, in which case, if $H = \{x \in \mathcal{V} : \langle p, x \rangle = p_0\}$, then we say that p exposes F or F is* exposed *by p.*

It should be pointed out that not all faces of a convex set are exposed faces. For example, consider the convex set S depicted in Fig. 1.4, where $S = \{x \in \mathbb{R}^2 : (x_1 - 1)^2 + x_2^2 \leq 1\} \cup \{x \in \mathbb{R}^2 : -1 \leq x_1 \leq 1, -1 \leq x_2 \leq 1\}$. Then the points $x = (1,1)$ and $y = (1,-1)$ are nonexposed faces of S.

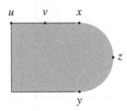

Fig. 1.4 A convex set with two nonexposed faces, namely $\{x\}$ and $\{y\}$

1.4.2 Separation Theorems

The notion of separation is crucial and very useful in convexity theory. A hyperplane $H = \{x \in \mathcal{V} : \langle p, x \rangle = p_0\}$ is said to *properly separate* convex sets S_1 and S_2 if

$$\langle p, x \rangle \geq p_0 \geq \langle p, y \rangle \text{ for all } x \in S_1 \text{ and } y \in S_2,$$

and $S_1 \cup S_2 \not\subseteq H$. On the other hand, H is said to *strongly separate* convex sets S_1 and S_2 if

$$\langle p,x \rangle > p_0 > \langle p,y \rangle \text{ for all } x \in S_1 \text{ and } y \in S_2.$$

The following two standard results are needed in the proofs of the separation theorems below.

Theorem 1.32 (Weierstrass) *Let S be a nonempty compact set in $\mathcal{V}$ and let f be a continuous real-valued function on S. Then f attains a maximum and a minimum on S.*

Theorem 1.33 (The Projection Theorem) *Let S be a nonempty closed convex set in $\mathcal{V}$ and let $\hat{x} \notin S$. Further, let $||x|| = \langle x,x \rangle^{1/2}$. Then there exists a unique x^* in S such that*

$$x^* = arg\ min_{x \in S} ||\hat{x} - x||.$$

x^* is said to be the projection of $\hat{x}$ on S. Moreover, x^* is the projection of $\hat{x}$ on S if and only if $\langle \hat{x} - x^*, x - x^* \rangle \leq 0$ for all $x \in S$.

Note that the existence of x^* follows from Weierstrass Theorem and its uniqueness follows from the convexity of S. Moreover, if S is an affine set, then x^* is the projection of $\hat{x}$ on S if and only if $\langle \hat{x} - x^*, x - x^* \rangle = 0$ for all $x \in S$.

The first separation theorem establishes the existence of a hyperplane strongly separating a point and a closed convex set.

Theorem 1.34 (The Strong Separation Theorem) *Let S be a nonempty closed convex set in $\mathcal{V}$ and let $\hat{x} \notin S$. Then there exists a hyperplane $H = \{x \in \mathcal{V} : \langle p,x \rangle = p_0\}$ such that*

$$\langle p,\hat{x} \rangle > p_0 \text{ and } \langle p,x \rangle \leq p_0 \text{ for all } x \in S.$$

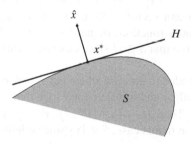

Proof. By the projection theorem, there exists x^* in S such that $\langle \hat{x} - x^*, x - x^* \rangle \leq 0$ for all $x \in S$. Let $p = \hat{x} - x^*$. Then $p \neq \mathbf{0}$ and thus $||p||^2 = \langle p, \hat{x} - x^* \rangle > 0$. Let $p_0 = \langle p,x^* \rangle$. Hence, $\langle p,\hat{x} \rangle > p_0$ and $\langle p,x \rangle \leq p_0$ for all $x \in S$.

$\square$

A remark is in order here. Since $\langle p, \hat{x} - x^* \rangle = ||p||^2$, i.e., $\langle p,\hat{x} \rangle = p_0 + ||p||^2$, it follows that $\langle p,\hat{x} \rangle > p_0 + ||p||^2/2$. Let $p_0' = p_0 + ||p||^2/2 = \langle p, x^* + p/2 \rangle$. Then by choosing the hyperplane $H = \{x : \langle p,x \rangle = p'_0\}$, i.e., by choosing H to pass through the point $x^* + p/2$ instead of x^*, we have $\langle p,\hat{x} \rangle > p'_0$ and $\langle p,x \rangle < p'_0$ for all $x \in S$.

A second version of the strong separation theorem is given next. It shows that two disjoint closed convex sets can be strongly separated provided that one of them is bounded.

Theorem 1.35 (The Strong Separation Theorem) *Let S_1 and S_2 be two nonempty disjoint closed convex set in $\mathcal{V}$ and assume that S_2 is bounded. Then there exists a hyperplane $H = \{x \in \mathcal{V} : \langle p,x \rangle = p_0\}$ such that*

$$\langle p,x \rangle < p_0 \text{ for all } x \in S_1, \text{ and } \langle p,y \rangle \geq p_0 \text{ for all } y \in S_2.$$

Proof. Let $S = S_1 - S_2$. Then S is a closed convex set and $\mathbf{0} \notin S$ since S_1 and S_2 are disjoint. Thus, there exist $p \neq 0$ and p'_0 such that $\langle p,\mathbf{0} \rangle > p'_0$ and $\langle p,x-y \rangle \leq p'_0$ for all $x \in S_1$ and $y \in S_2$. Thus, $p'_0 < 0$ and hence $\langle p,x \rangle < \langle p,y \rangle$ for all $x \in S_1$ and $y \in S_2$. The result follows by setting $p_0 = \min_{y \in S_2} \langle p,y \rangle$.   □

It is worthy of note that the condition S_2 is bounded cannot be dropped. For example, the two sets $S_1 = \{x \in \mathbb{R}^2 : x_2 \geq 1/x_1, x_1 > 0\}$ and $S_2 = \{x \in \mathbb{R}^2 : x_2 = 0, x_1 > 0\}$ cannot be strongly separated.

The following theorem shows that for any convex set S, there exists a supporting hyperplane to S at a boundary point.

Theorem 1.36 (The Supporting Hyperplane Theorem) *Let S be a convex set with a nonempty interior and let $y \in \partial S$. Then there exists a supporting hyperplane at y to S. That is, there exists $H = \{x \in \mathcal{V} : \langle p,x \rangle = p_0 = \langle p,y \rangle\}$ such that $\langle p,x \rangle \leq p_0\}$ for all $x \in S$ and $S \not\subset H$.*

Proof. Let $\{y^k\}$ be a sequence in $\mathcal{V} \setminus \text{cl}(S)$ that converges to y. Thus, for each k, there exists a unit p^k such that $\langle p^k,y^k \rangle > \langle p^k,x \rangle$ for all $x \in \text{cl}(S)$. Since p^k is in the compact set $\{x \in \mathcal{V} : ||x|| = 1\}$, it follows that there exists a subsequence $\{p^{k_i}\}$ that converges to a unit vector p. By taking the limit as $k \to \infty$ and since the inner product is a continuous function, we have $p_0 = \langle p,y \rangle \geq \langle p,x \rangle$ for all $x \in S$. To complete the proof, observe that $S \not\subset H$ since otherwise, $\text{int}(S) = \emptyset$, a contradiction.   □

This result can be extended to the case where $\text{int}(S) = \emptyset$ and $y \in \text{rbd}(S)$. Let $\mathcal{V}'$ be the subspace parallel to $\text{aff}(S)$ and translate set S such that $y = 0$. Take the sequence $\{y^k\}$ to be in $\mathcal{V}' \setminus \text{cl}(S)$. Then $H = \{x \in \mathcal{V}' : \langle p,x \rangle = p_0\} + \mathcal{V}'^\perp$ is a supporting hyperplane in $\mathcal{V}$. Note that in this case, $S \neq \{y\}$ since otherwise $y \in \text{relint}(S = \{y\})$. Consequently, $S \not\subset H$.

Theorem 1.37 (The Separation Theorem) *Let S_1 and S_2 be two nonempty convex sets in $\mathcal{V}$ such that $\text{relint}(S_1) \cap \text{relint}(S_2) = \emptyset$. Then there exists a hyperplane $H = \{x \in \mathcal{V} : \langle p,x \rangle = p_0\}$ such that*

$$\langle p,x \rangle \leq p_0 \text{ for all } x \in S_1, \text{ and } \langle p,y \rangle \geq p_0 \text{ for all } y \in S_2.$$

Moreover, $(S_1 \cup S_2) \not\subset H$.

Proof. Let $S = S_1 - S_2$. Then S is a convex set and $\mathbf{0} \notin \text{relint}(S)$ since $\text{relint}(S_1)$ and $\text{relint}(S_2)$ are disjoint. Thus, there exists a hyperplane $H = \{z : \langle p, z \rangle = 0\}$ such that $\langle p, z \rangle \leq 0$ for all $z \in S$ and $\langle p, \hat{z} \rangle < 0$ for some $\hat{z}$ in S. Consequently, $\langle p, x \rangle \leq \langle p, y \rangle$ for all $x \in S_1$ and $y \in S_2$; and $\langle p, \hat{x} \rangle < \langle p, \hat{y} \rangle$ for some $\hat{x}$ in S_1 and $\hat{y} \in S_2$. The result follows by setting $p_0 = \inf_{y \in S_2} \langle p, y \rangle$.

$\square$

The reverse of Theorem 1.37 is also true. Suppose that H properly separates S_1 and S_2 and assume to the contrary that $x \in (\text{relint}(S_1) \cap \text{relint}(S_2))$. Thus x must be in H. Let y be any point in S_2, $y \neq x$. Then there exists z in S_2 and $0 < \lambda < 1$ such that $x = \lambda y + (1 - \lambda)z$. Thus, both y and z are in H and hence $S_2 \subseteq H$. By a similar argument, $S_1 \subseteq H$. Thus, we have a contradiction since $(S_1 \cup S_2) \not\subseteq H$.

The sets S_1 and S_2 of Theorem 1.37 that are most relevant for our purposes are cones, linear subspaces, and affine sets. As a result, we have the following corollary of Theorem 1.37, which will be used in the proofs of the theorems of the alternative.

Corollary 1.4 *The assertion in Theorem 1.37 that $\langle p, x \rangle \leq p_0$ for all $x \in S_1$ reduces to*

1. $\langle p, x \rangle \leq 0$ *for all $x \in S_1$ if S_1 is a cone.*
2. $\langle p, x \rangle = 0$ *for all $x \in S_1$ if S_1 is a linear subspace.*
3. $\langle p, \hat{x} \rangle \leq p_0$ *and $\langle p, x \rangle = 0$ for all $x \in \mathscr{L}$ if S_1 is the affine set $\hat{x} + \mathscr{L}$, where $\hat{x}$ is a point and $\mathscr{L}$ is a subspace.*

Proof. We prove Statement 3, the proofs of the other two statements are similar. Assume that $S_1 = \hat{x} + \mathscr{L}$. Then $\langle p, \hat{x} \rangle \leq p_0$ since $\mathbf{0} \in \mathscr{L}$. Now, by way of contradiction, assume that $\langle p, \bar{x} \rangle \neq 0$ for some $\bar{x} \in \mathscr{L}$. Let α be a scalar. Then $\alpha \langle p, \bar{x} \rangle$ can be made large enough so that $\langle p, \hat{x} + \alpha \bar{x} \rangle = \langle p, \hat{x} \rangle + \alpha \langle p, \bar{x} \rangle > p_0$, a contradiction.

$\square$

1.4.3 Polar Cones

Let K be a cone, then the set

$$K^\circ = \{y \in \mathscr{V} : \langle y, x \rangle \leq 0 \text{ for all } x \in K\}. \tag{1.11}$$

is called the *polar* of K. As immediate consequences of this definition, we have that K° is a closed convex cone and $K \subseteq (K^\circ)^\circ$. Moreover, if $K_1 \subseteq K_2$, then, evidently, $K_2^\circ \subseteq (K_1)^\circ$. It is worth pointing out that if K is a subspace of $\mathscr{V}$, then K° is the orthogonal complement of K. The cone $(-K^\circ)$ is called the *dual* of K. Consequently, cone K is self-dual if $K^\circ = -K$. Next, we prove a few important properties of the polar cone.

Lemma 1.6 *Let K be a nonempty cone in $\mathscr{V}$. Then*

$$K^\circ = (cl(K))^\circ.$$

Proof. $(cl(K))^\circ \subseteq K^\circ$ since $K \subseteq cl(K)$. Now let $y \in K^\circ$ and let $\bar{x} \in cl(K)$. Then there exists a sequence $\{x^k\}$ in K that converges to $\bar{x}$. Thus, $\langle y, x^k \rangle \leq 0$ for all k and hence, $\langle y, \bar{x} \rangle \leq 0$. Therefore, $y \in (cl(K))^\circ$ since $\bar{x}$ was arbitrary. Consequently, $K^\circ \subseteq (cl(K))^\circ$ and the result follows.

$\square$

Theorem 1.38 *Let K be a nonempty convex cone in $\mathcal{V}$. Then*

$$(K^\circ)^\circ = cl(K).$$

Proof. As noted earlier, $K \subseteq (K^\circ)^\circ$. Now since $(K^\circ)^\circ$ is closed and since $cl(K)$ is the smallest closed set containing K, it follows that $cl(K) \subseteq (K^\circ)^\circ$. Therefore, it suffices to show that $cl(K) \supseteq (K^\circ)^\circ$. To this end, suppose to the contrary that there exists $x \in (K^\circ)^\circ$ and $x \notin cl(K)$. Thus, by the Strong Separation Theorem, there exist $p \neq \mathbf{0}$ and p_0 such that

$$\langle p, x \rangle > p_0 \geq \langle p, y \rangle \text{ for all } y \in cl(K).$$

But since $\mathbf{0} \in K$, it follows that $p_0 \geq 0$. Furthermore, by Corollary 1.4, we have

$$\langle p, x \rangle > 0 \text{ and } 0 \geq \langle p, y \rangle \text{ for all } y \in cl(K).$$

Hence, $p \in (cl(K))^\circ = K^\circ$. But $x \in (K^\circ)^\circ$. Therefore, $\langle p, x \rangle \leq 0$, a contradiction.

$\square$

Theorem 1.39 *Let K_1 and K_2 be two nonempty cones in $\mathcal{V}$. Then*

$$(K_1 + K_2)^\circ = K_1^\circ \cap K_2^\circ.$$

Proof. Let $y \in (K_1^\circ \cap K_2^\circ)$. Then $\langle y, x \rangle \leq 0$ for all $x \in K_1$ and $\langle y, z \rangle \leq 0$ for all $z \in K_2$. Hence, $\langle y, x + z \rangle \leq 0$ for all $x \in K_1$ and $z \in K_2$. Therefore, $y \in (K_1 + K_2)^\circ$ and thus $K_1^\circ \cap K_2^\circ \subseteq (K_1 + K_2)^\circ$.

On the other hand, let $y \in (K_1 + K_2)^\circ$. Then $\langle y, x + z \rangle \leq 0$ for all $x \in K_1$ and $z \in K_2$. But $\mathbf{0} \in K_2$. Thus $\langle y, x \rangle \leq 0$ for all $x \in K_1$. Similarly, $\langle y, z \rangle \leq 0$ for all $z \in K_2$. Hence, $(K_1 + K_2)^\circ \subseteq K_1^\circ \cap K_2^\circ$ and the result follows.

$\square$

Corollary 1.5 *Let K_1 and K_2 be two nonempty closed convex cones in $\mathcal{V}$. Then*

$$(K_1 \cap K_2)^\circ = cl(K_1^\circ + K_2^\circ).$$

Proof. Since K_1 and K_2 are closed and convex, we have $(K_1^\circ)^\circ = K_1$ and $(K_2^\circ)^\circ = K_2$. Thus $K_1 \cap K_2 = (K_1^\circ)^\circ \cap (K_2^\circ)^\circ = (K_1^\circ + K_2^\circ)^\circ$. Therefore, $(K_1 \cap K_2)^\circ = ((K_1^\circ + K_2^\circ)^\circ)^\circ = cl(K_1^\circ + K_2^\circ)$, where the last equality follows from Theorem 1.38.

$\square$

The need for the closure in the above corollary can be understood in light of the fact that $K_1^\circ + K_2^\circ$ need not be closed, while $(K_1 \cap K_2)^\circ$ is always closed.

For closed convex cones, the Projection Theorem specializes to the following:

Theorem 1.40 *Let K be a nonempty closed convex cone in $\mathscr{V}$ and let $\hat{x} \notin K$. Then x^* is the closest point in K to $\hat{x}$ if and only if $(\hat{x} - x^*) \in K^\circ$ and $\langle \hat{x} - x^*, x^* \rangle = 0$.*

Proof. Assume that x^* is the closest point in K to $\hat{x}$. Then $\langle \hat{x} - x^*, x - x^* \rangle \leq 0$ for all $x \in K$. This implies that $\langle \hat{x} - x^*, x^* \rangle = 0$. To see this, observe that this implication trivially follows if $x^* = \mathbf{0}$. On the other hand, if $x^* \neq \mathbf{0}$, then $\mathbf{0}$ and $2x^*$ are in K since K is a cone. Consequently, $\langle \hat{x} - x^*, x^* \rangle \geq 0$ and $\langle \hat{x} - x^*, x^* \rangle \leq 0$, and hence $\langle \hat{x} - x^*, x^* \rangle = 0$. As a result, we have that $\langle \hat{x} - x^*, x \rangle \leq 0$ for all $x \in K$; i.e., $(\hat{x} - x^*) \in K^\circ$.

To prove the other direction, note that if $\langle \hat{x} - x^*, x^* \rangle = 0$ and if $\langle \hat{x} - x^*, x \rangle \leq 0$ for all $x \in K$. Then it follows, trivially, that $\langle \hat{x} - x^*, x - x^* \rangle \leq 0$ for all $x \in K$. $\qquad \square$

We conclude this subsection with the following well-known important decomposition result.

Theorem 1.41 (Moreau [149]) *Let K be a nonempty closed convex cone in $\mathscr{V}$. Then the following two statements are equivalent:*

(i) *For every $\hat{x} \in \mathscr{V}$, there exist a unique $x^* \in K$ and a unique $y^* \in K^\circ$ such that:*

$$\hat{x} = x^* + y^* \text{ and } \langle x^*, y^* \rangle = 0,$$

(ii) *x^* and y^* are, respectively, the closest points in K and K° to $\hat{x}$.*

Proof. Assume that Statement (ii) holds. If $\hat{x}$ lies in K or K°, then Statement (i) trivially holds. Thus, assume that $\hat{x} \notin (K \cup K^\circ)$ and let $\bar{y} = \hat{x} - x^*$. Then Theorem 1.40 implies that $\bar{y} \in K^\circ$ and $\langle \hat{x} - \bar{y}, \bar{y} \rangle = \langle x^*, \bar{y} \rangle = \langle x^*, \hat{x} - x^* \rangle = 0$. Moreover, $(\hat{x} - \bar{y}) \in (K^\circ)^\circ$ since $x^* \in K = (K^\circ)^\circ$. Consequently, it follows from Theorem 1.40 that $\bar{y}$ is the closest point in K° to $\hat{x}$. Hence, $\bar{y} = y^*$ and $\langle x^*, y^* \rangle = 0$.

To prove the other direction, assume that Statement (i) holds. Then $(\hat{x} - x^*) = y^*$ lies in K° and $\langle \hat{x} - x^*, x^* \rangle = \langle y^*, x^* \rangle = 0$. Thus, it follows from Theorem 1.40 that x^* is the closest point in K to $\hat{x}$. By a similar argument, we have that y^* is the closest point in K° to $\hat{x}$. $\qquad \square$

1.4.4 The Boundary of Convex Sets

The boundary is of special interest in the study of convex sets. Let S be a nonempty closed convex set in $\mathscr{V}$ and let $\hat{x} \in S$. The *normal cone* of S at $\hat{x}$ is

$$N_S(\hat{x}) = \{c \in \mathscr{V} : \langle c, \hat{x} \rangle \geq \langle c, x \rangle \text{ for all } x \in S\}.$$

Three facts follow immediately from this definition. First, normal cones are closed and convex. Second, $N_S(\hat{x}) = \{\mathbf{0}\}$ if and only if $\hat{x}$ is an interior point of S. Third, the set $\hat{x} + N_S(\hat{x})$ is precisely the set of all points in $\mathscr{V}$ whose projection on S is $\hat{x}$.

Let d be a nonzero vector in $\mathcal{V}$, d is said to be a *feasible direction* of S at $\hat{x}$ if

$$\exists\, \delta > 0 : \hat{x} + \delta d \text{ lies in } S.$$

Let $F_S(\hat{x})$ be the set of all feasible directions of S at $\hat{x}$, and the origin $\mathbf{0}$. $F_S(\hat{x})$ is called the *cone of feasible directions* of S at $\hat{x}$. Let d^1 and d^2 be in $F_S(\hat{x})$ and let $\lambda : 0 \leq \lambda \leq 1$. Then $\hat{x} + \delta_1 d^1$ and $\hat{x} + \delta_2 d^2$ are in S for some $\delta_1 > 0$ and some $\delta_2 > 0$. Assume that $\delta_1 \leq \delta_2$. Since S is convex, it follows that $\hat{x} + \delta_1 d^2$ is also in S. Therefore, $\lambda(\hat{x} + \delta_1 d^1) + (1 - \lambda)(\hat{x} + \delta_1 d^2) = \hat{x} + \delta_1(\lambda d^1 + (1 - \lambda)d^2)$ lies in S. Hence, $\lambda d^1 + (1 - \lambda)d^2 \in F_S(\hat{x})$ and thus $F_S(\hat{x})$ is convex. However, $F_S(\hat{x})$ need not be closed. For example, consider the unit disk in the plane $S = \{x \in \mathbb{R}^2 : \|x\| \leq 1\}$ and let $\hat{x} = (0,1)$ (see Fig. 1.5). Then $N_S(\hat{x}) = \{\alpha \begin{bmatrix} 0 \\ 1 \end{bmatrix}$ where $\alpha \geq 0\}$ and $F_S(\hat{x}) = \{d = \begin{bmatrix} d_1 \\ d_2 \end{bmatrix} : d_2 < 0\}$. Thus $F_S(\hat{x})$ is not closed.

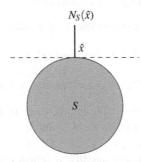

$N_S(\hat{x})$

$\hat{x}$

S

Fig. 1.5 A nonclosed cone of feasible directions $F_S(\hat{x})$

It is worth noting that $F_S(\hat{x}) = $ conic hull of $(S - \{\hat{x}\}) = \{\alpha(x - \hat{x}) : \alpha \geq 0, x \in S\}$ since S is convex. The *tangent cone* of S at $\hat{x}$, denoted by $T_S(\hat{x})$, is the closure of $F_S(\hat{x})$.

Theorem 1.42 *Let S be a nonempty closed convex set in $\mathcal{V}$ and let $\hat{x} \in S$. Then*

$$T_S(\hat{x}) = (N_S(\hat{x}))^\circ.$$

Proof.   Let $d \in F_S(\hat{x})$. Then $\hat{x} + \delta d$ lies in S for some $\delta > 0$. Let $c \in N_S(\hat{x})$. Then $\langle c, \hat{x} \rangle \geq \langle c, \hat{x} + \delta d \rangle$. Thus $\langle c, d \rangle \leq 0$. Hence, $d \in (N_S(\hat{x}))^\circ$ and thus $F_S(\hat{x}) \subseteq (N_S(\hat{x}))^\circ$. Therefore, $\mathrm{cl}(F_S(\hat{x})) = T_S(\hat{x}) \subseteq (N_S(\hat{x}))^\circ$ or $N_S(\hat{x}) \subseteq (T_S(\hat{x}))^\circ$.

To prove the other direction, let $c \in (T_S(\hat{x}))^\circ$. Then $\langle c, d \rangle \leq 0$ for all $d \in T_S(\hat{x})$. Let $x \neq \hat{x}$ be any point in S. Since S is convex, $\hat{x} + (x - \hat{x})/2$ lies in S. Thus, $x - \hat{x}$ lies in $T_S(\hat{x})$. Therefore, $\langle c, x - \hat{x} \rangle \leq 0$. Hence, $c \in N_S(\hat{x})$.

$\square$

Recall that an extreme point x^{**} of S is exposed if there exists a hyperplane H such that $\{x^{**}\} = H \cap S$; i.e., if there exists $p \in \mathcal{V}$, $p \neq \mathbf{0}$ such that $\langle p, x^{**} \rangle < \langle p, x \rangle$ for all $x \in S \setminus \{x^{**}\}$.

Lemma 1.7 *Let S be a nonempty convex compact set in $\mathcal{V}$ and let $\hat{x} \in \mathcal{V}$. Let x^{**} be a farthest point in S to $\hat{x}$; i.e., $x^{**} = \arg\max_{x \in S} ||\hat{x} - x||$. Then x^{**} is exposed.*

Proof.   Let $x \in S$. Then $||\hat{x} - x||^2 = ||\hat{x} - x^{**} + x^{**} - x||^2 = ||\hat{x} - x^{**}||^2 + ||x^{**} - x||^2 + 2\langle \hat{x} - x^{**}, x^{**} - x \rangle$. Hence, for any point $x \in S$, we have

$$||\hat{x} - x||^2 - ||\hat{x} - x^{**}||^2 = ||x^{**} - x||^2 + 2\langle \hat{x} - x^{**}, x^{**} - x \rangle \leq 0.$$

Let $p = \hat{x} - x^{**}$. Then

$$2\langle p, x^{**} - x \rangle \leq -||x^{**} - x||^2 \leq 0.$$

Therefore, $\langle p, x^{**} \rangle \leq \langle p, x \rangle$ for all $x \in S$. Moreover, $\langle p, x^{**} \rangle = \langle p, x \rangle$ if and only if $x = x^{**}$. To see this, note that if $x = x^{**}$, then it trivially follows that $\langle p, x^{**} \rangle = \langle p, x \rangle$. On the other hand, if $\langle p, x^{**} \rangle = \langle p, x \rangle$, then $||x^{**} - x||^2 = 0$. Let $H = \{x : \langle p, x \rangle = \langle p, x^{**} \rangle\}$. Then H is a supporting hyperplane to S at x^{**} and $H \cap S = \{x^{**}\}$. As a result, x^{**} is exposed.

$\square$

We should point out that while the closest point of a convex compact set S to $\hat{x}$ is unique, the farthest point of S to $\hat{x}$ need not be so.

Lemma 1.8 ([189]) *Let S be a nonempty convex compact set in $\mathcal{V}$. If there exists a hyperplane $H = \{x : \langle p, x \rangle = p_0\}$ such that $\langle p, \hat{x} \rangle > p_0$ for some $\hat{x} \in S$. Then there exists an exposed point x^{**} in S such that $\langle p, x^{**} \rangle > p_0$.*

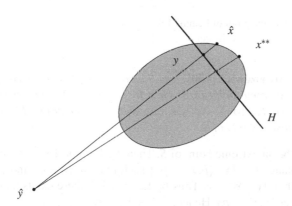

Fig. 1.6 Illustration of the proof of Lemma 1.8

Proof.   Let y be the closest point in H to $\hat{x}$ and wlog assume that $p = \hat{x} - y$. Let $\hat{y} = \hat{x} - \alpha p$ for some large scalar $\alpha > 0$ and let x^{**} be a farthest point in S to $\hat{y}$. Hence, $||\hat{y} - x^{**}||^2 \geq ||\hat{y} - \hat{x}||^2$; i.e.,

$$||\hat{x} - \alpha p - x^{**}||^2 = ||\hat{x} - x^{**}||^2 + \alpha^2 ||p||^2 - 2\alpha \langle p, \hat{x} - x^{**} \rangle \geq \alpha^2 ||p||^2.$$

Therefore,

$$2\alpha\langle p,\hat{x}-x^{**}\rangle \leq ||\hat{x}-x^{**}||^2. \tag{1.12}$$

Now if $\langle p,\hat{x}-x^{**}\rangle > 0$, then by taking α large enough, we get that $2\alpha\langle p,\hat{x}-x^{**}\rangle > ||\hat{x}-x^{**}||^2$, a contradiction to Eq. (1.12). Therefore, $\langle p,\hat{x}-x^{**}\rangle \leq 0$ and hence $\langle p,x^{**}\rangle \geq \langle p,\hat{x}\rangle > p_0$. Furthermore, by Lemma 1.7, x^{**} is exposed and the result follows.

$\square$

Lemma 1.9 *Let S be a nonempty convex compact set in $\mathcal{V}$ and let $\hat{x}$ be an extreme point of S. Then for each $r > 0$, there exists a hyperplane $H = \{x : \langle p,x\rangle = p_0\}$ such that $\langle p,\hat{x}\rangle > p_0$ and $\langle p,x\rangle \leq p_0$ for all $x \in S$ such that $||x - \hat{x}|| \geq r$.*

Proof. Let $S' = \mathrm{conv}(\{x \in S : ||x - \hat{x}|| \geq r\})$. Therefore, by Theorem 1.19, S' is compact. Thus, obviously, $\hat{x} \notin \{x \in S : ||x - \hat{x}|| \geq r\})$. Now assume that $\hat{x} \in S'$. Then $\hat{x}$ is a proper convex combination of points in $\{x \in S : ||x - \hat{x}|| \geq r\})$, a contradiction since $\hat{x}$ is an extreme point of S. Therefore, $\hat{x} \notin S'$ and the result follows from the Strong Separation Theorem.

$\square$

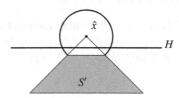

Fig. 1.7 Illustration of the proof of Lemma 1.9

Theorem 1.43 (Straszewicz [180]) *Let S be a nonempty convex compact set in $\mathcal{V}$ and let $\hat{x}$ be an extreme point of S. Then for each $r > 0$, there exists an exposed point x^{**} of S such that $||x^{**} - \hat{x}|| \leq r$; i.e., every extreme point of S is the limit of a sequence of exposed points of S.*

Proof. Let $\hat{x}$ be an extreme point of S. Then, by Lemma 1.9, for each $r > 0$, there exists a hyperplane $H = \{x : \langle p,x\rangle = p_0\}$ such that $\langle p,\hat{x}\rangle > p_0$ and $\langle p,x\rangle \leq p_0$ for all $x \in S$ such that $||x - \hat{x}|| \geq r$. Thus by Lemma 1.8, there exists an exposed point x^{**} of S such that $\langle p,\hat{x}\rangle > p_0$. Hence, $||x^{**} - \hat{x}|| < r$.

$\square$

It should be pointed out that Straszewicz Theorem applies not only to convex compact sets but to closed convex sets as well. Indeed, let S be a closed convex set and let $\hat{x}$ be an extreme point of S. Further, let $B = \{x : ||x - \hat{x}|| \leq \alpha\}$ for some $\alpha > 0$. Then there exists a sequence $\{x^k\}$ of exposed points of $S \cap B$ that converge to $\hat{x}$. Clearly, the tail of this sequence must lie the interior of B and hence the points in the tail are exposed points of S.

Chapter 2
Positive Semidefinite Matrices

Positive semidefinite (PSD) and positive definite (PD) matrices are closely connected with Euclidean distance matrices. Accordingly, they play a central role in this monograph. This chapter reviews some of the basic results concerning these matrices. Among the topics discussed are various characterizations of PSD and PD matrices, theorems of the alternative for the semidefinite cone, the facial structures of the semidefinite cone and spectrahedra, as well as the Borwein–Wolkowicz facial reduction scheme.

2.1 Definitions and Basic Results

Definition 2.1 *An $n \times n$ real symmetric matrix A is said to be* positive definite (PD) *if*
$$x^T A x > 0 \text{ for all } x \in \mathbb{R}^n, x \neq \mathbf{0}.$$

An immediate consequence of this definition is that PD matrices are nonsingular. For suppose that A is singular. Then there exists $x \neq \mathbf{0}$ such that $Ax = \mathbf{0}$. Hence, $x^T A x = 0$ and thus A is not PD.

Definition 2.2 *An $n \times n$ real symmetric matrix A is said to be* positive semidefinite (PSD) *if*
$$x^T A x \geq 0 \text{ for all } x \in \mathbb{R}^n.$$

We use the notation $A \succ \mathbf{0}$ ($A \succeq \mathbf{0}$) to denote that A is a real symmetric PD (PSD) matrix. Another easy consequence of the definition is that if $A \succ \mathbf{0}$ ($A \succeq \mathbf{0}$), then every diagonal entry of A is positive (nonnegative). Similarly, if A is a PD (PSD) block matrix, then every diagonal block of A is PD (PSD).

As always, $\mathscr{S}^n$ denotes the space of $n \times n$ real symmetric matrices endowed with the inner product $\langle A, B \rangle = \text{trace}(AB)$. $A \in \mathscr{S}^n$ is called *negative semidefinite* if $(-A)$ is positive semidefinite, and it is called *negative definite* if $(-A)$ is positive definite.

© Springer Nature Switzerland AG 2018
A.Y. Alfakih, *Euclidean Distance Matrices and Their Applications in Rigidity Theory*, https://doi.org/10.1007/978-3-319-97846-8_2

Let $\mathscr{S}_+^n$ and $\mathscr{S}_{++}^n$ denote, respectively, the sets of $n \times n$ real symmetric PSD and real symmetric PD matrices. Evidently, $\mathscr{S}_+^n$ and $\mathscr{S}_{++}^n$ are pointed convex cones in $\mathscr{S}^n$. Moreover, it is not hard to see that $\dim(\mathscr{S}_+^n) = \dim(\mathscr{S}_{++}^n) = n(n+1)/2$. Let $\hat{A} \in \mathscr{S}^n \backslash \mathscr{S}_+^n$, then $\hat{x}^T \hat{A} \hat{x} < 0$ for some unit vector $\hat{x}$ in $\mathbb{R}^n$. Hence, for any B in $\mathscr{S}^n$, there exists $\varepsilon > 0$ such that $\hat{x}^T \hat{A} \hat{x} - \varepsilon \hat{x}^T B \hat{x} < 0$; i.e., $\hat{A} - \varepsilon B$ is not in $\mathscr{S}_+^n$. Therefore, $\mathscr{S}^n \backslash \mathscr{S}_+^n$ is open and hence $\mathscr{S}_+^n$ is closed. On the other hand, let $A \in \mathscr{S}_{++}^n$ and let x be a unit vector in $\mathbb{R}^n$. Then, for any B in $\mathscr{S}^n$, there exists $\delta > 0$ such that $x^T(A - \delta B)x > 0$. Indeed, by Rayleigh–Ritz Theorem, $x^T A x - \delta x^T B x \geq \lambda_n(A) - \delta \lambda_1(B)$, where $\lambda_n(A) > 0$. Thus, if $\lambda_1(B) \leq 0$, i.e., if B is negative semidefinite, then $A - \delta B \succ 0$ for all $\delta > 0$. Otherwise, $A - \delta B \succ 0$ for $\delta : 0 < \delta < \lambda_n(A)/\lambda_1(B)$. Consequently,

$$\text{int}(\mathscr{S}_+^n) = \mathscr{S}_{++}^n \text{ and } \mathscr{S}_+^n = \text{cl}(\mathscr{S}_{++}^n).$$

Theorem 2.1 *Let A and S be two $n \times n$ real matrices and assume that $A \in \mathscr{S}^n$ and S is nonsingular. Then $SAS^T \succ 0\,(\succeq 0)$ if and only if $A \succ 0\,(\succeq 0)$.*

Proof.   Assume that $A \succ 0$ and let x be any nonzero vector in $\mathbb{R}^n$. Then $x^T SAS^T x > 0$ since $S^T x \neq \mathbf{0}$. Hence, $SAS^T \succ 0$. On the other hand, assume that $SAS^T \succ 0$ and let x be any nonzero vector in $\mathbb{R}^n$. Let $x = S^T y$, then $y \neq \mathbf{0}$. Therefore, $x^T A x = y^T SAS^T y > 0$. Hence, $A \succ 0$. The proof for the semidefinite case is similar. $\qquad\square$

Note that Theorem 2.1 is a special case of *Sylvester law of inertia*. The lemma that follows, known as Schur complement, plays an important role in the theory of semidefinite matrices and will be used repeatedly throughout the monograph.

Theorem 2.2 (Schur Complement) *Let $M \in \mathscr{S}^n$ and assume that M is partitioned as $M = \begin{bmatrix} A & B \\ B^T & C \end{bmatrix}$. Further, assume that $A \succ 0$. Then $M \succ 0\,(\succeq 0)$ if and only if*

$$C - B^T A^{-1} B \succ 0\,(\succeq 0).$$

The matrix $C - B^T A^{-1} B$ is called the Schur complement of A.

Proof.   Let $S = \begin{bmatrix} I & \mathbf{0} \\ -B^T A^{-1} & I \end{bmatrix}$. Then $SMS^T = \begin{bmatrix} A & \mathbf{0} \\ \mathbf{0} & C - B^T A^{-1} B \end{bmatrix}$. Thus, the result follows from Theorem 2.1 since S is obviously nonsingular. $\qquad\square$

It is worth pointing out that S in the proof of Schur complement is an elementary matrix. Thus, multiplying M by S from the left is equivalent to a block Gaussian elimination step. Furthermore, it is evident that $\det(M) = \det(A)\det(C - B^T A^{-1} B)$.

2.2 Characterizations

PD and PSD matrices can be characterized in terms of eigenvalues, principal minors, and Gram matrices. We begin first with the eigenvalue characterization.

2.2.1 Eigenvalues

Theorem 2.3 *Let $A \in \mathscr{S}^n$. Then A is positive definite if and only if all of its eigenvalues are positive.*

Proof. Assume that the eigenvalues of A are all positive. Let $A = Q\Lambda Q^T$ be the spectral decomposition of A and let x be any nonzero vector in $\mathbb{R}^n$. Further, let $y = Q^T x$. Thus $y \neq 0$. Therefore, $x^T A x = y^T \Lambda y = \sum_{i=1}^n \lambda_i (y_i)^2 > 0$. Hence, A is PD.

To prove the other direction, assume that one eigenvalue of A, say λ_1, is ≤ 0. Let x^1 be an eigenvector of A corresponding to λ_1. Then $(x^1)^T A x^1 = \lambda_1 (x^1)^T x^1 \leq 0$ and hence A is not PD.

$\square$

Corollary 2.1 *Let $A \in \mathscr{S}^n$. Then A is positive definite if and only if A^{-1} is positive definite.*

Similarly, we have

Theorem 2.4 *Let $A \in \mathscr{S}^n$. Then A is positive semidefinite if and only if all of its eigenvalues are nonnegative.*

Therefore, PD matrices are precisely the nonsingular PSD matrices. Next, we turn to the principal minor characterization.

2.2.2 Principal Minors

Theorem 2.5 *Let $A \subset \mathscr{S}^n$. Then A is positive definite if and only if all of its leading principal minors are positive.*

Proof. Assume that the kth leading principal minor of A is nonpositive and assume that A is partitioned as $A = \begin{bmatrix} A_k & B \\ B^T & C \end{bmatrix}$, where A_k is the kth leading principal submatrix of A. Thus $\det(A_k) \leq 0$. Hence, A_k has an eigenpair $(\hat{\lambda}, \hat{x})$ where $\hat{\lambda} \leq 0$. Let $x^T = [\hat{x}^T \ \ 0] \in \mathbb{R}^n$. Then $x^T A x = \hat{x}^T A_k \hat{x} \leq 0$ and hence A is not PD.

To prove the other direction, assume that the leading principal minors of A are all positive. We use induction on n to prove that A is PD. The result is obvious for $n = 1$. Thus, assume that the result is true for $n = k$. Let A_{k+1} be the $(k+1)$th leading principal submatrix of A, and let A_{k+1} be partitioned as $A_{k+1} = \begin{bmatrix} A_k & b \\ b^T & c \end{bmatrix}$, where A_k is the kth leading principal submatrix of A. Thus, the leading principal minors of A_k are all positive and hence, by the induction hypothesis, A_k is PD. Moreover, $\det(A_{k+1}) = \det(A_k)\det(c - b^T A_k^{-1} b) > 0$. Therefore, $c - b^T A_k^{-1} b > 0$ and hence, by Schur complement, A_{k+1} is PD.

$\square$

At this point, it would be tempting to conjecture that A is PSD if and only if all leading principal minors of A are nonnegative. Unfortunately, this is false [155]. Let

$A = \begin{bmatrix} 0 & 0 \\ 0 & -1 \end{bmatrix}$. Then, the two leading principal minors of A are nonnegative while, obviously, A is not PSD. In fact, in the semidefinite case, all principal minors, not just the leading ones, need to be nonnegative.

Theorem 2.6 *Let $A \in \mathscr{S}^n$. Then A is positive semidefinite if and only if all of its principal minors are nonnegative.*

Proof. Let A be partitioned as $A = \begin{bmatrix} A_k & B \\ B^T & C \end{bmatrix}$. Assume that one principal minor of A is negative and wlog assume that $\det(C) < 0$. Hence, C has an eigenpair $(\hat{\lambda}, \hat{x})$ where $\hat{\lambda} < 0$. Let $x^T = [\mathbf{0} \ \hat{x}^T] \in \mathbb{R}^n$. Then $x^T A x = \hat{x}^T C \hat{x} < 0$. Therefore, A is not PSD.

To prove the other direction, assume that all principal minors of A are nonnegative. Let $\chi_A(\lambda) = c_0 + c_1 \lambda + \cdots + c_k \lambda^k + \cdots + (-1)^n \lambda^n$ be the characteristic polynomial of A. Then by Theorem 1.8, for $k = 0, \ldots, n-1$, $c_k = (-1)^k \alpha_k$ where $\alpha_k > 0$. Now assume, to the contrary, that A has a negative eigenvalue, say λ_1. Then

$$\chi_A(\lambda_1) = \alpha_0 - \alpha_1 \lambda_1 + \alpha_2 (\lambda_1)^2 - \alpha_3 (\lambda_1)^3 + \cdots + (-\lambda_1)^n > 0$$

since each term is positive, a contradiction. Hence, all eigenvalues of A are nonnegative and thus A is PSD.

□

Finally, we turn to the Gram matrix characterization.

2.2.3 Gram Matrices

Let $p^1, \ldots, p^n$ be a point configuration in $\mathbb{R}^k$, and assume that these points are not contained in a proper hyperplane in $\mathbb{R}^k$. Then the $n \times n$ symmetric matrix $A = (a_{ij})$, where

$$a_{ij} = (p^i)^T p^j$$

is called the *Gram matrix* of this configuration. If P is the $n \times k$ matrix whose ith row is equal to $(p^i)^T$, then P has full column rank. Furthermore, $A = PP^T$ and hence $\text{rank}(A) = k$.

Theorem 2.7 *Let $A \in \mathscr{S}^n$. Then A is positive definite if and only if $A = PP^T$, where P is nonsingular.*

Proof. Assume that $A = PP^T$, where P is nonsingular. Then $x^T A x = ||P^T x||^2 > 0$ for all nonzero $x \in \mathbb{R}^n$ and hence A is PD. On the other hand, assume that A is PD and let $A = Q \Lambda Q^T$ be the spectral decomposition of A. Let $P = Q \Lambda^{1/2}$, where $(\Lambda^{1/2})_{ii} = \sqrt{\Lambda_{ii}}$. Then P is nonsingular and $A = PP^T$.

□

Similarly, we have

Corollary 2.2 *Let $A \in \mathscr{S}^n$. Then A is positive semidefinite of rank k if and only if $A = PP^T$, where P is an $n \times k$ matrix with full column rank.*

2.3 Miscellaneous Properties

Further properties of PSD matrices are given in this section. Recall that $\mathscr{S}^n_+$ denotes the set of $n \times n$ real symmetric PSD matrices.

Proposition 2.1 *Let $A \in \mathscr{S}^n_+$ and assume that $a_{ii} = 0$ for some i. Then all entries of A in the ith row (and consequently the ith column) are zeros.*

Proof.   Assume, to the contrary, that $a_{ik} \neq 0$ for some k. Then the principal minor of A induced by rows i and k is negative, a contradiction.

$\square$

An immediate consequence of Proposition 2.1 is that if $A \in \mathscr{S}^n_+$, then trace$(A) = 0$ if and only if $A = \mathbf{0}$.

Let $A \in \mathscr{S}^n_+$, then A has a unique positive semidefinite square root, denoted by $A^{1/2}$, such that $A = A^{1/2}A^{1/2}$. To this end, let $A = Q\Lambda Q^T$ be the spectral decomposition of A. Further let $\Lambda^{1/2}$ denote the positive square root of Λ. Then $A^{1/2} = Q\Lambda^{1/2}Q^T$. Notice that rank$(A^{1/2}) = $ rank(A).

Proposition 2.2 *Let A and B be in $\mathscr{S}^n_+$. Then trace$(AB) \geq 0$. Moreover, trace$(AB) = 0$ if and only if $AB = \mathbf{0}$.*

Proof.   Clearly, trace$(AB) = $ trace$(B^{1/2}A^{1/2}A^{1/2}B^{1/2}) = ||A^{1/2}B^{1/2}||^2_F \geq 0$. Now assume that trace$(AB) = 0$. Therefore, $A^{1/2}B^{1/2} = \mathbf{0}$ and hence $AB = \mathbf{0}$. The other direction is trivial.

$\square$

Proposition 2.3 *Let $A - PP^T$. Then null$(A) = $ null(P^T).*

Proof.   The fact that null$(P^T) \subseteq $ null(A) is obvious. Now let $x \in $ null(A), then $x^T A x = x^T PP^T x = ||P^T x||^2_F = 0$. Hence, $P^T x = \mathbf{0}$ and thus null$(A) \subseteq $ null(P^T).

$\square$

Let $A = PP^T$. Then the following two facts are immediate consequences of Proposition 2.3. First, rank$(A) = $ rank(P). Second, $x^T A x = 0$ iff $x \in $ null(A).

Proposition 2.4 *Let A and B be in $\mathscr{S}^n_+$ and let rank$(A) + $ rank$(B) \geq n+1$. Then trace$(AB) > 0$.*

Proof.   Assume that rank$(B) = r$ and let $B = PP^T$, where P is $n \times r$. Thus col$(P) \nsubseteq $ null(A) since dim null$(A) \leq r - 1$. Therefore, trace$(AB) = $ trace$(P^T AP) > 0$ since $P^T AP$ is a nonzero PSD matrix.

$\square$

Proposition 2.5 *Let A and B be in $\mathscr{S}_+^n$ and let $C = \lambda A + (1 - \lambda)B$ for some $\lambda : 0 < \lambda < 1$. Then*

$$null(C) = null(A) \cap null(B).$$

The fact that $(null(A) \cap null(B)) \subseteq null(C)$ is obvious. To prove the other inclusion, let $x \in null(C)$. Then $x^T C x = \lambda x^T A x + (1 - \lambda)x^T B x = 0$. Thus, $\lambda x^T A x = (1 - \lambda)x^T B x = 0$. But $\lambda > 0$ and $1 - \lambda > 0$. Therefore, $x^T A x = x^T B x = 0$ and hence $x \in (null(A) \cap null(B))$.

$\square$

Proposition 2.6 *Let $M \in \mathscr{S}^n$. Assume that M is partitioned as $M = \begin{bmatrix} A & B \\ B^T & C \end{bmatrix}$ and that A is nonsingular. If M is positive semidefinite, then*

$$null(C) \subseteq null(B).$$

Proof.   Clearly, $C \succeq 0$ and A is PD since it is nonsingular. Thus $B^T A^{-1} B \succeq 0$. By Schur's complement, $C - B^T A^{-1} B \succeq 0$. Let $x \in null(C)$. Then $x^T (C - B^T A^{-1} B)x = -x^T B^T A^{-1} B x \geq 0$. Therefore, $x^T B^T A^{-1} B x = 0$. Consequently, $Bx = 0$ since A^{-1} is PD, and thus $x \in null(B)$.

$\square$

Proposition 2.7 *Let $M(t) = \begin{bmatrix} A' + tA & tB \\ tB^T & tC \end{bmatrix}$, where $A' \succ 0$, $C \succeq 0, C \neq 0$ and t is a scalar. If $null(C) \subseteq null(B)$, then there exists $\hat{t} \neq 0$ such that $M(\hat{t})$ is positive semidefinite.*

Proof.   Let W and U be the matrices whose columns form orthonormal bases of $col(C)$ and $null(C)$, respectively. Thus, $BU = 0$ and $W^T C W = \Lambda$, where Λ is the diagonal matrix consisting of the positive eigenvalues of C. Moreover, the matrix $Q = \begin{bmatrix} I & 0 & 0 \\ 0 & W & U \end{bmatrix}$ is orthogonal. On the other hand, $M(t)$ is PSD if and only if $Q^T M(t) Q$ is PSD. Therefore, $M(t)$ is PSD if and only if

$$\begin{bmatrix} A' + tA & tBW \\ tW^T B^T & t\Lambda \end{bmatrix} \succeq 0. \tag{2.1}$$

Now by Schur complement, (2.1) holds iff $A' + t(A - BW\Lambda^{-1}W^T B^T) \succeq 0$. Moreover, since $A' \succ 0$, it follows that $A' + \hat{t}(A - BW\Lambda^{-1}W^T B^T) \succeq 0$ for some $\hat{t} \neq 0$. Thus the result holds.

$\square$

Let $f : \mathbb{R} \to \mathbb{R}$ and let $f[A] = (f(a_{ij}))$ denote the matrix obtained from matrix A by applying f to A entrywise. The following theorem is an immediate consequence of Schur Product Theorem.

Theorem 2.8 *Let $A = (a_{ij})$ be a real symmetric positive semidefinite matrix, then $exp[A]$ is also symmetric positive semidefinite.*

Proof.   The ijth entry of exp$[A]$ is given by

$$e^{a_{ij}} = 1 + a_{ij} + \frac{a_{ij}^2}{2!} + \frac{a_{ij}^3}{3!} + \cdots.$$

Thus,

$$\exp[A] = E + A + \frac{A \circ A}{2!} + \frac{A \circ A \circ A}{3!} + \cdots,$$

where E denotes the matrix of all 1's. But, by Schur Product Theorem, each term in this sum is PSD. Hence, exp$[A]$ is PSD.

$\square$

2.4 Theorems of the Alternative

Theorems of the alternative, the most famous of which is the celebrated Farkas lemma, are an indispensable tool in optimization theory. These theorems assert that exactly one of two given systems of linear inequalities or linear matrix inequalities has a solution. Thus, they underpin the duality theory of linear programming and semidefinite programming. Moreover, the theorems of the alternative are intimately connected with the separation theorems of convex sets. In this section, several theorems of the alternative for the semidefinite cone are presented.

Recall that the polar of $\mathscr{S}_+^n$ is

$$(\mathscr{S}_+^n)^\circ = \{Y \in \mathscr{S}^n : \text{trace}(YX) \leq 0 \text{ for all } X \succeq \mathbf{0}\},$$

and that the dual of cone K is $-K^\circ$.

Theorem 2.9 *The cone of symmetric positive semidefinite matrices is self-dual, i.e.,*

$$(\mathscr{S}_{++}^n)^\circ = (\mathscr{S}_+^n)^\circ = -\mathscr{S}_+^n.$$

Proof.   The first equality follows by applying Lemma 1.6 to the cone $\mathscr{S}_{++}^n$. To prove the second equality, assume that $Y \in (\mathscr{S}_+^n)^\circ$. Then trace$(YX) \leq 0$ for all $X \in \mathscr{S}_+^n$. Let $X = xx^T$, where x is any vector in $\mathbb{R}^n$. Thus trace$(YX) = x^T Y x \leq 0$. Therefore, $-Y \succeq \mathbf{0}$ and hence $(\mathscr{S}_+^n)^\circ \subseteq (-\mathscr{S}_+^n)$.

To prove the other inclusion, assume that $(-Y) \succeq \mathbf{0}$. Then by Proposition 2.2, trace$(YX) \leq 0$ for all $X \in \mathscr{S}_+^n$. Thus, $Y \in (\mathscr{S}_+^n)^\circ$ and hence $(-\mathscr{S}_+^n) \subseteq (\mathscr{S}_+^n)^\circ$.

$\square$

Next, we turn to the theorems of the alternative.

Theorem 2.10 (Homogeneous) *Let $A^1, \ldots, A^m$ be given matrices in $\mathscr{S}^n$. Then exactly one of the following two statements holds:*

(i) There exists $x \in \mathbb{R}^m$ such that $x_1 A^1 + \cdots + x_m A^m \succ \mathbf{0}$.
(ii) There exists $Y \succeq \mathbf{0}$, $Y \neq \mathbf{0}$ such that trace$(YA^i) = 0$ for $i = 1, \ldots, m$.

Proof.   Assume that both statements hold. Then $\text{trace}(Y\sum_{i=1}^{m}x_iA^i) > 0$. On the other hand, $\text{trace}(Y\sum_{i=1}^{m}x_iA^i) = \sum_{i=1}^{m}x_i\,\text{trace}(YA^i) = 0$, a contradiction.

Now assume that Statement (i) does not hold and let $\mathscr{L} = \text{span}\,\{A^1,\ldots,A^m\}$. Then $\mathscr{L}\cap \text{int}(\mathscr{S}_+^n) = \emptyset$. Therefore, by the Separation Theorem and Corollary 1.4, there exists $Y \in \mathscr{S}^n, Y \neq \mathbf{0}$ such that $\text{trace}(AY) = 0$ for all $A \in \mathscr{L}$ and $\text{trace}(YB) \geq 0$ for all $B \in \mathscr{S}_+^n$. Therefore, $\text{trace}(YA^i) = 0$ for all $i = 1,\ldots,m$, and by Theorem 2.9, Y belongs to $-(\mathscr{S}_+^n)^\circ = \mathscr{S}_+^n$.

$\square$

Theorem 2.11 (Nonhomogeneous) *Let $A^0,A^1,\ldots,A^m$ be given matrices in $\mathscr{S}^n$. Then exactly one of the following two statements holds:*

  (i) *There exists $x \in \mathbb{R}^m$ such that $\mathscr{A}(x) := A^0 + x_1A^1 + \cdots + x_mA^m \succ \mathbf{0}$.*
  (ii) *There exists $Y \succeq \mathbf{0}$, $Y \neq \mathbf{0}$ such that $\text{trace}(YA^0) \leq 0$ and $\text{trace}(YA^i) = 0$ for $i = 1,\ldots,m$.*

Proof.   Assume that both statements hold. Then $\text{trace}(Y\mathscr{A}(x)) > 0$. On the other hand, $\text{trace}(Y\mathscr{A}(x)) = \text{trace}(YA^0) + \sum_{i=1}^{k}x_i\text{trace}(YA^i) \leq 0$, a contradiction.

Now assume that Statement (i) does not hold and let $S_1 = \{\mathscr{A}(x) : x \in \mathbb{R}^m\}$. Then $S_1 \cap \text{int}(\mathscr{S}_+^n) = \emptyset$. Therefore, by the Separation Theorem, there exist $Y \in \mathscr{S}^n, Y \neq \mathbf{0}$, and scalar p_0 such that $\text{trace}(YA) \leq p_0$ for all $A \in S_1$ and $\text{trace}(YB) \geq p_0$ for all $B \in \mathscr{S}_+^n$. Hence, by Corollary 1.4, it follows that $\text{trace}(YA^0) \leq 0$ and $\text{trace}(YA^i) = 0$ for all $i = 1,\ldots,m$. It also follows that $\text{trace}(YB) \geq 0$ for all $B \succeq \mathbf{0}$ and hence $Y \succeq \mathbf{0}$.

$\square$

The case where $A^0 \succeq \mathbf{0}$ is of particular interest to us. Theorem 2.11, in this case, can be strengthened to the following easily proved corollary.

Corollary 2.3 *Let $A^0,A^1,\ldots,A^m$ be given matrices in $\mathscr{S}^n$ and let $\mathscr{A}(x) := A^0 + x_1A^1 + \cdots + x_mA^m$. Assume that $A^0 \succeq 0$. Then exactly one of the following two statements holds:*

  (i) *There exists x such that $\mathscr{A}(x) \succ \mathbf{0}$.*
  (ii) *There exists $Y \succeq \mathbf{0}, Y \neq \mathbf{0}$, such that $\text{trace}(YA^0) = 0$ and $\text{trace}(YA^i) = 0$ for $i = 1,\ldots,m$.*

Proof.   The result follows since $0 \leq \text{trace}(A^0 Y) \leq 0$.

$\square$

Since any affine set can also be represented as the intersection of hyperplanes, we obtain equivalent forms of Theorem 2.10, Theorem 2.11, and Corollary 2.3. To this end, let $S_1 = \{X \in \mathscr{S}^n : \text{trace}(A^iX) = b_i \text{ for } i = 1,\ldots,m\}$ and assume that $\hat{X} \in S_1$. Then $S_1 = \{X \in \mathscr{S}^n : \text{trace}(A^i(X - \hat{X})) = 0 \text{ for } i = 1,\ldots,m\}$. Hence, $S_1 = \hat{X} + x_1B^1 + \cdots + x_kB^k$ where $\{B^1,\ldots,B^k\}$ is a basis of the orthogonal complement of $\text{span}\,\{A^1,\ldots,A^m\}$ in $\mathscr{S}^n$. Now let $Y \succeq \mathbf{0}$, $Y \neq \mathbf{0}$ such that $\text{trace}(Y\hat{X}) \leq 0$ and $\text{trace}(YB^i) = 0$ for $i = 1,\ldots,k$. Then $Y = y_1A^1 + \cdots + y_mA^m$ for some scalars $y_1,\ldots,y_m$, and $\text{trace}(Y\hat{X}) = \sum_{i=1}^{m}y_i\,\text{trace}(A^i\hat{X}) = \sum_{i=1}^{m}y_ib_i$. Consequently, we have the following results.

Corollary 2.4 (Homogeneous) *Let $A^1,\ldots,A^m$ be given matrices in $\mathscr{S}^n$. Then exactly one of the following two statements holds:*

(i) *There exists $X \succ 0$ such that $trace(A^iX) = 0$ for $i = 1,\ldots,m$.*
(ii) *There exists $y \in \mathbb{R}^m$ such that $y_1A^1 + \cdots + y_mA^m \succeq 0, \neq 0$.*

Theorem 2.12 (Nonhomogeneous) *Let $A^1,\ldots,A^m$ be given matrices in $\mathscr{S}^n$ and let $b \in \mathbb{R}^m$. Assume that the set $\{X \in \mathscr{S}^n : trace(A^iX) = b_i \text{ for } i = 1,\ldots,m\}$ is not empty. Then exactly one of the following two statements holds:*

(i) *There exists $X \succ 0$ such that $trace(A^iX) = b_i$ for $i = 1,\ldots,m$.*
(ii) *There exists $y \in \mathbb{R}^m$ such that $y_1A^1 + \cdots + y_mA^m \succeq 0, \neq 0$ and $b^T y \leq 0$.*

Corollary 2.5 *Let $A^1,\ldots,A^m$ be given matrices in $\mathscr{S}^n$ and let $b \in \mathbb{R}^m$. Further, let $\mathscr{F} = \{X \in \mathscr{S}^n_+ : trace(A^iX) = b_i \text{ for } i = 1,\ldots,m\}$ and assume that $\mathscr{F} \neq \emptyset$. Then exactly one of the following two statements holds:*

(i) *There exists $X \succ 0$ such that $trace(A^iX) = b_i$ for $i = 1,\ldots,m$.*
(ii) *There exists $y \in \mathbb{R}^m$ such that $\Omega(y) = y_1A^1 + \cdots + y_mA^m \succeq 0, \neq 0$ and $b^T y = 0$.*

Proof. Let $\hat{X} \in \mathscr{F}$ and assume that Statement (ii) of Theorem 2.12 holds. Then $0 \geq b^T y = \sum_{i=1}^m y_i \, trace(A^i\hat{X}) = trace(\hat{X}\Omega(y)) \geq 0$ since $\hat{X} \succeq 0$. Therefore, $b^T y = 0$. □

The following simple result will be needed in the sequel.

Corollary 2.6 *Let $\mathscr{F}$ and $\Omega(y)$ be as in Corollary 2.5 and assume that Statement (ii) of Corollary 2.5 holds. Then $trace(\Omega(y)X) = 0$ for all $X \in \mathscr{F}$.*

Proof. Let $X \in \mathscr{F}$, then $trace(\Omega(y)X) = \sum_{i=1}^m y_i \, trace(A^iX) = b^T y = 0$. □

A remark is in order here. The affine set $\{x \in \mathbb{R}^m : A^0 + x_1A^1 + \cdots + x_mA^m\}$ is always nonempty, while the affine set $\{X \in \mathscr{S}^n : trace(A^iX) = b_i \text{ for } i = 1,\ldots,m\}$ may or may not be empty. Hence, the assumption that this set is not empty is made in Theorem 2.12.

2.5 Semidefinite Programming (SDP)

In this section, we present a few basic results concerning semidefinite programming (SDP) that will be needed in the monograph. Two excellent references on this subject are [198, 137]. As noted earlier, theorems of the alternative are the main ingredient in the proof of the strong duality theorem of SDP. Let $A^0, A^1,\ldots,A^m$ be given linearly independent matrices in $\mathscr{S}^n$ and let b be a given vector in $\mathbb{R}^m$. Then a primal SDP problem is of the form

$$(\text{P}) \quad \inf \quad f(x) = b^T x$$
$$\text{subject to } A^0 + x_1A^1 + \cdots + x_mA^m \succeq 0.$$

The dual problem of (P) is

$$
\begin{aligned}
\text{(D)} \quad & \sup \quad v(Y) = -\mathrm{trace}(A^0 Y) \\
& \text{subject to} \quad \mathrm{trace}(A^i Y) = b_i \quad \text{for } i = 1, \dots, m. \\
& \qquad\qquad\quad Y \succeq 0
\end{aligned}
$$

It should be pointed out that the dual of the dual problem is the primal problem. Thus, SDP problems come in dual pairs. Any solution that satisfies the constraints is called a *feasible solution*. The set of all feasible solutions is called the *feasible region*. Problem (P) satisfies *Slater's condition* if there exists $\hat{x} \in \mathbb{R}^m$ such that $A^0 + \hat{x}_1 A^1 + \cdots + \hat{x}_m A^m \succ 0$. Likewise, Problem (D) satisfies Slater's condition if there exists $\hat{Y} \succ 0$ which satisfies the constraints $\mathrm{trace}(A^i \hat{Y}) = b_i$ for $i = 1, \dots, m$. In other words, an SDP problem satisfies Slater's condition iff its feasible region intersects the interior of $\mathscr{S}_+^n$.

The following are the basic duality theorems of SDP.

Theorem 2.13 (Weak Duality) *Let (P) and (D) be a primal–dual pair of SDP problems as above. Then for any primal feasible solution x and any dual feasible solution Y, we have*

$$
f(x) \geq v(Y).
$$

Proof. $f(x) - v(Y) = \sum_{i=1}^m b_i x_i + \mathrm{trace}(A^0 Y) = \sum_{i=1}^m \mathrm{trace}(A^i Y) x_i + \mathrm{trace}(A^0 Y) = \mathrm{trace}((\sum_{i=1}^m x_i A^i + A^0) Y) \geq 0.$

□

Theorem 2.14 (Strong Duality) *Let (P) and (D) be a primal–dual pair of SDP problems as above. Assume that (P) and (D) are feasible and (P) satisfies Slater's condition. Then*

$$
f^* = v^*,
$$

where f^* and v^* are the respective optimal values. Moreover, the dual optimal value is attained.

Proof. By the definition of an optimal solution, there does not exist an x such that $A^0 + x_1 A^1 + \cdots + x_m A^m \succeq 0$ and $b^T x < f^*$. Therefore, the system

$$
\begin{bmatrix} f^* & 0 \\ 0 & A^0 \end{bmatrix} + x_1 \begin{bmatrix} -b_1 & 0 \\ 0 & A^1 \end{bmatrix} + \cdots + x_m \begin{bmatrix} -b_m & 0 \\ 0 & A^m \end{bmatrix} \succ 0.
$$

is infeasible. Thus, by Theorem 2.11, there exist $y \geq 0$ and $Y \succeq 0$, not both of which are zero, such that $-y b_i + \mathrm{trace}(A^i Y) = 0$ for all $i = 1, \dots, m$; and $y f^* + \mathrm{trace}(A^0 Y) \leq 0$. Now $y > 0$ since otherwise, Theorem 2.11 would imply that (P) does not satisfy Slater's condition. Therefore, Y/y is a feasible solution of (D) with objective value $v(Y/y) = -\mathrm{trace}(A^0 Y/y) \geq f^* \geq v^*$. Hence, $v(Y/y) = f^* = v^*$.

□

We remark here that in the absence of Slater's condition, an SDP problem may have a finite duality gap and/or the optimal value may not be attained.

Assume that Y is a feasible solution of (D) with rank s, and let W' and U' be the matrices whose columns form orthonormal bases of $\mathrm{col}(Y)$ and $\mathrm{null}(Y)$, respectively. Further [26, 154], let

$$\mathscr{L} = \text{span}\{A^1, \dots, A^m\} \tag{2.2}$$

$$\mathscr{T}_Y = \left\{ C \in \mathscr{S}^n : C = [W' \ U'] \begin{bmatrix} \Phi_1 & \Phi_2 \\ \Phi_2^T & 0 \end{bmatrix} \begin{bmatrix} W'^T \\ U'^T \end{bmatrix} \right\}, \tag{2.3}$$

where Φ_1 is a symmetric matrix of order s. $\mathscr{T}_Y$ is called the *tangent space* at Y to the set of symmetric matrices of rank s. Y is said to be *nondegenerate* if

$$\mathscr{T}_Y + \mathscr{L}^\perp = \mathscr{S}^n. \tag{2.4}$$

Otherwise, Y is said to be *degenerate*. Note that Eq. (2.4) is equivalent to

$$\mathscr{T}_Y^\perp \cap \mathscr{L} = \{0\}, \tag{2.5}$$

where $\mathscr{T}_Y^\perp$, the orthogonal complement of $\mathscr{T}_Y$, is given by

$$\mathscr{T}_Y^\perp = \{C \in \mathscr{S}^n : YC = 0\} = \{C \in \mathscr{S}^n : C = U'\Phi U'^T\},$$

where Φ is a symmetric matrix of order $n - s$.

Theorem 2.15 (Alizadeh et al. [26]) *Let x and Y be optimal solutions of (P) and (D), respectively. If Y is nondegenerate, then x is unique.*

Proof. Since x is an optimal solution of (P), it follows that $\mathscr{X}(x) = A^0 + \sum_i x_i A^i \succeq 0$ and $b^T x = -\text{trace}(A^0 Y)$, i.e., $\text{trace}(\mathscr{X}(x)Y) = 0$. Hence, $\mathscr{X}(x)Y = 0$ since both $\mathscr{X}(x)$ and Y are PSD. Now assume that x' is an optimal solution of (P). Then $\mathscr{X}(x')Y = 0$ and hence $(\mathscr{X}(x) - \mathscr{X}(x'))Y = 0$. Consequently, $\sum_i (x_i - x_i')A^i$ lies in $\mathscr{T}_Y^\perp$. Therefore, (2.5) implies that $\sum_i (x_i' - x_i)A^i = 0$ and hence $x' = x$ since $A^1, \dots, A^m$ are linearly independent. $\qquad\square$

2.6 The Facial Structure of $\mathscr{S}_+^n$

It is well known [165, 35, 34, 36, 108, 154] that the faces of the positive semidefinite cone $\mathscr{S}_+^n$ can be characterized either in terms of the null space or in terms of the column space.

Lemma 2.1 *Let A and B be two matrices in $\mathscr{S}_+^n$ and let F be a face of $\mathscr{S}_+^n$ containing A. If $\text{null}(A) \subseteq \text{null}(B)$, then $B \in F$.*

Proof. Assume that $\text{null}(A) \subseteq \text{null}(B)$. Now if $A = 0$, then $\text{null}(A) = \mathbb{R}^n$ and thus $B = 0$. Hence, the result follows trivially. Therefore, assume that $A \neq 0$ and let U be the matrix whose columns form an orthonormal basis of $\text{null}(A)$. Further, let $A = W\Lambda W^T$ be the spectral decomposition of A, where Λ is the diagonal matrix consisting of the positive eigenvalues of A. Thus the matrix $Q = [W \ U]$ is orthogonal. Let t be a scalar. Then $A - t(B - A) \succeq 0$ if and only if $Q^T(A - t(B - A))Q \succeq 0$.

But

$$Q^T(A - t(B - A))Q = \begin{bmatrix} \Lambda - tW^T(B-A)W & 0 \\ 0 & 0 \end{bmatrix}.$$

Hence, there exists $\hat{t} > 0$ such that $A - \hat{t}(B - A) \succeq 0$. Let $C = A - \hat{t}(B - A)$. Then

$$A = \frac{1}{1+\hat{t}}C + \frac{\hat{t}}{1+\hat{t}}B.$$

Therefore, $B \in F$.

□

Example 2.1 Let $A = \begin{bmatrix} 1 & 1 & 0 \\ 1 & 0 & 1 \\ 1 & -1 & 0 \end{bmatrix}$, $B = \begin{bmatrix} 1 & 1 & 1 \\ 1 & 0 & 0 \\ 1 & -1 & -1 \end{bmatrix}$ and $C = \begin{bmatrix} 1 & 1 & 1 \\ 1 & 1 & 1 \\ 1 & 1 & 1 \end{bmatrix}$. Then

$null(A) = \{0\}$, $null(B) = col(\begin{bmatrix} 0 \\ 1 \\ -1 \end{bmatrix})$ and $null(C) = col(\begin{bmatrix} 1 & 0 \\ 0 & 1 \\ -1 & -1 \end{bmatrix})$. Hence,

$null(A) \subset null(B) \subset null(C)$. Accordingly, any face of $\mathscr{S}_+^n$ containing A must contain B and C, and any face of $\mathscr{S}_+^n$ containing B must contain C.

Recall that face$(A, \mathscr{S}_+^n)$ denotes the minimal face of $\mathscr{S}_+^n$ containing A. Also, recall that, by Theorem 1.28, A lies in the relative interior of face$(A, \mathscr{S}_+^n)$.

Theorem 2.16 (Barker and Carlson [35]) Let $A \in \mathscr{S}_+^n$. Then

$$face(A, \mathscr{S}_+^n) = \{B \in \mathscr{S}_+^n : null(A) \subseteq null(B)\}. \tag{2.6}$$
$$= \{B \in \mathscr{S}_+^n : col(B) \subseteq col(A)\}. \tag{2.7}$$

Proof.   Let $S = \{B \in \mathscr{S}_+^n : null(A) \subseteq null(B)\}$. Then, by Lemma 2.1, $S \subseteq$ face$(A, \mathscr{S}_+^n)$. Now if face$(A, \mathscr{S}_+^n) = \{A\}$, then face$(A, \mathscr{S}_+^n) \subseteq S$ and we are done. Therefore, let $B \in$ face$(A, \mathscr{S}_+^n)$ where $B \neq A$. Since A lies in relint(face$(A, \mathscr{S}_+^n)$), there exist $C \in$ face$(A, \mathscr{S}_+^n)$ and $0 < \lambda < 1$, such that $A = \lambda B + (1 - \lambda)C$. Hence, by Proposition 2.5, $null(A) \subseteq null(B)$. Thus $B \in S$ and therefore face$(A, \mathscr{S}_+^n) \subseteq S$.

□

It is an easy observation that face$(I, \mathscr{S}_+^n) = \mathscr{S}_+^n$ and face$(0, \mathscr{S}_+^n) = 0$. Moreover, the following corollaries are immediate.

Corollary 2.7 Let $A \in \mathscr{S}_+^n$. Then

$$relint(face(A, \mathscr{S}_+^n)) = \{B \in \mathscr{S}_+^n : null(A) = null(B)\},$$
$$= \{B \in \mathscr{S}_+^n : col(B) = col(A)\}.$$

Corollary 2.8 Let A and B be in $\mathscr{S}_+^n$. Then

$$face(A, \mathscr{S}_+^n) \subseteq face(B, \mathscr{S}_+^n) \quad if \ and \ only \ if \ col(A) \subseteq col(B),$$
$$face(A, \mathscr{S}_+^n) = face(B, \mathscr{S}_+^n) \quad if \ and \ only \ if \ col(A) = col(B).$$

Corollary 2.9 *Let $A \in \mathscr{S}_+^n$ and let rank$(A) = r$, $r \leq n - 1$. Further, let W be the $n \times r$ matrix whose columns form an orthonormal basis of col(A). Then*

$$face(A, \mathscr{S}_+^n) = \{W\Phi W^T \text{ for some } \Phi \in \mathscr{S}_+^r\},$$
$$relint(face(A, \mathscr{S}_+^n)) = \{W\Phi W^T \text{ for some } \Phi \in \mathscr{S}_{++}^r\}.$$

Corollary 2.10 *Let F be a face of $\mathscr{S}_+^n$. Then*

$$F = \{B \in \mathscr{S}_+^n : \mathscr{L} \subseteq \text{null}(B) \text{ for some subspace } \mathscr{L} \text{ of } \mathbb{R}^n\}.$$

Note that Corollary 2.9 follows in part from Proposition 1.1. As a result, the faces of $\mathscr{S}_+^n$ are isomorphic to positive semidefinite cones of lower dimensions, and they are in a one-to-one correspondence with the subspaces of $\mathbb{R}^n$. As the following theorem shows, all faces of $\mathscr{S}_+^n$ are exposed.

Theorem 2.17 *Let $A \in \mathscr{S}_+^n$ and let rank$(A) = r$, $r \leq n - 1$. Further, let U be the $n \times (n - r)$ matrix whose columns form an orthonormal basis of null(A); and let $H = \{X \in \mathscr{S}^n : trace(UU^T X) = 0\}$. Then*

$$face(A, \mathscr{S}_+^n) = H \cap \mathscr{S}_+^n. \tag{2.8}$$

That is, all faces of $\mathscr{S}_+^n$ are exposed.

Proof. H is a supporting hyperplane of $\mathscr{S}_+^n$ at A since trace$(UU^T X) \geq 0$ for all X in $\mathscr{S}_+^n$. Let $X \in$ face$(A, \mathscr{S}_+^n)$. Then $XU = 0$. Hence, $X \in (H \cap \mathscr{S}_+^n)$ and thus face$(A, \mathscr{S}_+^n) \subseteq (H \cap \mathscr{S}_+^n)$. To prove the other inclusion, let $X \in (H \cap \mathscr{S}_+^n)$. Then trace$(U^T XU) = 0$. But $U^T XU \succeq 0$. Hence, $U^T XU = 0$. Thus, $XU = 0$ and hence $X \in$ face$(A, \mathscr{S}_+^n)$. Therefore, $(H \cap \mathscr{S}_+^n) \subseteq$ face$(A, \mathscr{S}_+^n)$.

$\square$

2.7 The Facial Structure of Spectrahedra

Of particular interest to us are sets formed by the intersection of the positive semidefinite cone $\mathscr{S}_+^n$ with an affine set. Such sets are called *spectrahedra*. Evidently, a spectrahedron is a closed convex set. A spectrahedron $\mathscr{F}$ has two equivalent representations depending on the representation of the affine set. Each of these representations has its own advantages. In this section, it is advantageous to use the following description of $\mathscr{F}$. Let $A^0, A^1, \ldots, A^m$ be linearly independent matrices in $\mathscr{S}^n$. Then $\mathscr{F}$ can be parameterized as

$$\mathscr{F} = \{x \in \mathbb{R}^m : \mathscr{A}(x) \succeq 0\},$$

where

$$\mathscr{A}(x) := A^0 + x_1 A^1 + \cdots + x_m A^m.$$

Assume that there exists $\hat{x} \in \mathbb{R}^m$ such that $\mathscr{A}(\hat{x}) \succ \mathbf{0}$, i.e., assume that $\mathscr{A}(x)$ satisfies Slater's condition. Then we may assume, wlog, that $A^0 \succ \mathbf{0}$ since $\mathscr{F}$ can be expressed as $\mathscr{F} = \{x \in \mathbb{R}^m : \mathscr{A}(\hat{x}) + \sum_{i=1}^m (x_i - \hat{x}_i)A^i \succeq \mathbf{0}\}$. On the other hand, if $A^0 \succ \mathbf{0}$, then for any $x \in \mathbb{R}^m$, there exists $\varepsilon > 0$ such that $A^0 + \varepsilon \sum_{i=1}^m x_i A^i \succeq \mathbf{0}$. Hence, $\mathrm{int}(\mathscr{F}) \neq \emptyset$. The reverse statement is not true; i.e., if a spectrahedron has a nonempty interior, then it may or may not satisfy Slater's condition.

Example 2.2 *Let* $A^0 = \begin{bmatrix} 1 & 1 \\ 1 & 1 \end{bmatrix}$, $A^1 = \begin{bmatrix} 1 & 0 \\ 0 & -1 \end{bmatrix}$ *and* $A^2 = \begin{bmatrix} 0 & -1 \\ -1 & 0 \end{bmatrix}$. *Then the spec-trahedron* $\mathscr{F} = \{x \in \mathbb{R}^2 : \mathscr{A}(x) = A^0 + x_1 A^1 + x_2 A^2 \succeq \mathbf{0}\}$ *is clearly given by the unit disk centered at* $(0, 1)$. *Notice that* $\mathscr{A}(x) \succ \mathbf{0}$ *for all x in the interior of this disk. Hence,* $\mathscr{A}(x)$ *satisfies Slater's condition. However,* $\mathscr{F}$ *can also be represented as* $\mathscr{F} = \{x \in \mathbb{R}^2 : \mathscr{A}'(x) = A'^0 + x_1 A'^1 + x_2 A'^2 \succeq \mathbf{0}\}$, *where*

$$
A'^0 = \begin{bmatrix} 1 & 1 & -2 \\ 1 & 1 & -2 \\ -2 & -2 & 4 \end{bmatrix}, A'^1 = \begin{bmatrix} 1 & 0 & -1 \\ 0 & -1 & 1 \\ -1 & 1 & 0 \end{bmatrix} \text{ and } A'^2 = \begin{bmatrix} 0 & -1 & 1 \\ -1 & 0 & 1 \\ 1 & 1 & -2 \end{bmatrix}.
$$

Notice that the three principal minors of $\mathscr{A}'(x)$ *of order 2 are equal. Also, notice that* $\det(\mathscr{A}'(x)) = 0$ *for all x since the 3rd row of* $\mathscr{A}'(x)$ *is a linear combination of its first two rows. Hence* $\mathscr{A}'(x) \not\succ \mathbf{0}$ *for all x, and thus* $\mathscr{A}'(x)$ *does not satisfy Slater's condition.*

In this monograph we will be interested in minimal faces of $\mathscr{F}$ as well as those of $\mathscr{S}_+^n$.

Theorem 2.18 (Ramana and Goldman [156]) *Let* $\mathscr{F} = \{x \in \mathbb{R}^m : \mathscr{A}(x) \succeq \mathbf{0}\}$ *and let* $x \in \mathscr{F}$. *Then*

$$
face(x, \mathscr{F}) = \{y \in \mathscr{F} : null(\mathscr{A}(x)) \subseteq null(\mathscr{A}(y))\}. \tag{2.9}
$$

Proof. Let $S = \{y \in \mathscr{F} : null(\mathscr{A}(x)) \subseteq null(\mathscr{A}(y))\}$ and let $y \in S$. Then by an argument similar to that in the proof of Lemma 2.1, there exists $\hat{t} > 0$ such that $\mathscr{A}(x) - \hat{t}(\mathscr{A}(y) - \mathscr{A}(x)) \succeq \mathbf{0}$. Let $z = x - \hat{t}(y - x)$. Then $\mathscr{A}(z) = \mathscr{A}(x) - \hat{t}(\mathscr{A}(y) - \mathscr{A}(x))$ and thus $z \in \mathscr{F}$. Hence,

$$
x = \frac{1}{1+\hat{t}}z + \frac{\hat{t}}{1+\hat{t}}y.
$$

Therefore, $y \in face(x, \mathscr{F})$ and hence $S \subseteq face(x, \mathscr{F})$.

To prove the other inclusion, note that if $face(x, \mathscr{F}) = \{x\}$, then $face(x, \mathscr{F}) \subseteq S$ and we are done. Therefore, let $y \in face(x, \mathscr{F})$ where $y \neq x$. Since x lies in $\mathrm{relint}(face(x, \mathscr{F}))$, there exist $z \in face(x, \mathscr{F})$ and $0 < \lambda < 1$ such that $x = \lambda y + (1 - \lambda)z$. Then $\mathscr{A}(x) = \lambda \mathscr{A}(y) + (1 - \lambda)\mathscr{A}(z)$. By an argument similar to that in the proof of Theorem 2.16, it follows that $null(\mathscr{A}(x)) \subseteq null(\mathscr{A}(y))$ and thus $y \in S$. Therefore, $face(x, \mathscr{F}) \subseteq S$.   □

The following corollary is immediate.

Corollary 2.11 *Let* $\mathscr{F} = \{x \in \mathbb{R}^m : \mathscr{A}(x) \succeq 0\}$ *and let* $x \in \mathscr{F}$. *Then*

$$relint(face(x, \mathscr{F})) = \{y \in \mathscr{F} : null(\mathscr{A}(x)) = null(\mathscr{A}(y))\}. \qquad (2.10)$$

It is worth pointing out that, for a face F of $\mathscr{F}$, all matrices $\{\mathscr{A}(x) : x \in relint(F)\}$ have the same rank. That is, the rank is constant over the relative interior of a face. Let S be a convex subset of $\mathscr{F}$. Then it easily follows from Theorem 1.28 and Corollary 1.2 that face$(S, \mathscr{F})$, the minimal face of $\mathscr{F}$ containing S, is the face of $\mathscr{F}$ whose relative interior intersects relint(S) [154].

Example 2.3 *Let* $\mathscr{F} = \{x \in \mathbb{R}^2 : A^0 + x_1 A^1 + x_2 A^2 \succeq 0\}$ *be the spectrahedron of Example 2.2. Then* $null(\mathscr{A}(0)) = null(A^0) = col(\begin{bmatrix} 1 \\ -1 \end{bmatrix})$. *Hence, face*$(0, \mathscr{F}) = \{x \in \mathscr{F} : x_1 + x_2 = 0 = x_1 - x_2\} = \{0\}$.

As was the case for the faces of $\mathscr{S}_+^n$, all faces of a spectrahedron are exposed.

Theorem 2.19 (Ramana and Goldman [156]) *Let* $\mathscr{F} = \{y \in \mathbb{R}^m : \mathscr{A}(y) \succeq 0\}$ *and let* $x \in \mathscr{F}$. *If* $\mathscr{A}(x) \succ 0$, *then face*$(x, \mathscr{F}) = \mathscr{F}$. *Otherwise, let* U *be the matrix whose columns form an orthonormal basis of* $null(\mathscr{A}(x))$; *and let* $H = \{y \in \mathbb{R}^m : trace(U^T \mathscr{A}(y)U) = 0\}$. *Then*

$$face(x, \mathscr{F}) = \mathscr{F} \cap H, \qquad (2.11)$$

i.e., all faces of $\mathscr{F}$ *are exposed.*

Proof. H is a supporting hyperplane to $\mathscr{F}$ since trace$(UU^T \mathscr{A}(y)) \geq 0$ for all $y \in \mathscr{F}$. Now let $y \in (\mathscr{F} \cap H)$. Therefore, $U^T \mathscr{A}(y)U = 0$ and hence $\mathscr{A}(y)U = 0$. Consequently, y lies in face$(x, \mathscr{F})$ and thus $(\mathscr{F} \cap H) \subseteq$ face$(x, \mathscr{F})$.

To prove the other inclusion, let $y \in$ face$(x, \mathscr{F})$. Then $\mathscr{A}(y)U = 0$. Consequently, $y \in (\mathscr{F} \cap H)$ and hence face$(x, \mathscr{F}) \subseteq (\mathscr{F} \cap H)$.

$\square$

Theorem 2.20 *Let* $\mathscr{F} = \{x \in \mathbb{R}^m : \mathscr{A}(x) \succeq 0\}$ *and let* S *be a convex subset of* $\mathscr{F}$. *Further, let* $\hat{x} \in S$. *Then the following statements are equivalent:*

(i) rank $\mathscr{A}(\hat{x}) \geq$ *rank* $\mathscr{A}(x)$ *for all* $x \in S$.
(ii) face$(\hat{x}, \mathscr{F}) =$ *face*$(S, \mathscr{F})$.
(iii) $\hat{x}$ *lies in relint*(S).

Proof. (i) $\Rightarrow$ (ii)

Since $\hat{x} \in S$, it follows that face$(\hat{x}, \mathscr{F}) \subseteq$ face$(S, \mathscr{F})$. To prove the other inclusion, let x be any point in S and let $y = \lambda \hat{x} + (1 - \lambda)x$ for some $0 < \lambda < 1$. Then $y \in S$ since $\mathscr{F}$ is convex. Moreover, $\mathscr{A}(y) = \lambda \mathscr{A}(\hat{x}) + (1 - \lambda)\mathscr{A}(x)$. Thus, by proposition 2.5, null$(\mathscr{A}(y)) \subseteq$ null$(\mathscr{A}(\hat{x}))$ and null$(\mathscr{A}(y)) \subseteq$ null$(\mathscr{A}(x))$. Hence, rank $\mathscr{A}(y) \geq$ rank $\mathscr{A}(\hat{x})$. But by assumption, rank $\mathscr{A}(y) \leq$ rank $\mathscr{A}(\hat{x})$. Therefore, rank $\mathscr{A}(y) =$ rank $\mathscr{A}(\hat{x})$ and consequently null$(\mathscr{A}(y)) =$ null$(\mathscr{A}(\hat{x}))$. Hence, null$(\mathscr{A}(\hat{x})) \subseteq$ null$(\mathscr{A}(x))$. Therefore, $x \in$ face$(\hat{x}, \mathscr{F})$ and thus $S \subseteq$ face$(\hat{x}, \mathscr{F})$. As a result, face$(S, \mathscr{F}) \subseteq$ face$(\hat{x}, \mathscr{F})$.

(ii) $\Rightarrow$ (iii)

Assume that face($\hat{x}, \mathscr{F}$) = face($S, \mathscr{F}$) and assume, to the contrary, that $\hat{x} \notin$ relint(S). Then $\hat{x}$ lies in the relative boundary of S. Hence, by the Supporting Hyperplane Theorem, there exists a hyperplane H containing $\hat{x}$ but not S. Therefore, $H \cap$ face($S, \mathscr{F}$) is a face of $\mathscr{F}$ containing $\hat{x}$ of a smaller dimension than face($S, \mathscr{F}$). This contradicts the definition of a minimal face.

(iii) $\Rightarrow$ (i)

Assume that $\hat{x} \in$ relint(S) and let y be any point in S. Then there exist $z \in S$ and λ : $0 < \lambda < 1$ such that $\hat{x} = \lambda y + (1 - \lambda)z$. Therefore, $\mathscr{A}(\hat{x}) = \lambda \mathscr{A}(y) + (1 - \lambda)\mathscr{A}(z)$. Then, by proposition 2.5, null($\mathscr{A}(\hat{x})$) $\subseteq$ null($\mathscr{A}(y)$). Consequently, rank $\mathscr{A}(\hat{x}) \geq$ rank $\mathscr{A}(y)$.

$\square$

The following theorem gives a representation of the affine hull of the minimal face of $\mathscr{F}$ containing x.

Theorem 2.21 *Let* $\mathscr{F} = \{x \in \mathbb{R}^m : \mathscr{A}(x) \succeq 0\}$ *and let* $x \in \mathscr{F}$. *Further, let* U *be the matrix whose columns form an orthonormal basis of* null($\mathscr{A}(x)$). *Then the affine hull of* face($x, \mathscr{F}$) *is given by*

$$aff(face(x, \mathscr{F})) = \{y \in \mathbb{R}^m : \mathscr{A}(y)U = 0\}. \tag{2.12}$$

Proof. Let $\mathscr{L} = \{y \in \mathbb{R}^m : \mathscr{A}(y)U = 0\}$ and let y be a point in aff(face($x, \mathscr{F}$)). Then $y = \lambda w + (1 - \lambda)v$ for some points w and v in face($x, \mathscr{F}$) and some scalar λ. Thus, $\mathscr{A}(w)U = \mathscr{A}(v)U = 0$ and hence $\mathscr{A}(y)U = \lambda \mathscr{A}(w)U + (1 - \lambda)\mathscr{A}(v)U = 0$. Consequently, $y \in \mathscr{L}$ and thus aff(face($x, \mathscr{F}$)) $\subseteq \mathscr{L}$.

To prove the other inclusion, note that if $\mathscr{L} = \{x\}$, then we are done. Therefore, let $y \in \mathscr{L}$, $y \neq x$. Let W be the matrix whose columns form an orthonormal basis for col($\mathscr{A}(x)$). Thus, $Q = [U \ W]$ is orthogonal and $W^T \mathscr{A}(x)W \succ 0$. Hence, there exists $t > 0$ such that $W^T \mathscr{A}(x)W - t(W^T \mathscr{A}(y)W - W^T \mathscr{A}(x)W) \succeq 0$. Let $z = x - t(y - x)$. Therefore, $W^T \mathscr{A}(z)W \succeq 0$ and $\mathscr{A}(z)U = 0$. Moreover, $Q^T \mathscr{A}(z)Q \succeq 0$ and hence, $\mathscr{A}(z) \succeq 0$. Therefore, $z \in$ face($x, \mathscr{F}$). Moreover, since $y = (1 + 1/t)x - z/t$, it follows that y belongs to aff(face($x, \mathscr{F}$)) and hence $\mathscr{L} \subseteq$ aff(face($x, \mathscr{F}$)).

$\square$

2.8 Facial Reduction

As we saw earlier, Slater's condition is sufficient for SDP strong duality and its absence can result in theoretical and numerical problems. Even though it holds generically [76], Slater's condition fails in many interesting instances of the SDP problem since the feasible regions of these problems are contained in the boundary of $\mathscr{S}_+^n$, and thus do not intersect the interior of $\mathscr{S}_+^n$. Borwein and Wolkowicz [49, 50] devised a facial reduction algorithm to regularize such problems and to turn the absence of Slater's condition to our advantage, see, e.g., [54, 74].

Lemma 2.2 *An SDP problem satisfies Slater's condition if and only if the minimal face of its feasible region is $\mathscr{S}_+^n$.*

Proof. Assume that an SDP problem satisfies Slater's condition and let $\mathscr{F}$ be its feasible region. Then $\mathscr{F} \cap \mathscr{S}_{++}^n \neq \emptyset$ and thus let $A \in (\mathscr{F} \cap \mathscr{S}_{++}^n)$. Hence face$(A, \mathscr{S}_+^n)$ $\subseteq$ face$(\mathscr{F}, \mathscr{S}_+^n) \subseteq \mathscr{S}_+^n$. But, face$(A, \mathscr{S}_+^n) = \mathscr{S}_+^n$ since A has a trivial null space. Therefore, face$(\mathscr{F}, \mathscr{S}_+^n) = \mathscr{S}_+^n$.

To prove the other direction, assume that face$(\mathscr{F}, \mathscr{S}_+^n) = \mathscr{S}_+^n = $ face$(I, \mathscr{S}_+^n)$ and let $A \in$ relint$(\mathscr{F})$. Then face$(A, \mathscr{S}_+^n) = $ face$(\mathscr{F}, \mathscr{S}_+^n) = $ face$(I, \mathscr{S}_+^n)$ and thus A is PD. Therefore, Slater's condition holds. $\qquad \square$

Another characterization of Slater's condition is given in Lemma 2.4 below. Consequently, in the absence of Slater's condition, the Borwein–Wolkowicz facial reduction algorithm aims at finding the minimal face of the feasible region by generating a sequence of faces of $\mathscr{S}_+^n$ containing $\mathscr{F}$, each of which is a proper subset of the previous one. In other words, this algorithm generates matrices $\mathscr{U}_1, \ldots, \mathscr{U}_{k+1}$ such that

$$\text{face}(\mathscr{F}, \mathscr{S}_+^n) = \text{face}(\mathscr{U}_{k+1}\mathscr{U}_{k+1}^T, \mathscr{S}_+^n) \subset \cdots \subset \text{face}(\mathscr{U}_1\mathscr{U}_1^T, \mathscr{S}_+^n) \subset \mathscr{S}_+^n, \quad (2.13)$$

where $\text{col}(\mathscr{U}_{k+1}) \subset \cdots \subset \text{col}(\mathscr{U}_1) \subset \text{col}(I) = \mathbb{R}^n$. An important point to bear in mind is that $\hat{X}$ lies in relint$(\mathscr{F})$ iff face$(\mathscr{F}, \mathscr{S}_+^n) = $ face$(\hat{X}, \mathscr{S}_+^n)$ iff $\text{col}(\hat{X}) = \text{col}(\mathscr{U}_{k+1})$. It is worth noting that face$(\mathscr{F}, \mathscr{S}_+^n)$ is isomorphic to the cone $\mathscr{S}_+^r$ for some $r \leq n$. As a result, Slater's condition holds if $\mathscr{F}$ is embedded in $\mathscr{S}_+^r$ instead of $\mathscr{S}_+^n$; i.e., if $\mathscr{F}$ is embedded in the smallest possible space.

Notice that theorems of the alternative presented above do not involve the rank. The Borwein–Wolkowicz facial reduction scheme is used to establish the following theorem of the alternative involving the rank (see also [136]). For the purposes of this section, it is advantageous to describe a spectrahedron as $\mathscr{F} = \{X \in \mathscr{S}_+^n : \text{trace}(A^l X) = b_l \text{ for } l = 1, \ldots, m\}$.

Theorem 2.22 (Alfakih [14]) *Let $A^1, \ldots, A^m$ be given linearly independent matrices in $\mathscr{S}^n$ and let b be a given nonzero vector in $\mathbb{R}^m$. Further, let*

$$\mathscr{F} = \{X \in \mathscr{S}_+^n : \text{trace}(A^i X) = b_i \text{ for } i = 1, \ldots, m\}.$$

Let X^* be a matrix in $\mathscr{F}$ such that rank$(X^*) = r$, $r \leq n - 1$. Then exactly one of the following two statements holds:

 (i) *There exists an $X \in \mathscr{F}$ such that rank$(X) \geq r + 1$.*
 (ii) *There exist nonzero matrices $\Omega^0, \Omega^1, \ldots, \Omega^k$, for some $k \leq n - r - 1$ such that:*

  a. *$\Omega^j = \sum_{i=1}^m x_i^j A^i (j = 0, 1, \ldots, k)$, for some scalars x_i^j's.*
  b. *$\Omega^0 \succeq 0$, $\mathscr{U}_1^T \Omega^1 \mathscr{U}_1 \succeq 0, \ldots, \mathscr{U}_k^T \Omega^k \mathscr{U}_k \succeq 0$.*
  c. *trace$(\Omega^j X^*) = 0$ for $j = 0, 1, \ldots, k$.*
  d. *rank$(\Omega^0) + $ rank$(\mathscr{U}_1^T \Omega^1 \mathscr{U}_1) + \cdots + $ rank$(\mathscr{U}_k^T \Omega^k \mathscr{U}_k) = n - r$.*

Here, $\mathcal{U}_1,\ldots,\mathcal{U}_{k+1}$ and $\mathcal{W}_0,\mathcal{W}_1,\ldots,\mathcal{W}_k$ are full column rank matrices defined as follows: For $i = 0,1,\ldots,k$, $col(\mathcal{W}_i) = null(\mathcal{U}_i^T \Omega^i \mathcal{U}_i)$ and $\mathcal{U}_{i+1} = \mathcal{U}_i \mathcal{W}_i$, where $\mathcal{U}_0 = I_n$.

Two remarks concerning Theorem 2.22 are in order. First, if $r = n-1$, then this theorem reduces to Corollary 2.5. Second, the integer k in this theorem, i.e., the number of steps needed in the facial reduction procedure, depends on the choice of the matrix $\mathcal{U}_i^T \Omega^i \mathcal{U}_i$ in each step. The minimum number of such steps is called the *singularity degree* of $\mathcal{F}$ [181]. For example, if $k = 0$, then the singularity degree is one and the facial reduction procedure terminates in one step.

The following lemma is the crucial ingredient in the proof of Theorem 2.22.

Lemma 2.3 *Let $\mathcal{F}$, X^*, $\mathcal{U}_j$, $\mathcal{U}_{j+1}$ and $\mathcal{W}_j$ be as in Theorem 2.22. Then*

$$\mathcal{F} \subseteq face(\mathcal{U}_{j+1}\mathcal{U}_{j+1}^T, \mathscr{S}_+^n) \subset face(\mathcal{U}_j\mathcal{U}_j^T, \mathscr{S}_+^n), \tag{2.14}$$

if the following two conditions hold:

1. $\mathcal{F} \subset face(\mathcal{U}_j\mathcal{U}_j^T, \mathscr{S}_+^n)$,
2. $\exists$ *a nonzero* $\Omega^j = \sum_{i=1}^m x_i^j A^i$ *for some scalars x_i^j's such that $\mathcal{U}_j^T \Omega^j \mathcal{U}_j \succeq 0$ and* $trace(\Omega^j X^*) = 0$.

Proof. Assume that Conditions 1 and 2 hold. Then since $\mathcal{F} \subset face(\mathcal{U}_j\mathcal{U}_j^T, \mathscr{S}_+^n)$, Corollary 2.9 implies that

$$\mathcal{F} = \{X = \mathcal{U}_j Y_j \mathcal{U}_j^T : Y_j \succeq 0, trace(A^i X) = b_i \text{ for } i = 1,\ldots,m\}.$$

Now for any $X \in \mathcal{F}$, we have

$$trace(\Omega^j X) = \sum_{i=1}^m x_i^j \, trace(A^i X) = \sum_{i=1}^m x_i^j b_i = \sum_{i=1}^m x_i^j \, trace(A^i X^*) = 0.$$

But $trace(\Omega^j X) = trace(\mathcal{U}_j^T \Omega^j \mathcal{U}_j Y_j)$. Consequently, $\mathcal{U}_j^T \Omega^j \mathcal{U}_j Y_j = 0$ since both Y_j and $\mathcal{U}_j^T \Omega^j \mathcal{U}_j$ are PSD. Therefore, by Proposition 1.1, $Y_j = \mathcal{W}_j Y_{j+1} \mathcal{W}_j^T$ for some $Y_{j+1} \succeq 0$. Hence, $\mathcal{F} = \{X : X = \mathcal{U}_{j+1} Y_{j+1} \mathcal{U}_{j+1}^T : Y_{j+1} \succeq 0, trace(A^i X) = b_i$ for $i = 1,\ldots,m\}$, i.e., $\mathcal{F} \subset face(\mathcal{U}_{j+1}\mathcal{U}_{j+1}^T, \mathscr{S}_+^n)$. Moreover, $face(\mathcal{U}_{j+1}\mathcal{U}_{j+1}^T, \mathscr{S}_+^n) \subset face(\mathcal{U}_j\mathcal{U}_j^T, \mathscr{S}_+^n)$ since $col(\mathcal{U}_{j+1}) \subset col(\mathcal{U}_j)$. □

The following observation concerning Lemma 2.3 is worth pointing out. Suppose that $\mathcal{U}_j$ is $n \times s$. Then $face(\mathcal{U}_j\mathcal{U}_j^T, \mathscr{S}_+^n)$ is isomorphic to $\mathscr{S}_+^s$. Hence, if $rank(\mathcal{U}_j^T \Omega^j \mathcal{U}_j) = \delta$, then $\mathcal{W}_j$ is $s \times (s-\delta)$ and therefore, $\mathcal{U}_{j+1}$ is $n \times (s-\delta)$. As a result, $face(\mathcal{U}_{j+1}\mathcal{U}_{j+1}^T, \mathscr{S}_+^n)$ is isomorphic to $\mathscr{S}_+^{s-\delta}$. Accordingly, the higher the rank of $\mathcal{U}_j^T \Omega^j \mathcal{U}_j$ is, the greater the difference between the dimensions of $face(\mathcal{U}_j\mathcal{U}_j^T, \mathscr{S}_+^n)$ and $face(\mathcal{U}_{j+1}\mathcal{U}_{j+1}^T, \mathscr{S}_+^n)$ will be. Consequently, the higher the ranks of the matrices $\mathcal{U}_j^T \Omega^j \mathcal{U}_j$'s are, the fewer steps the facial reduction scheme will need.

Proof of Theorem 2.22. All minimal faces in this proof are faces of $\mathscr{S}^n_+$ and thus we will drop the second argument of face. We first prove that if Statement (i) does not hold, then Statement (ii) holds. Therefore, assume that there does not exist $X \in \mathscr{F}$ such that $\mathrm{rank}(X) \geq r+1$; i.e., assume that $\mathrm{face}(\mathscr{F}) = \mathrm{face}(X^*)$. Then there does not exist $X \succ 0$ such that $\mathrm{trace}(A^i X) = b_i$ for $i = 1, \ldots, m$. Hence, by Corollary 2.5, there exists a nonzero $\Omega^0 \succeq 0$ such that $\Omega^0 = \sum_{i=1}^m x_i^0 A^i$ and $\mathrm{trace}(\Omega^0 X^*) = 0$. Now if $\mathrm{rank}(\Omega^0) = n - r$, then we are done and $k = 0$ in the theorem. Therefore, assume that $\mathrm{rank}(\Omega^0) = n - r - \delta_1$ for some $\delta_1 \geq 1$, and let $\mathscr{W}_0$ be an $n \times (r + \delta_1)$ full column rank matrix such that $\mathrm{col}(\mathscr{W}_0) = \mathrm{null}(\Omega^0)$. Since $\mathscr{F} \subset \mathrm{face}(I) = \mathscr{S}^n_+$, it follows from Lemma 2.3 that

$$\mathscr{F} \subseteq \mathrm{face}(\mathscr{U}_1 \mathscr{U}_1^T) \subset \mathrm{face}(I_n) = \mathscr{S}^n_+,$$

where $\mathscr{U}_1 = I_n \mathscr{W}_0$ is $n \times (r + \delta_1)$ with full column rank. Moreover, $\mathrm{face}(\mathscr{F}) = \mathrm{face}(X^*) \neq \mathrm{face}(\mathscr{U}_1 \mathscr{U}_1^T)$. That is, $X^* = \mathscr{U}_1 Y_1^* \mathscr{U}_1^T$, where Y_1^* is a singular PSD matrix. Furthermore, there does not exist $(r + \delta_1) \times (r + \delta_1)$ matrix $Y_1 \succ 0$ such that $\mathrm{trace}(A^i \mathscr{U}_1 Y_1 \mathscr{U}_1^T) = b_i$ for $i = 1, \ldots, m$. Hence, by Corollary 2.5, there exists a nonzero $\Omega^1 = \sum_{i=1}^m x_i^1 A^i$ such that $\mathscr{U}_1^T \Omega^1 \mathscr{U}_1 \succeq 0$ and $\mathrm{trace}(\mathscr{U}_1^T \Omega^1 \mathscr{U}_1 Y_1^*) = \mathrm{trace}(\Omega^1 X^*) = 0$. If $\mathrm{rank}(\mathscr{U}_1^T \Omega^1 \mathscr{U}_1) = \delta_1$, then we are done and $k = 1$ in the theorem since $\mathrm{rank}(\Omega^0) + \mathrm{rank}(\mathscr{U}_1^T \Omega^1 \mathscr{U}_1) = n - r$. Thus, assume that $\mathrm{rank}(\mathscr{U}_1^T \Omega^1 \mathscr{U}_1) = \delta_1 - \delta_2$, where $1 \leq \delta_2 \leq \delta_1 - 1$. Let $\mathscr{W}_1$ be an $(r + \delta_1) \times (r + \delta_2)$ full column rank matrix such that $\mathrm{col}(\mathscr{W}_1) = \mathrm{null}(\mathscr{U}_1^T \Omega^1 \mathscr{U}_1)$. Since $\mathscr{F} \subset \mathrm{face}(\mathscr{U}_1 \mathscr{U}_1^T)$, it follows from Lemma 2.3 that

$$\mathscr{F} \subseteq \mathrm{face}(\mathscr{U}_2 \mathscr{U}_2^T) \subset \mathrm{face}(\mathscr{U}_1 \mathscr{U}_1^T) \subset \mathrm{face}(I_n) = \mathscr{S}^n_+,$$

where $\mathscr{U}_2 = \mathscr{U}_1 \mathscr{W}_1$ is $n \times (r + \delta_2)$ with full column rank.

Observe that, at each step, a lower dimensional face containing $\mathscr{F}$ is obtained. Also, $\delta_1, \delta_2, \ldots$ is a strictly decreasing sequence of positive integers bounded above by $n - r - 1$. Thus after at most $n - r$ steps, we must have some k, $k \leq n - r - 1$, such that $\mathscr{U}_k$ is $n \times (r + \delta_k)$ and $\mathrm{rank}(\mathscr{U}_k^T \Omega^k \mathscr{U}_k) = \delta_k$. But, $\mathrm{rank}(\Omega^0) = n - r - \delta_1$, $\mathrm{rank}(\mathscr{U}_1^T \Omega^1 \mathscr{U}_1) = \delta_1 - \delta_2$, $\ldots$, $\mathrm{rank}(\mathscr{U}_{k-1}^T \Omega^{k-1} \mathscr{U}_{k-1}) = \delta_{k-1} - \delta_k$. Therefore, Statement (ii) holds.

Second, we prove that if Statement (ii) holds then Statement (i) does not hold. Thus, for $k \geq 1$, assume that $\mathrm{rank}(\Omega^0) = n - r - \delta_1$,

$$\mathrm{rank}(\mathscr{U}_1^T \Omega^1 \mathscr{U}_1) = \delta_1 - \delta_2, \ldots, \mathrm{rank}(\mathscr{U}_{k-1}^T \Omega^{k-1} \mathscr{U}_{k-1}) = \delta_{k-1} - \delta_k,$$

and $\mathrm{rank}(\mathscr{U}_k^T \Omega^k \mathscr{U}_k) = \delta_k$. Therefore, $\mathscr{W}_0$ is $n \times (r + \delta_1)$, $\mathscr{W}_1$ is $(r + \delta_1) \times (r + \delta_2)$, $\ldots$, $\mathscr{W}_{k-1}$ is $(r + \delta_{k-1}) \times (r + \delta_k)$ and $\mathscr{W}_k$ is $(r + \delta_k) \times r$. Moreover, $\mathscr{U}_1 = \mathscr{W}_0$, $\mathscr{U}_2 = \mathscr{W}_0 \mathscr{W}_1$, $\ldots$, $\mathscr{U}_{k+1} = \mathscr{W}_0 \cdots \mathscr{W}_k$. Hence, $\mathscr{U}_{k+1}$ is $n \times r$.

Now successive application of Lemma 2.3 yields

$$\mathscr{F} \subseteq \mathrm{face}(\mathscr{U}_{k+1} \mathscr{U}_{k+1}^T) \subset \mathrm{face}(\mathscr{U}_k \mathscr{U}_k^T) \subset \cdots \subset \mathrm{face}(\mathscr{U}_1 \mathscr{U}_1^T) \subset \mathscr{S}^n_+.$$

But $X^* \in \mathscr{F}$. Thus, face($\mathscr{F}$) is a face of $\mathscr{S}_+^n$ containing X^*. Hence, by the definition of a minimal face, it follows that face(X^*) $\subseteq$ face($\mathscr{F}$). Similarly, face($\mathscr{F}$) $\subseteq$ face($\mathscr{U}_{k+1} \mathscr{U}_{k+1}^T$). Thus, face($X^*$) $\subseteq$ face($\mathscr{U}_{k+1} \mathscr{U}_{k+1}^T$) or col($X^*$) $\subseteq$ col($\mathscr{U}_{k+1}$). Therefore, face(X^*) = face($\mathscr{U}_{k+1} \mathscr{U}_{k+1}^T$) since rank($X^*$) = r and $\mathscr{U}_k$ is $n \times r$. Consequently, face(X^*) = face($\mathscr{F}$) and thus Statement (i) does not hold.

Now if $k = 0$, i.e., if rank(Ω^0) = $n - r$, then $\mathscr{W}_0$ is $n \times r$ and hence $\mathscr{U}_1 = I_n \mathscr{W}_0$ is also $n \times r$. By applying Lemma 2.3 we get

$$\text{face}(X^*) \subseteq \mathscr{F} \subseteq \text{face}(\mathscr{U}_1 \mathscr{U}_1^T) \subset \mathscr{S}_+^n.$$

Hence, col(X^*) $\subseteq$ col($\mathscr{U}_1$) and thus col(X^*) = col($\mathscr{U}_1$). Therefore, face(X^*) = face($\mathscr{F}$).

$\square$

Example 2.4 *As an illustration of the facial reduction algorithm and the proof of Theorem 2.22, consider the spectrahedron $\mathscr{F} = \{X \succeq 0 : trace(XA^i) = b_i \text{ for } i = 1, \ldots, 4\}$, where*

$$A^1 = \frac{1}{9} \begin{bmatrix} 16 & 4 & 4 \\ 4 & 1 & 1 \\ 4 & 1 & 1 \end{bmatrix}, A^2 = \frac{1}{9} \begin{bmatrix} 1 & 4 & 1 \\ 4 & 16 & 4 \\ 1 & 4 & 1 \end{bmatrix}, A^3 = \begin{bmatrix} 0 & 0 & 0 \\ 0 & 1 & -1 \\ 0 & -1 & 1 \end{bmatrix},$$

$$A^4 = \begin{bmatrix} 1 & 0 & -1 \\ 0 & 0 & 0 \\ -1 & 0 & 1 \end{bmatrix} \text{ and } b = \begin{bmatrix} 0 \\ 1 \\ 1 \\ 4 \end{bmatrix}. \text{ Then } \mathscr{F} = \{X^* = \frac{1}{4} \begin{bmatrix} 1 & -1 & -3 \\ -1 & 1 & 3 \\ -3 & 3 & 9 \end{bmatrix}\}.$$

That is, $\mathscr{F}$ is a singleton and rank(X^*) = r = 1. Therefore, $\mathscr{F}$ has empty interior and hence, by Corollary 2.5, there exists a nonzero $\Omega^0 = \sum_{i=1}^4 x_i^0 A^i \succeq 0$ such that $trace(\Omega^0 X^*) = b^T x^0 = 0$. It is easy to see that, by setting $x_1^0 = 1, x_2^0 = x_3^0 = x_4^0 = 0$,

we get $\Omega^0 = A^1 \succeq 0$. Thus, $\mathscr{W}_0 = \begin{bmatrix} -1 & -1 \\ 4 & 0 \\ 0 & 4 \end{bmatrix}$. Notice that rank($\Omega^0$) = 1. Hence,

$\delta_1 = 1$ *since $n - r = 2$. Therefore, $\mathscr{U}_1 = \mathscr{W}_0$ and hence, $\mathscr{F} \subset face(\mathscr{U}_1 \mathscr{U}_1^T, \mathscr{S}_+^3)$, i.e.,*

$$\mathscr{F} = \{X : X = \mathscr{U}_1 Y_1 \mathscr{U}_1^T \succeq 0 \text{ and } trace(XA^i) = b_i \text{ for } i = 1, \ldots, 4\},$$
$$= \{Y_1 \succeq 0 : trace(Y_1 \mathscr{U}_1^T A^i \mathscr{U}_1) = b_i \text{ for } i = 1, \ldots, 4\},$$
$$= \{Y_1^* = \frac{1}{64} \begin{bmatrix} 1 & 3 \\ 3 & 9 \end{bmatrix}\},$$

since $\mathscr{U}_1^T A^1 \mathscr{U}_1 = 0$, $\mathscr{U}_1^T A^2 \mathscr{U}_1 = \begin{bmatrix} 25 & 5 \\ 5 & 1 \end{bmatrix}$, $\mathscr{U}_1^T A^3 \mathscr{U}_1 = \begin{bmatrix} 16 & -16 \\ -16 & 16 \end{bmatrix}$ and $\mathscr{U}_1^T A^4 \mathscr{U}_1 = \begin{bmatrix} 1 & 5 \\ 5 & 25 \end{bmatrix}$.

Again, by Corollary 2.5, there exists a nonzero $\Omega^1 = \sum_{i=2}^4 x_i^1 \mathscr{U}_1^T A^i \mathscr{U}_1 \succeq 0$ such that $trace(\mathscr{U}_1^T \Omega^1 \mathscr{U}_1 Y_1^*) = trace(\Omega^1 X^*) = b^T x^1 = 0$, or $x_2^1 + x_3^1 + 4x_4^1 = 0$. Elimi-

nating x_3^1 yields

$$\mathcal{U}_1^T \Omega^1 \mathcal{U}_1 = 3 \begin{bmatrix} 3x_2^1 - 21x_4^1 & 7x_2^1 + 23x_4^1 \\ 7x_2^1 + 23x_4^1 & -5x_2^1 - 13x_4^1 \end{bmatrix} \succeq \mathbf{0}.$$

Hence, $\det(\mathcal{U}_1^T \Omega^1 \mathcal{U}_1) = -192(x_2^1 + 2x_4^1)^2 \geq 0$. Therefore, $x_2^1 = -2x_4^1$. Thus, by setting $x_2^1 = 2$, $x_3^1 = 2$ and $x_4^1 = -1$, we get

$$\Omega^1 = \frac{1}{9} \begin{bmatrix} -7 & 8 & 11 \\ 8 & 50 & -10 \\ 11 & -10 & 11 \end{bmatrix} \text{ and } \mathcal{U}_1^T \Omega^1 \mathcal{U}_1 = 9 \begin{bmatrix} 9 & -3 \\ -3 & 1 \end{bmatrix} \succeq \mathbf{0}.$$

Notice that $\operatorname{rank}(\mathcal{U}_1^T \Omega^1 \mathcal{U}_1) = \delta_1 = 1$. Thus, $\operatorname{rank}(\Omega^0) + \operatorname{rank}(\mathcal{U}_1^T \Omega^1 \mathcal{U}_1) = 2 = n - r$. Therefore, $k = 1$ in Theorem 2.22, i.e., the singularity degree of $\mathscr{F}$ is 2.

Now $\mathcal{W}_1 = \begin{bmatrix} 1 \\ 3 \end{bmatrix}$. Hence, $\mathcal{U}_2 = \mathcal{U}_1 \mathcal{W}_1 = \begin{bmatrix} -4 \\ 4 \\ 12 \end{bmatrix}$. Thus, $\mathscr{F} \subseteq \operatorname{face}(\mathcal{U}_2 \mathcal{U}_2^T, \mathscr{S}_+^3)$ and more precisely, $\operatorname{face}(\mathscr{F}, \mathscr{S}_+^3) = \operatorname{face}(\mathcal{U}_2 \mathcal{U}_2^T, \mathscr{S}_+^3)$. Note that $X^* = \mathcal{U}_2 \mathcal{U}_2^T / 64$.

We saw in the preceding discussion that $\operatorname{face}(\mathscr{F}, \mathscr{S}_+^n) = \operatorname{face}(\mathcal{U}_{k+1} \mathcal{U}_{k+1}^T, \mathscr{S}_+^n)$. As we noted earlier, $\operatorname{face}(\mathscr{F}, \mathscr{S}_+^n)$, as well as any other face of $\mathscr{S}_+^n$, is exposed. To be more precise, let $\overline{\mathcal{U}}_{k+1}$ be the $n \times (n - r)$ full column rank matrix such that $\operatorname{col}(\overline{\mathcal{U}}_{k+1}) = \operatorname{null}(\mathcal{U}_{k+1}^T)$. Then $\operatorname{face}(\mathscr{F}, \mathscr{S}_+^n) = H \cap \mathscr{S}_+^n$, where $H = \{X \in \mathscr{S}^n : \operatorname{trace}(\overline{\mathcal{U}}_{k+1} \overline{\mathcal{U}}_{k+1}^T X) = 0\}$. Now, the exposing matrix $\overline{\mathcal{U}}_{k+1} \overline{\mathcal{U}}_{k+1}^T$ can be found in one step if and only if the singularity degree of $\mathscr{F}$ is one. Let $\mathcal{M} : \mathscr{S}^n \to \mathbb{R}^m$ be the linear transformation where $\mathcal{M}_i(X) = \operatorname{trace}(A^i X)$. As it turns out, the minimal faces of $\mathcal{M}(\mathscr{S}_+^n)$, unlike those of $\mathscr{S}_+^n$, may or may not be exposed. Drusvyatskiy et al. [75] proved that the singularity degree of $\mathscr{F}$ is one iff $\operatorname{face}(b, \mathcal{M}(\mathscr{S}_+^n))$ is exposed. Before presenting their result, we provide another characterization of Slater's condition.

Lemma 2.4 ([74]) *Let $\mathscr{F} = \{X \in \mathscr{S}_+^n : \mathcal{M}(X) = b\}$ and assume that $\mathscr{F} \neq \emptyset$. Then $\mathscr{F}$ satisfies Slater's condition if and only if b lies in $\operatorname{relint}(\mathcal{M}(\mathscr{S}_+^n))$.*

Proof. We have, by Theorem 1.21, that $\operatorname{relint}(\mathcal{M}(\mathscr{S}_+^n)) = \mathcal{M}(\operatorname{relint}(\mathscr{S}_+^n))$. Now, obviously, there exists $X \succ \mathbf{0} : \mathcal{M}(X) = b$ iff b lies in $\mathcal{M}(\operatorname{relint}(\mathscr{S}_+^n))$ and hence the result follows.

$\square$

Consequently, in the absence of Slater's condition, we have the following theorem.

Theorem 2.23 (Drusvyatskiy et al. [75]) *Let $\mathcal{M} : \mathscr{S}^n \to \mathbb{R}^m$ be a linear transformation and let $\mathscr{F} = \{X \in \mathscr{S}_+^n : \mathcal{M}(X) = b\}$. Further, let $X^* \in \operatorname{relint}(\mathscr{F})$. Assume that $\operatorname{rank}(X^*) = r \leq n - 1$ and that $\mathcal{M}(\mathscr{S}_+^n)$ is closed. Then the following statements are equivalent:*

(i) The singularity degree of $\mathscr{F}$ is 1.

(ii) w exposes face$(b, \mathcal{M}(\mathcal{S}^n_+))$.
(iii) $\mathcal{M}^*(w)$ exposes face$(\mathcal{F}, \mathcal{S}^n_+)$.
(iv) $\mathcal{M}^*(w) \succeq \mathbf{0}$, $b^T w = 0$ and rank $\mathcal{M}^*(w) = n - r$.

Theorem 2.23 has interesting implications for dimensional rigidity, which we study in Chap. 10. Next, we prove the following theorem which is used in the proof of Theorem 2.23, and which is interesting in its own right.

Theorem 2.24 Let $\mathcal{M} : \mathcal{S}^n \to \mathbb{R}^m$ be a linear transformation and let $\mathcal{F} = \{X \in \mathcal{S}^n_+ : \mathcal{M}(X) = b\}$. Assume that $\mathcal{M}(\mathcal{S}^n_+)$ is closed and let $X^* \in relint(\mathcal{F})$. Then

(i) $\mathcal{M}(face(X, \mathcal{S}^n_+)) \subseteq face(b, \mathcal{M}(\mathcal{S}^n_+))$ for any $X \in \mathcal{F}$.
(ii) $\mathcal{M}(face(X^*, \mathcal{S}^n_+)) = \mathcal{M}(face(\mathcal{F}, \mathcal{S}^n_+)) = face(b, \mathcal{M}(\mathcal{S}^n_+))$.

Proof.   Let $X \in \mathcal{F}$. Then obviously, X lies in relint(face$(X, \mathcal{S}^n_+)$). Hence, by Theorem 1.21, it follows that $\mathcal{M}(X) = b$ lies in relint($\mathcal{M}(face(X, \mathcal{S}^n_+))$). Note that $\mathcal{M}(face(X, \mathcal{S}^n_+))$ is not necessarily a face of $\mathcal{M}(\mathcal{S}^n_+)$. However, obviously b lies in relint(face$(b, \mathcal{M}(\mathcal{S}^n_+))$). Thus, Theorem 1.28 implies that $\mathcal{M}(face(X, \mathcal{S}^n_+)) \subseteq$ face$(b, \mathcal{M}(\mathcal{S}^n_+))$.

To prove Statement (ii), let $F = \mathcal{S}^n_+ \cap \mathcal{M}^{-1}(face(b, \mathcal{M}(\mathcal{S}^n_+)))$ and note that $\mathcal{M}(F) = face(b, \mathcal{M}(\mathcal{S}^n_+))$ and $\mathcal{F} = \mathcal{S}^n_+ \cap \mathcal{M}^{-1}(b)$. Then obviously, $\mathcal{F} \subseteq F$. Moreover, by Theorem 1.29, F is a face of $\mathcal{S}^n_+$. But b lies in relint(face$(b, \mathcal{M}(\mathcal{S}^n_+))$). Thus, Theorem 1.21 implies that there exists X in relint(F) such that $\mathcal{M}(X) = b$, i.e., $\mathcal{F} \cap relint(F) \neq \emptyset$. Therefore, by Corollary 1.2, face$(\mathcal{F}, \mathcal{S}^n_+) = F$ and thus, Statement (ii) holds since face$(\mathcal{F}, \mathcal{S}^n_+) = face(X^*, \mathcal{S}^n_+)$.

$\square$

Example 2.5 To illustrate the previous theorem, let $\mathcal{M} : \mathcal{S}^2 \to \mathbb{R}^2$, where

$$\mathcal{M}(\begin{bmatrix} x & z \\ z & y \end{bmatrix}) = \begin{bmatrix} x \\ y \end{bmatrix}.$$

Then $\mathcal{M}(\mathcal{S}^2_+) = \mathbb{R}^2_+ = \{x \in \mathbb{R}^2 : x_1 \geq 0, x_2 \geq 0\}$. Let $b = e$. Then $\mathcal{M}^{-1}(b) = \{\begin{bmatrix} 1 & z \\ z & 1 \end{bmatrix} : z \in \mathbb{R}\}$ and thus $\mathcal{F} = \mathcal{S}^2_+ \cap \mathcal{M}^{-1}(b) = \{\begin{bmatrix} 1 & z \\ z & 1 \end{bmatrix} : z^2 \leq 1\}$. As a result, face$(b, \mathcal{M}(\mathcal{S}^2_+)) = \mathbb{R}^2_+$ since b lies in the relative interior of $\mathcal{M}(\mathcal{S}^2_+)$. Notice that $\mathcal{F}$ satisfies Slater's condition. Now let $X^* = I$. Then $X^* \in int(\mathcal{F})$ and thus face$(X^*, \mathcal{S}^n_+) = \mathcal{S}^n_+$. Therefore, $\mathcal{M}(face(X^*, \mathcal{S}^n_+)) = \mathbb{R}^2_+ = face(b, \mathcal{M}(\mathcal{S}^2_+))$.

On the other hand, let $\bar{X} = ee^T$. Then $\bar{X}$ lies in the boundary of $\mathcal{F}$ and face$(\bar{X}, \mathcal{S}^n_+) = \{zee^T : z \geq 0\}$. Thus, $\mathcal{M}(face(\bar{X}, \mathcal{S}^n_+)) = \mathbb{R}^2_+ \cap \{x \in \mathbb{R}^2 : x_1 = x_2\}$.

Chapter 3
Euclidean Distance Matrices (EDMs)

This chapter provides an introduction to Euclidean distance matrices (EDMs). Our primary focus is on various characterizations and basic properties of EDMs. The chapter also discusses methods to construct new EDMs from old ones, and presents some EDM necessary and sufficient inequalities. It also provides a discussion of the Cayley–Menger matrix and Schoenberg Transformations.

An $n \times n$ matrix $D = (d_{ij})$ is called a *Euclidean distance matrix (EDM)* if there exist points $p^1, \ldots, p^n$ in some Euclidean space such that

$$d_{ij} = ||p^i - p^j||^2 \text{ for } i, j = 1, \ldots, n.$$

The dimension of the affine span of these points is called the *embedding dimension* of D. If the embedding dimension of D is r, we always assume that $p^1, \ldots, p^n$ are points in $\mathbb{R}^r$. Of particular interest throughout the monograph is the EDM associated with the standard simplex, i.e., $\Delta = E - I$, where E is the $n \times n$ matrix of all 1's and I is the identity matrix of order n. Clearly, the embedding dimension of Δ is $n - 1$. Observe that if $d_{ij} = 0$ for some $i \neq j$, then the ith and the jth columns (rows) of D are identical. Conversely, if the ith and the jth columns (rows) of D are identical, then $d_{ij} = 0$ since $d_{ii} = 0$. Consequently, D has no repeated columns (rows) iff the off-diagonal entries of D are all nonzero iff no two of the generating points of D coincide.

Let the embedding dimension of D be r, then the $n \times r$ full column rank matrix

$$P = \begin{bmatrix} (p^1)^T \\ \vdots \\ (p^n)^T \end{bmatrix} \tag{3.1}$$

is called a *configuration matrix* of D. Consequently, the Gram matrix of points $p^1, \ldots, p^n$ is the $n \times n$ matrix

$$B = PP^T. \tag{3.2}$$

Thus, B is positive semidefinite (PSD) and of rank r. Moreover, the entries of D can be expressed in terms of the entries of B as follows.

© Springer Nature Switzerland AG 2018
A.Y. Alfakih, *Euclidean Distance Matrices and Their Applications in Rigidity Theory*, https://doi.org/10.1007/978-3-319-97846-8_3

$$d_{ij} = ||p^i - p^j||^2$$
$$= (p^i)^T p^i + (p^j)^T p^j - 2(p^i)^T p^j$$
$$= b_{ii} + b_{jj} - 2b_{ij}.$$

Denote the vector of all 1's by e, and the vector consisting of the diagonal entries of a matrix A by $\text{diag}(A)$. Define the linear operator $\mathcal{K} : \mathcal{S}^n \to \mathcal{S}^n$ by

$$\mathcal{K}(A) = \text{diag}(A)e^T + e(\text{diag}(A))^T - 2A. \tag{3.3}$$

Then, it is immediate that an EDM D can be expressed in terms of the Gram matrix of its generating points as

$$D = \mathcal{K}(B). \tag{3.4}$$

Gram matrix B is invariant under orthogonal transformations since for any $r \times r$ orthogonal matrix Q, it follows that $(PQ)(PQ)^T = PP^T$. However, B is not invariant under translations. Let a be a nonzero vector in $\mathbb{R}^r$ and let P' and B' be, respectively, the images of P and B under a translation along a. Then $P' = P - ea^T$ and hence $B' = B + a^T a ee^T - Pae^T - ea^T P^T$. Note that $\mathcal{K}(a^T a ee^T - Pae^T - ea^T P^T) = \mathbf{0}$. Placing the origin at a particular point removes these r translational degrees of freedom. This amounts to requiring the configuration matrix P, and hence the Gram matrix B, to satisfy $P^T s = \mathbf{0}$ or $Bs = \mathbf{0}$ for some nonzero vector s in $\mathbb{R}^n$. For example, if $s = e/n$, then $Bs = \mathbf{0}$ is equivalent to placing the origin at the centroid of the generating points $p^1, \ldots, p^n$. On the other hand, if $s = e^i$, where e^i is the ith standard unit vector in $\mathbb{R}^n$, then $Bs = \mathbf{0}$ is equivalent to placing the origin at point p^i. A remark is in order here regarding the possible choices of such a vector s. Suppose that for some given s, $P^T s \neq \mathbf{0}$. Then there exists a translation such that $B's = \mathbf{0}$ only if $e^T s \neq 0$ since $P'^T = P^T - ae^T$. This fact will be manifest in the definition of the linear operator $\mathcal{T}$ in the next section.

Example 3.1 *Consider the EDM* $D = \begin{bmatrix} 0 & 18 & 36 \\ 18 & 0 & 18 \\ 36 & 18 & 0 \end{bmatrix}$ *generated by points lying on a hypersphere of radius 3.*

Let $s^1 = [1/2 \ 0 \ 1/2]^T$, *then D has a configuration matrix* $P = \begin{bmatrix} -3 & 0 \\ 0 & 3 \\ 3 & 0 \end{bmatrix}$ *satisfy-ing $P^T s^1 = \mathbf{0}$. In this case, the hypersphere is centered at the origin since $||p^1|| = ||p^2|| = ||p^3|| = 3$, while the centroid of the generating points is $P^T e/3 = [0 \ 1]^T$.*

Now let $s^2 = e/3$, *then D also has a configuration matrix* $P = \begin{bmatrix} -3 & -1 \\ 0 & 2 \\ 3 & -1 \end{bmatrix}$ *satis-fying $P^T s^2 = \mathbf{0}$. In this case, the hypersphere is centered at $a = [0 \ -1]^T$, while the centroid of the generating points coincides with the origin.*

Example 3.2 *Consider the EDM* $D = \begin{bmatrix} 0 & 10 & 10 & 4 \\ 10 & 0 & 4 & 2 \\ 10 & 4 & 0 & 2 \\ 4 & 2 & 2 & 0 \end{bmatrix}$. *Let* $s^1 = e/4$. *Then* D *has a*

configuration matrix $P = \begin{bmatrix} 0 & -2 \\ -1 & 1 \\ 1 & 1 \\ 0 & 0 \end{bmatrix}$ *satisfying* $P^T s^1 = \mathbf{0}$. *Now let* $s^2 = [1 \ 1 \ 1 \ -3]^T$.

Notice that $e^T s^2 = 0$ *and* $P^T s^2$ *happened to be* $\mathbf{0}$. *However, if* $P^T s \neq 0$ *for some s perpendicular to e, then D has no configuration matrix* P' *satisfying* $P'^T s = 0$.

3.1 The Basic Characterization of EDMs

In this section we focus on the basic characterization of EDMs. Other characterizations will be discussed later in this chapter.

Let $e^{\perp}$ denote the orthogonal complement of e in $\mathbb{R}^n$; i.e.,

$$e^{\perp} = \{x \in \mathbb{R}^n : e^T x = 0\}.$$

Two vectors are of particular importance in the theory of EDMs. First, vector e plays a fundamental role as it is part of the definition of $\mathcal{K}$. Second, vector s, where $e^T s = 1$, also plays an important role since it used to eliminate the Gram matrix translational degrees of freedom. In most of this monograph, we will be interested in the case where $s = e/n$. In this case, one should keep in mind that e is playing a dual role.

Let $Q = I - se^T$, where $e^T s = 1$. Then, obviously, Q is a projection matrix. More precisely, Q is the projection matrix onto $e^{\perp}$ along s. This follows since $Qs = 0$ and $e^T Qx = 0$ for any x; i.e., s and e are, respectively, in the null space and the left null space of Q. Evidently, if $s = e/n$, then Q is the orthogonal projection matrix onto $e^{\perp}$.

Define the linear operator $\mathcal{T} : \mathcal{S}^n \to \mathcal{S}^n$ by

$$\mathcal{T}(A) = -\frac{1}{2}(I - es^T)A(I - se^T), \tag{3.5}$$

where $s^T e = 1$. The motivation behind this definition is given in the next lemma, which establishes the connection between operators $\mathcal{K}$ and $\mathcal{T}$. The operators $\mathcal{K}$ and $\mathcal{T}$ were introduced by Critchley in [67], where several of their properties are discussed. Let $\mathcal{S}_h^n$ and $\mathcal{S}_s^n$ be the two subspaces of $\mathcal{S}^n$ defined as follows:

$$\mathcal{S}_h^n = \{A \in \mathcal{S}^n : \text{diag}(A) = \mathbf{0}\} \tag{3.6}$$
$$\mathcal{S}_s^n = \{A \in \mathcal{S}^n : As = \mathbf{0}\}. \tag{3.7}$$

Lemma 3.1 (Critchley [67]) *The operator $\mathscr{T}$, restricted to $\mathscr{S}_h^n$, and the operator $\mathscr{K}$, restricted to $\mathscr{S}_s^n$ are mutually inverse; i.e.,*

$$\mathscr{T}|_{\mathscr{S}_h^n} = (\mathscr{K}|_{\mathscr{S}_s^n})^{-1} \text{ and } \mathscr{K}|_{\mathscr{S}_s^n} = (\mathscr{T}|_{\mathscr{S}_h^n})^{-1}.$$

Proof. Let $A \in \mathscr{S}_h^n$, then $\mathscr{T}(A) = -(A - Ase^T - es^T A + s^T As\, ee^T)/2$. Hence, $\text{diag}(\mathscr{T}(A)) = As - (s^T As/2)\, e$. Thus, $\text{diag}(\mathscr{T}(A))e^T + e(\text{diag}(\mathscr{T}(A))^T = 2\mathscr{T}(A) + A$. Therefore, $\mathscr{K}(\mathscr{T}(A)) = A$.

On the other hand, let $A \in \mathscr{S}_s^n$. Then $\mathscr{K}(A)s = \text{diag}(A) + (s^T \text{diag}(A))e$ since $As = 0$ and $e^T s = 1$. Consequently, $s^T \mathscr{K}(A)s = 2s^T \text{diag}(A)$. Moreover, $\mathscr{K}(A)se^T = \text{diag}(A)e^T + s^T \text{diag}(A)\, ee^T$. Therefore,

$$-2\mathscr{T}(\mathscr{K}(A)) = \mathscr{K}(A) - \text{diag}(A)e^T - e(\text{diag}(A))^T = -2A$$

or $\mathscr{T}(\mathscr{K}(A)) = A$.

$\square$

It is worth noting that $\dim \mathscr{S}_h^n = \dim \mathscr{S}_s^n = n(n-1)/2$. The following two lemmas are easy consequences of the definitions of $\mathscr{K}$ and $\mathscr{T}$ [25].

Lemma 3.2 *The image of $\mathscr{K} = \mathscr{S}_h^n$ and the kernel of $\mathscr{K} = \{ae^T + ea^T : a \in \mathbb{R}^n\}$.*

Lemma 3.3 *The image of $\mathscr{T} = \mathscr{S}_s^n$ and the kernel of $\mathscr{T} = \{ae^T + ea^T : a \in \mathbb{R}^n\}$.*

Schoenberg [167] and Young and Householder [200] established the basic characterization of EDMs. The following theorem is a restatement, due to Gower [93], of their result.

Theorem 3.1 (Schoenberg, Young and Householder) *Let D be an $n \times n$ real symmetric matrix whose diagonal entries are all 0's. Then D is an EDM if and only if D is negative semidefinite on $e^\perp$; i.e., if and only if*

$$\mathscr{T}(D) \succeq 0.$$

Moreover, the embedding dimension of D is given by the rank of $\mathscr{T}(D)$.

Proof. Assume that D is an EDM of embedding dimension r. Then $D = \mathscr{K}(B)$ where B, the Gram matrix of the generating points of D, satisfies $Bs = \mathbf{0}$. Thus, $\mathscr{T}(D) = \mathscr{T}(\mathscr{K}(B)) = B$ is PSD of rank r.

On the other hand, assume that $B = \mathscr{T}(D)$ is PSD of rank r. Then, it follows from (3.5) that $Bs = \mathbf{0}$. Moreover, B can be decomposed as $B = PP^T$, where P is an $n \times r$ full column rank matrix. Therefore, $D = \mathscr{K}(\mathscr{T}(D)) = \mathscr{K}(PP^T)$, and hence D is an EDM generated by the points $p^1, \ldots, p^n$, where $(p^i)^T$ is the ith row of P.

$\square$

We should point out that if D has a negative entry, then obviously D is not an EDM. To see how this follows from Theorem 3.1, assume wlog that $d_{12} < 0$ and let $x^T = [1\ -1\ \mathbf{0}^T]$. Then clearly $x \in e^\perp$ and $x^T Dx = -2d_{12} > 0$, and hence D is not negative semidefinite on $e^\perp$.

The following two observations made in [92] shed more light on vector s in $\mathcal{T}$. First, a nonzero D is never a PSD matrix since diag$(D) = \mathbf{0}$. Hence, $B = \mathcal{T}(D)$ is PSD only if $I - se^T$ is singular. But $\det(I - se^T) = 0$ iff $e^T s = 1$. Second, $Ds \neq \mathbf{0}$. To see this, assume to the contrary that $s \in \text{null}(D)$. Then $e^T s = 0$ since $e \in \text{col}(D)$ (see Theorem 3.9 below).

In the next subsection, Theorem 3.1 is refined by exploiting the facial structure of the positive semidefinite cone $\mathscr{S}_+^n$.

3.1.1 The Orthogonal Projection on $e^\perp$

Different choices of projection matrices onto $e^\perp$ amount to different choices of the origin. In most situations, we will find it convenient to place the origin at the centroid of the generating points of D; i.e., to set $s = e/n$. Accordingly, let J denote the orthogonal projection matrix onto $e^\perp$; i.e.,

$$J = I - \frac{ee^T}{n}. \tag{3.8}$$

Throughout this monograph we make the following assumption:

Assumption 3.1 *Unless otherwise stated, we assume that the Gram matrix B of an EDM D satisfies Be = $\mathbf{0}$; i.e., the origin coincides with the centroid of the generating points $p^1, \ldots, p^n$.*

Under this assumption, we have

$$\mathcal{T}(D) = -\frac{1}{2}JDJ. \tag{3.9}$$

Different choices of a basis for $e^\perp$ give rise to different factorizations of J. Next, we present two such factorizations. The first one corresponds to a sparse, albeit nonorthogonal, basis of $e^\perp$, while the other one corresponds to an orthonormal, albeit dense, basis.

To obtain the first factorization, let [25]

$$U = \begin{bmatrix} -e_{n-1}^T \\ I_{n-1} \end{bmatrix}. \tag{3.10}$$

Then, clearly, the columns of U form a (nonorthogonal) basis of $e^\perp$. Consequently, J can be factorized as $J = UU^\dagger$, where $U^\dagger$ is the Moore–Penrose inverse of U. That is, $J = U(U^T U)^{-1}U^T$. As will be shown later, this factorization is particularly useful for pencil-and-paper computations.

To obtain the second factorization, let V be the $n \times (n-1)$ matrix whose columns form an orthonormal basis of $e^\perp$. In other words, V satisfies

$$V^T e = \mathbf{0} \text{ and } V^T V = I_{n-1}. \tag{3.11}$$

Hence, $J = VV^T$. This factorization of J is most useful for theoretical purposes. As an immediate consequence of (3.11), we have the following fact which will be used later in the monograph. Every $(n-1) \times (n-1)$ submatrix of V is nonsingular. To see this, let $\bar{V}$ denote the submatrix of V obtained by deleting, say, its nth row, and assume that $\bar{V}$ is singular. Then there exists a nonzero $\xi \in \mathbb{R}^{n-1}$ such that $\xi^T \bar{V} = \mathbf{0}$. Let $\rho^T = [\xi^T \ 0]$. Then $\rho^T V = 0$ and thus $\text{rank}(V) \leq n-2$ since $\rho \neq e$, a contradiction.

Obviously, V is not unique and many choices of V are possible. One such choice [21] is

$$V = \begin{bmatrix} y e_{n-1}^T \\ I_{n-1} + x e_{n-1} e_{n-1}^T \end{bmatrix}, \text{ where } y = -\frac{1}{\sqrt{n}} \text{ and } x = -\frac{1}{n + \sqrt{n}}. \tag{3.12}$$

This particular choice of V is related to Householder matrices as will become clear below. It is worth noting that

$$V = U(I_{n-1} + x e_{n-1} e_{n-1}^T) \tag{3.13}$$

and

$$U = V(I_{n-1} + \frac{x}{y} e_{n-1} e_{n-1}^T). \tag{3.14}$$

Let V_1 and V_2 be two $n_1 \times (n_1 - 1)$ and $n_2 \times (n_2 - 1)$ matrices, respectively, as defined in (3.11). When dealing with 2×2 block matrices, we are often faced with the problem of constructing an $n \times (n-1)$ matrix satisfying (3.11) out of V_1 and V_2, where $n = n_1 + n_2$ (see, e.g., Theorem 3.20 below). To this end, let

$$V = \begin{bmatrix} V_1 & \mathbf{0} & \alpha e_{n_1} \\ \mathbf{0} & V_2 & \beta e_{n_2} \end{bmatrix}, \tag{3.15}$$

where α and β are scalars. Hence, V satisfies (3.11) iff $\alpha n_1 + \beta n_2 = 0$ and $\alpha^2 n_1 + \beta^2 n_2 = 1$. Therefore, V satisfies (3.11) if $\alpha = \sqrt{\frac{n_2}{nn_1}}$ and $\beta = -\sqrt{\frac{n_1}{nn_2}}$.

Recall that the translational degrees of freedom can be eliminated by requiring the Gram matrix B to satisfy the constraint $Be = \mathbf{0}$. As we show next, this constraint can be dropped if instead of B, we use the corresponding projected Gram matrix.

3.1.2 The Projected Gram Matrix

It is an immediate consequence of Corollary 2.10 that set $F = \{A \in \mathcal{S}_+^n : Ae = \mathbf{0}\}$ is a maximal proper face of $\mathcal{S}_+^n$. Let $\mathcal{D}^n$ denote the set of $n \times n$ EDMs. Then $\mathcal{D}^n$ is the image of F under the linear operator $\mathcal{K}$. Moreover, F, by Corollary 2.9, can be expressed as $F = \{A : A = VXV^T, \text{ where } X \in \mathcal{S}_+^{n-1}\}$, where V is as defined in (3.11). This motivates the introduction [21] of the following two linear transformations: $\mathcal{K}_V : \mathcal{S}^{n-1} \to \mathcal{S}_h^n$ and $\mathcal{T}_V : \mathcal{S}_h^n \to \mathcal{S}^{n-1}$ defined by $\mathcal{K}_V(X) = \mathcal{K}(VXV^T)$ and $\mathcal{T}_V(A) = V^T \mathcal{T}(A)V$. That is,

$$\mathscr{K}_V(X) = \operatorname{diag}(VXV^T)e^T + e(\operatorname{diag}(VXV^T))^T - 2VXV^T \qquad (3.16)$$

$$\mathscr{T}_V(A) = -\frac{1}{2}V^T AV. \qquad (3.17)$$

Lemma 3.4 $\mathscr{T}(A) = V\mathscr{T}_V(A)V^T$ *for all* $A \in \mathscr{S}^n$.

Proof. Let $A \in \mathscr{S}^n$, then $\mathscr{T}(A)e = \mathbf{0}$. Thus, $V\mathscr{T}_V(A)V^T = VV^T\mathscr{T}(A)VV^T = \mathscr{T}(A)$ since $VV^T = J$ and since $\mathscr{T}(A)e = \mathbf{0}$.

$\square$

The next lemma is especially useful in the study of the geometry of the EDMs.

Lemma 3.5 ([21]) *The adjoint of* $\mathscr{K}_V$ *is given by*

$$\mathscr{K}_V^*(A) = 2V^T(Diag(Ae) - A)V.$$

Proof. Let $A \in \mathscr{S}^n$. Then $\operatorname{trace}(A\mathscr{K}_V(X)) = 2\operatorname{trace}(V^T(Diag(Ae) - A)VX)$ and the result follows.

$\square$

Note that $\mathscr{K}_V$ has a trivial kernel since the kernel of $\mathscr{K} = \{ae^T + ea^T : a \in \mathbb{R}^n\}$. The following lemma is analogous to Lemma 3.1.

Lemma 3.6 ([21]) *The transformation* $\mathscr{T}_V$, *restricted to* $\mathscr{S}_h^n$, *is the inverse of the transformation* $\mathscr{K}_V$; *i.e.,*

$$\mathscr{T}_V|_{\mathscr{S}_h^n} = (\mathscr{K}_V)^{-1} \text{ and } \mathscr{K}_V = (\mathscr{T}_V|_{\mathscr{S}_h^n})^{-1}.$$

Proof. Let $X \in \mathscr{S}^{n-1}$, then $\mathscr{T}_V(\mathscr{K}_V(X)) = V^T\mathscr{T}(\mathscr{K}(VXV^T))V = X$. On the other hand, let $A \in \mathscr{S}_h^n$ then $\mathscr{T}(A)e = 0$. Thus $\mathscr{K}_V(\mathscr{T}_V(A)) = \mathscr{K}(V\mathscr{T}_V(A)V^T) = \mathscr{K}(\mathscr{T}(A)) = A$.

$\square$

An immediate consequence of the definition of $\mathscr{T}$ and Lemma 3.4 is the fact that if $A \in \mathscr{S}_h^n$, then $\operatorname{rank}(\mathscr{T}(A)) = \operatorname{rank}(\mathscr{T}_V(A))$ and $\mathscr{T}(A) \succeq \mathbf{0}$ iff $\mathscr{T}_V(A) \succeq \mathbf{0}$. As a result, the following theorem is a restatement of Theorem 3.1.

Theorem 3.2 ([24, 21]) *Let D be an $n \times n$ real symmetric matrix whose diagonal entries are all 0's. Then D is an EDM if and only if*

$$\mathscr{T}_V(D) \succeq 0.$$

Moreover, the embedding dimension of D is given by the rank of $\mathscr{T}_V(D)$.

As a result, $\mathscr{D}^n$ is the image of $\mathscr{S}_+^{n-1}$ under the linear transformation $\mathscr{K}_V$; and $\mathscr{S}^{n-1}$ is the image of $\mathscr{D}^n$ under the linear transformation $\mathscr{T}_V$. Furthermore, if D is an EDM, then $\mathscr{T}(D)$ is the Gram matrix of D. Accordingly, the matrix $\mathscr{T}_V(D)$ is called the *projected Gram matrix* of D. As we will see below, projected Gram matrices have a geometric interpretation in terms of the volume of the simplex defined by the generating points of D.

The following characterization, which follows as a corollary of Theorem 3.2, is particularly useful for pencil-and-paper computation for small n.

Theorem 3.3 ([25]) *Let* $D = \begin{bmatrix} 0 & d^T \\ d & \bar{D} \end{bmatrix}$ *be an* $n \times n$ *real symmetric matrix whose diagonal entries are all 0's, where* $d \in \mathbb{R}^{n-1}$. *Then D is an EDM if and only if*

$$de_{n-1}^T + e_{n-1}d^T - \bar{D} \succeq 0. \tag{3.18}$$

Moreover, the embedding dimension of D is given by $rank(de_{n-1}^T + e_{n-1}d^T - \bar{D})$.

Proof. D is an EDM iff $(-V^T DV) \succeq 0$. But $V = UA$ where U is as defined in (3.10) and A is the nonsingular matrix given in (3.13). Thus, D is an EDM iff $-U^T DU \succeq 0$. The result follows since $U^T DU = -de_{n-1}^T - e_{n-1}d^T + \bar{D}$ and since $rank(U^T DU) = rank(V^T DV)$.

□

As an application of Theorem 3.3, we present a proof of the obvious fact that if D is an EDM, then the square roots of its entries satisfy the triangular inequality.

Theorem 3.4 *Let* $n \geq 3$ *and let D be an* $n \times n$ *EDM. Then for any distinct indices* i, j, k, *we have*

$$|\sqrt{d_{ij}} - \sqrt{d_{jk}}| \leq \sqrt{d_{ik}} \leq \sqrt{d_{ij}} + \sqrt{d_{jk}}.$$

Proof. Wlog assume that $i = 1, j = 2, z = 3$, and that $d_{12} = a, d_{13} = b$ and $d_{23} = c$. Then it follows from Theorem 3.3 that

$$\begin{bmatrix} 2a & a+b-c \\ a+b-c & 2b \end{bmatrix} \succeq 0.$$

Thus,

$$4ab - (a+b-c)^2 = (2\sqrt{a}\sqrt{b} - a - b + c)(2\sqrt{a}\sqrt{b} + a + b - c) \geq 0.$$

Hence, we have two possibilities. The first one is:

$$(2\sqrt{a}\sqrt{b} - a - b + c) \geq 0 \text{ and } (2\sqrt{a}\sqrt{b} + a + b - c) \geq 0$$

or $(\sqrt{a} - \sqrt{b})^2 \leq c \leq (\sqrt{a} + \sqrt{b})^2$.
    The second possibility is $(2\sqrt{a}\sqrt{b} - a - b + c) \leq 0$ and $(2\sqrt{a}\sqrt{b} + a + b - c) \leq 0$; or $(\sqrt{a} + \sqrt{b})^2 \leq c \leq (\sqrt{a} - \sqrt{b})^2$, a contradiction.

□

As the following example shows, the converse of Theorem 3.4 is false for $n \geq 4$.

Example 3.3 *Consider the matrix* $A = \begin{bmatrix} 0 & 1 & 1 & \alpha \\ 1 & 0 & 1 & \alpha \\ 1 & 1 & 0 & \alpha \\ \alpha & \alpha & \alpha & 0 \end{bmatrix}$. *The square roots of the entries of A satisfy the triangular inequality for* $\alpha = 0.3$ *since* $\sqrt{0.3} > 0.54$, *i.e.,* $\sqrt{\alpha} > 1/2$. *However, A is not an EDM. In fact, A is an EDM iff* $\alpha \geq 1/3$. *Moreover, the embedding dimension of A is 2 iff* $\alpha = 1/3$.

The following example shows the advantage of Theorem 3.3 for pencil-and-paper computations.

Example 3.4 *Let* $D = \begin{bmatrix} 0 & 10 & 10 & 4 \\ 10 & 0 & 4 & 2 \\ 10 & 4 & 0 & 2 \\ 4 & 2 & 2 & 0 \end{bmatrix}$. *Then* D *is an EDM of embedding dimension*

2 since $de_3^T + e_3 d^T - \bar{D} = \begin{bmatrix} 20 & 16 & 12 \\ 16 & 20 & 12 \\ 12 & 12 & 8 \end{bmatrix}$ *is PSD and of rank 2.*

As noted earlier, the choice of V in (3.12) is related to Householder matrices. This fact, which is established next, leads to yet another characterization of EDMs. Let Q be the $n \times n$ Householder matrix

$$Q = I - 2\frac{vv^T}{v^Tv}, \quad \text{where } v = \begin{bmatrix} 1 + \sqrt{n} \\ e_{n-1} \end{bmatrix}. \tag{3.19}$$

Then, $v^Tv = 2(n + \sqrt{n})$ and

$$2\frac{vv^T}{v^Tv} = \begin{bmatrix} 1 + 1/\sqrt{n} & e_{n-1}^T/\sqrt{n} \\ e_{n-1}/\sqrt{n} & e_{n-1}e_{n-1}^T/(n+\sqrt{n}) \end{bmatrix}.$$

Hence,

$$Q = \begin{bmatrix} y & ye_{n-1}^T \\ ye_{n-1} & I_{n-1} + xe_{n-1}e_{n-1}^T \end{bmatrix} = \begin{bmatrix} ye^T \\ V^T \end{bmatrix}.$$

Hence, the following theorem is a simple corollary of Theorem 3.2. One should keep in mind that $Q = Q^T$.

Theorem 3.5 (Hayden and Wells [102]) *Let* D *be an* $n \times n$ *real symmetric matrix whose diagonal entries are all 0's, and let* Q *be the householder matrix defined in (3.19). Then* D *is an EDM if and only if the submatrix of* $(-QDQ^T)$ *obtained by deleting the first row and the first column is positive semidefinite. Moreover, the embedding dimension of* D *is given by the rank of this submatrix.*

3.2 The Gale Matrix Z

The notion of Gale transform, which plays a key role in the theory of polytopes [84, 99], also plays an important role in the theory of EDMs. The *Gale space* of an $n \times n$ EDM D of embedding dimension r, denoted by $\text{gal}(D)$, is defined as

$$\text{gal}(D) = \text{null}(\begin{bmatrix} P^T \\ e^T \end{bmatrix}) = \text{null}(P^T) \cap \text{null}(e^T), \tag{3.20}$$

where P is a configuration matrix of D. Accordingly, the dimension of $\mathrm{gal}(D)$ is given by

$$\bar{r} = n - 1 - r. \tag{3.21}$$

Let P' be a configuration matrix of D obtained from P by a translation along a. Thus $P' = P - ea^T$. Hence, $\mathrm{null}(P'^T) \cap \mathrm{null}(e^T) = \mathrm{null}(P^T) \cap \mathrm{null}(e^T)$. Consequently, $\mathrm{gal}(D)$ is uniquely determined by D. Gale space can also be defined in terms of the null space of the Gram matrix $B = \mathcal{T}(D)$. More precisely, $\mathrm{gal}(D) = \mathrm{null}\left(\begin{bmatrix} B \\ e^T \end{bmatrix} \right)$ since $\mathrm{null}(P^T) = \mathrm{null}(B)$.

Any $n \times \bar{r}$ matrix Z whose columns form a basis of $\mathrm{gal}(D)$ is called a *Gale matrix* of D. Let $(z^i)^T$ be the ith row of Z, i.e., let

$$Z = \begin{bmatrix} (z^1)^T \\ \vdots \\ (z^n)^T \end{bmatrix}. \tag{3.22}$$

Then $z^1, \ldots, z^n$ are called *Gale transforms* of $p^1, \ldots, p^n$, respectively. As we will see in this monograph, in some cases it is more convenient to use Gale matrix Z, whereas in other cases, using Gale transforms $z^1, \ldots, z^n$ is more convenient.

Observe that the columns of Z encode the affine dependency of the generating points of D. As a result, Gale matrices are particularly useful in characterizing points in general position. Points $p^1, \ldots, p^n$ in $\mathbb{R}^r$ are said to be in *general position* in $\mathbb{R}^r$ if every $r + 1$ of these points is affinely independent; i.e., if every $r + 1$ of these points affinely spans $\mathbb{R}^r$. For instance, points in the plane are in general position if no three of them are collinear since three collinear points affinely span a straight line. An immediate consequence of this definition is that if n points are in general position in $\mathbb{R}^r$ and if $n \geq r + 2$, then any $n - 1$ of these points will continue to affinely span $\mathbb{R}^r$. In other words, deleting one point from a configuration of n ($n \geq r + 2$) points in general position in $\mathbb{R}^r$ does not decrease the dimension of the affine hull of the remaining points.

An EDM D of embedding dimension r is said to be in general position if its generating points are in general position in $\mathbb{R}^r$. The following lemma relates the affine dependence of a point configuration to the linear dependence of the corresponding Gale transforms.

Lemma 3.7 *Let $z^1, \ldots, z^n \in \mathbb{R}^{n-r-1}$ be Gale transforms of $p^1, \ldots, p^n \in \mathbb{R}^r$, respectively. Let $\mathscr{I}$ be a subset of $\{1, \ldots, n\}$ of cardinality $r + 1$. Then $\{p^i : i \in \mathscr{I}\}$ are affinely dependent if and only if $\{z^i : i \in \bar{\mathscr{I}}\}$ are linearly dependent, where $\bar{\mathscr{I}} = \{1, \ldots, n\} \setminus \mathscr{I}$.*

Proof. Wlog assume that $\mathscr{I} = \{1, \ldots, r+1\}$, i.e., assume that $p^1, \ldots, p^{r+1}$ are affinely dependent. Then there exists a nonzero $\lambda = (\lambda_i) \in \mathbb{R}^{r+1}$ such that $\sum_{i=1}^{r+1} \lambda_i p^i = \mathbf{0}$ and $\sum_{i=1}^{r+1} \lambda_i = 0$. Let x be the vector in $\mathbb{R}^n$ such that $x = \begin{bmatrix} \lambda \\ \mathbf{0} \end{bmatrix}$. Then $\sum_{i=1}^{n} x_i p^i = \mathbf{0}$ and $\sum_{i=1}^{n} x_i = 0$; i.e., $x \in \mathrm{gal}(D)$. As a result, if Gale matrix Z is partitioned as $Z = \begin{bmatrix} Z_1 \\ Z_2 \end{bmatrix}$, where Z_1 is $(r+1) \times (n-r-1)$. Then

$$x = \begin{bmatrix} \lambda \\ 0 \end{bmatrix} = \begin{bmatrix} Z_1 \\ Z_2 \end{bmatrix} \xi,$$

for some nonzero $\xi \in \mathbb{R}^{n-r-1}$. Consequently, the square matrix Z_2 is singular. Therefore, the rows of Z_2, i.e., $z^{r+2}, \ldots, z^n$ are linearly dependent. The result follows since each of the above steps is reversible.

$\square$

As an immediate corollary of Lemma 3.7, we have the following characterization of point configurations in general position.

Corollary 3.1 ([6]) *Let D be an $n \times n$ EDM of embedding dimension r, $r \leq n-2$, and let Z be a Gale matrix of D. Then D is in general position if and only if every submatrix of Z of order $(n-r-1)$ is nonsingular.*

More useful properties of Gale matrices are given next. Assume that $p^1, \ldots, p^{r+1}$ are affinely independent and that Gale matrix Z is partitioned as $Z = \begin{bmatrix} Z_1 \\ Z_2 \end{bmatrix}$, where Z_2 is $(n-1-r) \times (n-1-r)$. Then Z_2 is nonsingular. Moreover, the matrix obtained by multiplying Z from the right with Z_2^{-1} is also a Gale matrix. Consequently, by relabelling the nodes if necessary, we always have a Gale matrix of the form $Z = \begin{bmatrix} \bar{Z} \\ I_{n-1-r} \end{bmatrix}$.

Another useful property of Gale space is its connection with the null space of projected Gram matrices. This connection will be used repeatedly in this monograph.

Lemma 3.8 ([2]) *Let D be an $n \times n$ EDM of embedding dimension $r \leq n-2$, and let $X = \mathcal{T}_V(D)$ be the projected Gram matrix of D. Let Z and P, $P^T e = 0$, be a Gale matrix and a configuration matrix of D. Further, let U and W be the matrices whose columns form orthonormal bases of $\text{null}(X)$ and $\text{col}(X)$, respectively. Then*

1. *$VU = ZA$ for some nonsingular matrix A; i. e., VU is a Gale matrix of D.*
2. *$VW = PA'$ for some nonsingular matrix A'.*

Proof. $X = V^T P P^T V$. Thus $XU = 0$ iff $P^T VU = 0$. But $e^T VU = 0$. Hence, the columns of VU form a basis of $\text{gal}(D)$ and thus Statement 1 follows. Statement 2 follows since $Z^T VW = A^{-T} U^T V^T VW = 0$ and $e^T VW = 0$.

$\square$

More properties are given in Chap. 7, where Gale transform is revisited.

3.3 Basic Properties of EDMs

Several properties of EDMs follow from their characterizations in the previous section. Theorems 3.1 and 1.5 imply that an $n \times n$ EDM D has at least $n-1$ nonpositive eigenvalues. But $\text{diag}(D) = \mathbf{0}$ and hence $\text{trace}(D) = 0$. Consequently, a nonzero EDM has exactly one positive eigenvalue. A real symmetric matrix with exactly one simple positive eigenvalue is called *elliptic*. Moreover, an elliptic matrix C is said to be *special elliptic* [80] if $\text{diag}(C) = \mathbf{0}$. For example,

$$C = \begin{bmatrix} 0 & 1 \\ 1 & 0 \end{bmatrix} \tag{3.23}$$

is a special elliptic matrix. Accordingly, the set of nonzero EDMs is a proper subset of the set of nonnegative special elliptic matrices [80]. To see that not every nonnegative special elliptic matrix is an EDM, consider the matrix

$$C' = \begin{bmatrix} C & \mathbf{0} \\ \mathbf{0} & 0 \end{bmatrix},$$

where C is the matrix in (3.23). Obviously, C' is a nonnegative special elliptic matrix, but it is not an EDM.

The sign of the determinant of a nonsingular elliptic matrix is easily determined.

Lemma 3.9 *Let A be an $n \times n$ nonsingular elliptic matrix. Then the determinant of A has sign $(-1)^{n-1}$.*

Proof. A has exactly $n - 1$ negative eigenvalues since A is nonsingular and has exactly one positive eigenvalue. Therefore, sign $\det(A) = (-1)^{n-1}$ since the determinant of A is the product of its eigenvalues.

$\square$

The set of special elliptic matrices is characterized in the following theorem.

Theorem 3.6 (Fiedler [80]) *The set of $n \times n$, $n \geq 2$, special elliptic matrices is given by*

$$\{C \in \mathscr{S}^n : C = aa^T - A \neq \mathbf{0}, \text{ where } A \succeq \mathbf{0}, A \neq \mathbf{0} \text{ and } \text{diag}(C) = \mathbf{0}\}.$$

Proof. Let C be a special elliptic matrix and let λ be its positive eigenvalue with corresponding normalized eigenvector x. Further, let $C = \lambda xx^T - W \Lambda W^T$ be the spectral decomposition of C, where $(-\Lambda)$ is the diagonal matrix consisting of the nonpositive eigenvalues of C. Therefore, $C = aa^T - A$ where $A = W \Lambda W^T \succeq \mathbf{0}$ and $a = \sqrt{\lambda}\, x$.

On the other hand, assume that $C = aa^T - A \neq \mathbf{0}$, where $A \succeq \mathbf{0}$, $A \neq \mathbf{0}$ and $\text{diag}(C) = \mathbf{0}$. Then since $\text{trace}(C) = 0$ and $C \neq \mathbf{0}$, it follows that C has at least one positive eigenvalue. Now let $\mathscr{L} = a^\perp$. Then $\dim(\mathscr{L}) = n - 1$. Moreover, for each $y \in \mathscr{L}$, we have $y^T C y = -y^T A y \leq 0$. Therefore, by Theorem 1.5, C has at least $n - 1$ nonpositive eigenvalues; i.e., C has at most one positive eigenvalue. Consequently, C has exactly one positive eigenvalue and the result follows.

$\square$

It should be pointed out that the set of $n \times n$ special elliptic matrices, for $n \geq 4$, is not convex [80] since the matrix

$$\frac{1}{2}\begin{bmatrix} C & 0 \\ 0 & 0 \end{bmatrix} + \frac{1}{2}\begin{bmatrix} 0 & 0 \\ 0 & C \end{bmatrix},$$

where C is the matrix in (3.23), is not elliptic.

Next, we establish the connection between Gale matrices and EDMs.

Lemma 3.10 *Let D be a nonzero EDM and let Z be a Gale matrix of D. Further, let $B = \mathscr{T}(D)$ be the Gram matrix of D. Then*

$$DZ = e\xi^T,$$

where $\xi = Z^T diag(B)$.

Proof. This follows directly from the definition of $\mathscr{K}$ in (3.3) since $D = \mathscr{K}(B)$.
$\square$

Theorem 3.7 *Let D be a nonzero $n \times n$ EDM and let $gal(D)$ be its Gale space. Then*

$$null(D) \subseteq gal(D).$$

Proof. Let $x \in null(D)$ and let $B = \mathscr{T}(D)$. Then it follows from the definition of $\mathscr{T}$ in (3.9) that $2x^T Bx = -(e^T x)^2 e^T De/n^2 \le 0$. But since B is PSD and since $e^T De > 0$, it follows that $e^T x = 0$ and $x^T Bx = 0$. Consequently, $x \in gal(D)$.
$\square$

An immediate consequence of Theorem 3.7 is that the rank of an $n \times n$ nonzero EDM can assume only two values, and that these values are independent of n.

Theorem 3.8 (Gower [93]) *Let D be a nonzero $n \times n$ EDM of embedding dimension r. Then*

$$rank(D) = r + 1 \text{ or } rank(D) = r + 2.$$

Proof. On the one hand, it follows from Eq. (1.7) and the definition of $\mathscr{K}$ that $rank(D) \le r + 2$. On the other hand, by Theorem 3.7, $rank(D) \ge r + 1$ since $\dim gal(D) = n - r - 1$. Thus the result follows.
$\square$

Another consequence of Theorem 3.7 is that e is always in the column space of a nonzero EDM.

Theorem 3.9 (Gower [93]) *Let D be a nonzero $n \times n$ EDM. Then*

$$e \text{ lies in } col(D) \text{ or } DD^\dagger e = e,$$

where $D^\dagger$ is the Moore–Penrose inverse of D.

Proof. This follows from Theorem 3.7 since $gal(D)^\perp \subseteq col(D)$ and since $e \in gal(D)^\perp$. Recall that $DD^\dagger$ is the orthogonal projection on $col(D)$.
$\square$

The following theorem exploits the freedom to choose an origin.

Theorem 3.10 *Let D be a nonzero EDM and let $Dw = e$, then there exists a configuration matrix P of D such that $P^T w = \mathbf{0}$.*

Proof. The existence of w follows from Theorem 3.9. If $e^T w \neq 0$, let the Gram matrix be $B = -\frac{1}{2}(I - ew^T/(e^T w))D(I - we^T/(e^T w))$. Then $Bw = \mathbf{0}$ and hence $P^T w = \mathbf{0}$. In this case, the centroid of the generating points of D is $P^T e/n$.

On the other hand, if $e^T w = 0$, then let $B = -(I - es^T)D(I - se^T)/2$ for some s such that $e^T s = 1$. Thus, $Bw = \mathbf{0}$ since $D(I - se^T)w = e$. Hence $P^T w = \mathbf{0}$.   □

Example 3.5 *Consider the EDM* $D = \begin{bmatrix} 0 & 1 & 9 & 16 \\ 1 & 0 & 4 & 9 \\ 9 & 4 & 0 & 1 \\ 16 & 9 & 1 & 0 \end{bmatrix}$. *Then $Dw = e$ yields $w =$*

$\frac{1}{6}[1 \ -1 \ -1 \ 1]^T$ *and hence $e^T w = 0$. Let $B = -JDJ/2$ then the configuration*

matrix is $P = \begin{bmatrix} -2 \\ -1 \\ 1 \\ 2 \end{bmatrix}$. *Note that $P^T w = \mathbf{0}$ as well as $P^T e = \mathbf{0}$. In this case, the*

centroid of the generating points coincides with the origin.

The following theorem presents a necessary and sufficient condition for a special elliptic matrix to be an EDM.

Theorem 3.11 (Crouzeix and Ferland [68]) *Let D be a nonzero real symmetric matrix whose diagonal entries are all zeros. Assume that D has exactly one positive eigenvalue. Then D is an EDM if and only if there exists $w \in \mathbb{R}^n$ such that $Dw = e$ and $w^T e \geq 0$.*

Proof. Assume that D is an EDM. Then by Theorem 3.9, there exists w such that $Dw = e$. If $e^T w = 0$, there is nothing to prove. Therefore, assume that $e^T w \neq 0$. Thus, we can assume that the origin is chosen such that $P^T w = \mathbf{0}$. Then it follows from the definition of $\mathcal{K}$ that $Dw = e^T w \operatorname{diag}(B) + \operatorname{diag}(B)^T w \, e = e$. Hence,

$$e^T w \operatorname{diag}(B) = (1 - w^T \operatorname{diag}(B))e. \tag{3.24}$$

Thus, $2e^T w \, w^T \operatorname{diag}(B) = w^T e$ and hence $w^T \operatorname{diag}(B) = 1/2$. Hence, it follows from (3.24) that $e^T w > 0$ since $\operatorname{diag}(B) \geq 0$. Therefore, if D is an EDM, then either $e^T w = 0$ or $e^T w > 0$.

To prove the reverse direction, assume that there exists w such that $Dw = e$ and $e^T w \geq 0$. We consider two cases.

Case 1: $w^T e > 0$. Then the matrix $S = [w \ V]$ is nonsingular, where V is as defined in (3.11). Moreover,

$$S^T D S = \begin{bmatrix} w^T e & \mathbf{0} \\ \mathbf{0} & V^T D V \end{bmatrix}.$$

Therefore, since D has exactly one positive eigenvalue, it follows from *Sylvester law of inertia* that $S^T D S$ has exactly one positive eigenvalue, namely $e^T w$. Hence, $V^T D V$ is negative semidefinite. Consequently, D is an EDM since $\mathcal{T}_V(D)$ is PSD.

Case 2: $w^T e = 0$. Let $V = [w \ \bar{V}]$ and thus the matrix $S' = [e \ w \ \bar{V}]$ is nonsingu-

lar. Moreover, $S'^T DS' = \begin{bmatrix} e^T De & n & e^T D\bar{V} \\ n & 0 & \mathbf{0} \\ \bar{V}^T De & \mathbf{0} & \bar{V}^T D\bar{V} \end{bmatrix}$. Note that the Schur complement of

$\begin{bmatrix} e^T De & n \\ n & 0 \end{bmatrix}$ is

$$\bar{V}^T D\bar{V} - [\bar{V}^T De \ \mathbf{0}] \begin{bmatrix} 0 & 1/n \\ 1/n & -e^T De/n^2 \end{bmatrix} \begin{bmatrix} e^T D\bar{V} \\ \mathbf{0} \end{bmatrix} = \bar{V}^T D\bar{V}.$$

Therefore, let

$$E = \begin{bmatrix} 1 & 0 & \mathbf{0} \\ 0 & 1 & \mathbf{0} \\ \mathbf{0} & -\bar{V}^T De/n & I \end{bmatrix}. \text{ Then } E(S'^T DS')E^T = \begin{bmatrix} e^T De & n & \mathbf{0} \\ n & 0 & \mathbf{0} \\ \mathbf{0} & \mathbf{0} & \bar{V}^T D\bar{V} \end{bmatrix}.$$

But $\begin{bmatrix} e^T De & n \\ n & 0 \end{bmatrix}$ has one positive and one negative eigenvalue since its determinant is negative. Therefore, it follows from *Sylvester law of inertia* and the fact that D has exactly one positive eigenvalue that $\bar{V}^T D\bar{V}$ is negative semidefinite and hence $V^T DV = \begin{bmatrix} 0 & \mathbf{0} \\ \mathbf{0} & \bar{V}^T D\bar{V} \end{bmatrix}$ is negative semidefinite.

$\square$

It should be pointed out that if D is an EDM and if $Dw = e$, then whether $e^T w = 0$ or $e^T w > 0$ has a geometric significance. This issue will be investigated in great detail in Chap. 4. We conclude this section with the following theorem which extends the notion of the polynomial of a graph to EDMs [110].

Theorem 3.12 *Let D be a nonzero $n \times n$ EDM. Then there exists a polynomial $g(D)$ such that*

$$g(D) = \gamma x x^T, \tag{3.25}$$

where $x \in \mathbb{R}^n$ is the Perron eigenvector of D and γ is a scalar.

Proof. Let the distinct eigenvalues of D be $\lambda > -\mu_1 > \cdots > -\mu_k$. Therefore, $(D - \lambda I)x = \mathbf{0}$. Since D is symmetric, the minimal polynomial of D implies that

$$m(D) = (D - \lambda I)(D + \mu_1 I) \cdots (D + \mu_k I) = (D - \lambda I) \prod_{i=1}^{k} (D + \mu_i I) = 0.$$

Therefore, $\prod_{i=1}^{k} (D + \mu_i I) = xy^T$ for some vector y. But since D is symmetric, this implies that

$$g(D) = \prod_{i=1}^{k} (D + \mu_i I) = \gamma x x^T$$

for some scalar γ.

$\square$

3.4 The Cayley–Menger Matrix

The Cayley–Menger determinant [52, 143, 144, 46, 45, 66] is used to compute the
volume of a simplex. As will be shown in this section, this volume can also be
computed using the corresponding projected Gram matrix. Moreover, the Cayley–
Menger matrix and the Cayley–Menger determinant provide yet another characterization of EDMs.

Let D be an EDM, then $M = \begin{bmatrix} 0 & e^T \\ e & D \end{bmatrix}$ is called the *Cayley–Menger matrix* of D,
and $\det(M)$ is called the *Cayley–Menger determinant* of D. It should be pointed out
that M, not D, is what Menger [143] calls a distance matrix. As it is shown next, the
Cayley–Menger determinant is independent of the labeling of the generating points
of D.

Theorem 3.13 *Let D be an $n \times n$ EDM and let Q be an $n \times n$ permutation matrix.
Then*

$$\det\left(\begin{bmatrix} 0 & e^T \\ e & Q^T D Q \end{bmatrix}\right) = \det\left(\begin{bmatrix} 0 & e^T \\ e & D \end{bmatrix}\right).$$

Proof.   By definition of Q, $Qe = e$ and $Q^T e = e$. Therefore,

$$\begin{bmatrix} 1 & 0 \\ 0 & Q^T \end{bmatrix} \begin{bmatrix} 0 & e^T \\ e & D \end{bmatrix} \begin{bmatrix} 1 & 0 \\ 0 & Q \end{bmatrix} = \begin{bmatrix} 0 & e^T \\ e & Q^T D Q \end{bmatrix},$$

and the result follows.

$\square$

Suppose that D is nonsingular and let $Dw = e$. Further, assume that $e^T w = 1/(2\rho^2) > 0$. Then the inverse of the Cayley–Menger matrix M is given by

$$\begin{bmatrix} 0 & e^T \\ e & D \end{bmatrix}^{-1} = 2\rho^2 \begin{bmatrix} -1 & w^T \\ w & D^{-1}/(2\rho^2) - ww^T \end{bmatrix}. \tag{3.26}$$

The entries of M^{-1} have an interesting geometric interpretation which is discussed
in the next chapter.

The following theorem is a special case of a more general result of Chabrillac
and Crouzeix [53]. It provides a characterization of EDMs in terms of M.

Theorem 3.14 (Hayden and Wells [102] and Fiedler [80]) *Let D be a nonzero
real symmetric $n \times n$ matrix whose diagonal entries are all zeros and let
$M = \begin{bmatrix} 0 & e^T \\ e & D \end{bmatrix}$. Then D is an EDM if and only if M has exactly one positive eigenvalue, in which case, $\text{rank}(M) = r + 2$, where r is the embedding dimension of D.*

Proof. Let $Q = \begin{bmatrix} 0 & 1 & 0 \\ e/\sqrt{n} & 0 & V \end{bmatrix}$, then $Q^T M Q = \begin{bmatrix} \frac{1}{n} e^T D e & \sqrt{n} & \frac{1}{\sqrt{n}} e^T D V \\ \sqrt{n} & 0 & 0 \\ \frac{1}{\sqrt{n}} V^T D e & 0 & V^T D V \end{bmatrix}$. Note

that Q is orthogonal and that the Schur complement of $\begin{bmatrix} e^T D e/n & \sqrt{n} \\ \sqrt{n} & 0 \end{bmatrix}$ is

$$V^T D V - [\frac{1}{\sqrt{n}} V^T D e \;\; 0] \begin{bmatrix} 0 & 1/\sqrt{n} \\ 1/\sqrt{n} & -e^T D e/n^2 \end{bmatrix} \begin{bmatrix} e^T D V/\sqrt{n} \\ 0 \end{bmatrix} = V^T D V.$$

Thus, let

$$E = \begin{bmatrix} 1 & 0 & 0 \\ 0 & 1 & 0 \\ 0 & -V^T D e/n & I \end{bmatrix}. \text{ Then } E(Q^T M Q)E^T = \begin{bmatrix} e^T D e/n & \sqrt{n} & 0 \\ \sqrt{n} & 0 & 0 \\ 0 & 0 & V^T D V \end{bmatrix}.$$

Now, $\begin{bmatrix} e^T D e/n & \sqrt{n} \\ \sqrt{n} & 0 \end{bmatrix}$ has one positive and one negative eigenvalue since its de-

terminant is negative. Therefore, it follows from *Sylvester law of inertia* that M has exactly one positive eigenvalue if and only if $V^T D V$ is negative semidefinite. Observe that $\text{rank}(M) = \text{rank}(V^T D V) + 2$. $\qquad\square$

It should be noted that Theorem 3.14 still holds if e in M is replaced by $(-e)$.

The volume of a simplex can be computed in terms of its corresponding Cayley–Menger determinant. The area of a triangle of vertices at p^1, p^2, and p^3 in $\mathbb{R}^2$ is given by

$$\frac{1}{2!} \det\left(\begin{bmatrix} 1 & 1 & 1 \\ p^1 & p^2 & p^3 \end{bmatrix} \right).$$

This formula generalizes to simplices in higher dimensions. Let $p^1, \ldots, p^n$ be in $\mathbb{R}^{n-1}$ and let $V(p^1, \ldots, p^n)$ denote the volume of the simplex whose vertices are at $p^1, \ldots, p^n$. Then

$$V(p^1, \ldots, p^n) = \frac{1}{(n-1)!} \det\left(\begin{bmatrix} 1 & 1 & \cdots & 1 \\ p^1 & p^2 & \cdots & p^n \end{bmatrix} \right) = \frac{1}{(n-1)!} \det\left(\begin{bmatrix} e^T \\ P^T \end{bmatrix} \right).$$

Theorem 3.15 (Menger [144]) *Let D be an $n \times n$ EDM of embedding dimension $n-1$ and let $V(p^1, \ldots, p^n)$ denote the volume of the simplex defined by the generating points of D. Then*

$$V^2(p^1, \ldots, p^n) = \frac{(-1)^n}{2^{n-1}((n-1)!)^2} \det\left(\begin{bmatrix} 0 & e^T \\ e & D \end{bmatrix} \right).$$

Proof. Using the fact that $\det(A^T) = \det(A)$, we obtain that

$$((n-1)!)^2 V^2(p^1, \ldots, p^n) = \det([e \; P] \begin{bmatrix} e^T \\ P^T \end{bmatrix}) = \det([ee^T + PP^T]).$$

But this last determinant is obviously equal to $\det(\begin{bmatrix} 1 & e^T \\ \mathbf{0} & ee^T + PP^T \end{bmatrix})$, which in turn

is equal to $\det(\begin{bmatrix} 1 & e^T \\ -e & B \end{bmatrix})$. This follows by subtracting row 1 from rows 2 to n. Now

this last determinant is equal to $\det(\begin{bmatrix} 0 & e^T \\ -e & B \end{bmatrix})$ since its $(1,1)$ cofactor is $\det(B) = 0$.
But,

$$\det(\begin{bmatrix} 0 & e^T \\ -e & B \end{bmatrix}) = \frac{1}{(-2)^n} \det(\begin{bmatrix} 0 & e^T \\ 2e & -2B \end{bmatrix}) = \frac{2}{(-2)^n} \det(\begin{bmatrix} 0 & e^T \\ e & -2B \end{bmatrix}),$$

where the first equality follows by multiplying rows 2 to $n+1$ with -2; and the
second equality follows by factoring 2 out of the first column. Moreover,

$$\det(\begin{bmatrix} 0 & e^T \\ e & -2B \end{bmatrix}) = \det(\begin{bmatrix} 0 & e^T \\ e & \text{diag}(B)e^T + e(\text{diag}(B))^T - 2B \end{bmatrix}) = \det(\begin{bmatrix} 0 & e^T \\ e & D \end{bmatrix}),$$

where the first equality follows by adding to row i, $i = 2,\ldots,n+1$, $b_{i-1\,i-1}$ times
row 1; and by adding to column j, $j = 2,\ldots,n+1$, $b_{j-1\,j-1}$ times column 1. Thus
the result follows.

□

Observe that $(-1)^n \det(M)$ is positive. This follows as a simple consequence of
Lemma 3.9 and Theorem 3.14.

The volume of a simplex can be, equivalently, expressed in terms of its corresponding projected Gram matrix as shown by the following two corollaries.

Corollary 3.2 *Let D be an $n \times n$ EDM of embedding dimension $n-1$ and let
$V(p^1,\ldots,p^n)$ denote the volume of the simplex defined by the generating points of
D. Then*

$$V^2(p^1,\ldots,p^n) = \frac{n}{((n-1)!)^2} \det(\mathscr{T}_V(D)).$$

Proof. The proof of Theorem 3.14 implies that $\det(M) = -n\,\det(V^T DV) = (-1)^n 2^{n-1} n\,\det(\mathscr{T}_V(D))$.

□

Corollary 3.3 *Let $D = \begin{bmatrix} 0 & d^T \\ d & \bar{D} \end{bmatrix}$ be an $n \times n$ EDM of embedding dimension $n-1$;
and let $V(p^1,\ldots,p^n)$ denote the volume of the simplex defined by the generating
points of D. Then*

$$V^2(p^1,\ldots,p^n) = \frac{1}{2^{n-1}((n-1)!)^2} \det(de_{n-1}^T + e_{n-1}d^T - \bar{D}).$$

Proof. Recall from Eq. (3.13) that $V = UA$ where $A = I_{n-1} + xe_{n-1}e_{n-1}^T$.
Hence, $\det(A) = 1 + x(n-1) = 1/\sqrt{n}$. Moreover, $\det(\mathscr{T}_V(D)) = 2^{-n+1}\,(\det(A))^2\,\det(-U^T DU)$.

□

Example 3.6 *Consider the simplex with vertices at $p^1 = e^1$, $p^2 = e^2$, $p^3 = e^3$, and $p^4 = 0$, where e^1, e^2, and e^3 are the standard unit vectors in $\mathbb{R}^3$. The EDM generated by this configuration and the corresponding projected Gram matrix are*

$$D = \begin{bmatrix} 0 & 2 & 2 & 1 \\ 2 & 0 & 2 & 1 \\ 2 & 2 & 0 & 1 \\ 1 & 1 & 1 & 0 \end{bmatrix} \text{ and } X = \mathscr{T}_V(D) = \frac{1}{36}\begin{bmatrix} 35 & -1 & 5 \\ -1 & 35 & 5 \\ 5 & 5 & 11 \end{bmatrix}, \text{ respectively, where we used } V$$

as in (3.12). Then $det(M) = 8$. Hence, the volume of this simplex is $1/6$. Moreover, $det(X) = 1/4$. Hence, the volume of this simplex is also $1/6$.

$$Now\ X' = de^T + ed^T - \bar{D} = \begin{bmatrix} 4 & 2 & 2 \\ 2 & 4 & 2 \\ 2 & 2 & 2 \end{bmatrix}. \text{ Thus } det(X') = 8. \text{ Hence, again, the vol-}$$

ume of this simplex is $1/6$.

As an application of Corollary 3.3 we derive Heron's formula for the area of a triangle. The square of the area of a triangle with side lengths of a, b, and c is thus

$$\frac{1}{16} det(\begin{bmatrix} 2a^2 & a^2 + b^2 - c^2 \\ a^2 + b^2 - c^2 & 2b^2 \end{bmatrix}) = \frac{1}{16}(4a^2b^2 - (a^2 + b^2 - c^2)^2).$$

As we noted earlier, the Cayley–Menger determinant provides yet another characterization of EDMs.

Theorem 3.16 (Blumenthal [45]) *Let D be a nonzero $n \times n$ symmetric real matrix whose diagonal entries are all 0's. Let $M = \begin{bmatrix} 0 & e^T \\ e & D \end{bmatrix}$ and let Δ_i be the ith leading principal minor of M. Then the following two statements are equivalent:*

(i) *D is an EDM of embedding dimension $r \le n - 1$, where the first $r + 1$ of the generating points are affinely independent.*
(ii) *$(-1)^{i-1}\Delta_i > 0$ for $i = 3, \ldots, r+2$; and for each i, j: $r + 3 \le i < j \le n+1$, we have*

    (a) *The principal minor of M induced by $[1, \ldots, r+2, i]$ is zero,*
    (b) *The principal minor of M induced by $[1, \ldots, r+2, j]$ is zero,*
    (c) *The principal minor of M induced by $[1, \ldots, r+2, i, j]$ is zero,*

Proof. Assume that Statement (i) holds, then since the first $r + 1$ of the generating points are affinely independent, it follows that $V(p^1, \ldots, p^{i-1}) \ne 0$ for $i = 3, \ldots, r + 2$. Hence, it follows from Theorem 3.15 that for $i = 3, \ldots, r+2$, $(-1)^{i-1}\Delta_i > 0$ since $V^2(p^1, \ldots, p^{i-1}) > 0$. Moreover, $V(p^1, \ldots, p^{r+1}, p^{i-1}) = 0$ for each $i = r + 3, \ldots, n + 1$; and $V(p^1, \ldots, p^{r+1}, p^{i-1}, p^{j-1}) = 0$ for each $r + 3 \le i < j \le n+1$. Thus, Statement (ii) follows from Theorem 3.15 since the principal minors of M induced by $[1, \ldots, r+2, i]$ and $[1, \ldots, r+2, i, j]$ are proportional, respectively, to $V^2(p^1, \ldots, p^{r+1}, p^{i-1})$ and $V^2(p^1, \ldots, p^{r+1}, p^{i-1}, p^{j-1})$.

To prove the reverse direction, assume that Statement (ii) holds. Let $D = \begin{bmatrix} 0 & d^T \\ d & \bar{D} \end{bmatrix}$

and let $S = \begin{bmatrix} 1 & 0 & 0 \\ 0 & 1 & 0 \\ -d & -e & I \end{bmatrix}$. Then $SMS^T = \begin{bmatrix} 0 & 1 & 0 \\ 1 & 0 & 0 \\ 0 & 0 & U^T DU \end{bmatrix}$, where U is as defined in

(3.10). Let Δ_i' denote the ith leading principal minor of $U^T DU$. Then $\Delta_i = -\Delta_{i-2}'$ for
$i = 3, \ldots, n+1$. Hence, $(-1)^i \Delta_i' > 0$ for $i = 1, \ldots, r$; i.e., the first r leading principal
minors of $U^T DU$ are nonzero. Therefore, there exists elementary matrix S' such that

$S'(U^T DU)S'^T = \begin{bmatrix} \Gamma & 0 \\ 0 & A \end{bmatrix}$, where $\Gamma = \mathrm{Diag}(\gamma_1, \ldots, \gamma_r)$, and $A = (a_{ij})$ is $(n-r-1) \times$

$(n-r-1)$. Note that for $i = 1, \ldots, r$, we have $\Delta_i' = \gamma_1 \cdots \gamma_i$. Hence, $\gamma_1, \ldots, \gamma_r$ are all
negative and hence Γ is negative definite.

Now let $r+3 \le i < j \le n+1$, then the principal minors of M induced by
$[1, \ldots, r+2, i]$ and $[1, \ldots, r+2, i, j]$ are, respectively, equal to

$$\det\left(\begin{bmatrix} 0 & 1 & 0 & 0 \\ 1 & 0 & 0 & 0 \\ 0 & 0 & \Gamma & 0 \\ 0 & 0 & 0 & a_{ii} \end{bmatrix}\right) \text{ and } \det\left(\begin{bmatrix} 0 & 1 & 0 & 0 & 0 \\ 1 & 0 & 0 & 0 & 0 \\ 0 & 0 & \Gamma & 0 & 0 \\ 0 & 0 & 0 & a_{ii} & a_{ij} \\ 0 & 0 & 0 & a_{ji} & a_{jj} \end{bmatrix}\right).$$

Therefore, $a_{ii} = 0$ for all $i = 1, \ldots, n-r-1$. Consequently, $a_{ij} = 0$ for all
$i, j = 1, \ldots, n-r-1$. Hence $A = 0$. Therefore, $U^T DU$ is negative semidefinite
and of rank r. Thus Statement (i) holds.

$\square$

Example 3.7 *To illustrate Theorem 3.16, consider the matrix* $D = \begin{bmatrix} 0 & 1 & 4 & 9 \\ 1 & 0 & 1 & 4 \\ 4 & 1 & 0 & t \\ 9 & 4 & t & 0 \end{bmatrix}$ *and*

*let $r = 1$. Then Δ_3, the third leading principal minor of M is equal to 2. Moreover,
the principal minors of M induced by $[1, 2, 3, 4]$, $[1, 2, 3, 5]$ and $[1, 2, 3, 4, 5]$ are:*

$$\det\left(\begin{bmatrix} 0 & 1 & 1 & 1 \\ 1 & 0 & 1 & 4 \\ 1 & 1 & 0 & 1 \\ 1 & 4 & 1 & 0 \end{bmatrix}\right) = \det\left(\begin{bmatrix} 0 & 1 & 1 & 1 \\ 1 & 0 & 1 & 9 \\ 1 & 1 & 0 & 4 \\ 1 & 9 & 4 & 0 \end{bmatrix}\right) = 0 \text{ and } \det(M) = -2(t-1)^2.$$

Thus, D is an EDM of embedding dimension 1 iff $t = 1$.

3.5 Constructing New EDMs from Old Ones

In this section we show how to construct new EDMs from old ones. In particular, we
show how to construct a new EDM D' from an EDM D_1 and its Perron eigenvector,
and from two EDMs D_1 and D_2 and their two Perron eigenvectors. We, also, show

how to use Kronecker products to construct new EDMs. We begin with the following useful lemmas.

Lemma 3.11 *Let $\Lambda \succeq 0$, $a \in \mathbb{R}^n$ and let $\sigma > 0$ be a scalar. Then*

$$\Lambda^{1/2}(I - \sigma\Lambda^{1/2}aa^T\Lambda^{1/2})\Lambda^{1/2} \succeq 0 \text{ if and only if } 1 - \sigma a^T\Lambda a \geq 0.$$

Proof. If $a \in \text{null}(\Lambda)$, the result follows trivially. Thus assume that $\Lambda a \neq 0$. The sufficiency part is obvious. To prove the necessity part assume, to the contrary, that $1 - \sigma a^T\Lambda a < 0$. Then $a^T\Lambda^{1/2}(I - \sigma\Lambda^{1/2}aa^T\Lambda^{1/2})\Lambda^{1/2}a = (1 - \sigma a^T\Lambda a)a^T\Lambda a < 0$, a contradiction.

□

Lemma 3.12 *Let D be an $n \times n$ EDM and let $Dw = e$. Further, let (λ, x) be the Perron eigenpair of D and assume that x is normalized. Then*

$$(e^Tx)^2 \geq \lambda e^T w,$$

with equality holding if and only if $x = e/\sqrt{n}$.

Proof. Let $D = \lambda xx^T - W\Lambda W^T$ be the spectral decomposition of D, where $\Lambda \succeq 0$. Then

$$w^TW\Lambda W^Tw = \lambda(x^Tw)^2 - w^TDw = \frac{(x^TDw)^2}{\lambda} - e^Tw = \frac{(e^Tx)^2}{\lambda} - e^Tw \geq 0.$$

Now if $x = e/\sqrt{n}$, then $w = e/\lambda$ since $De = \lambda e$. Thus $(e^Tx)^2 = n = \lambda e^Tw$. On the other hand, assume that $(e^Tx)^2/\lambda - e^Tw = 0$. Then $w^TW\Lambda W^Tw = 0$ and hence $\Lambda W^Tw = 0$. Moreover, $\lambda x^Tw = x^TDw = x^Te$. Consequently, $Dw = \lambda(x^Tw)x = e^Txx = e$. Hence, $x = e/e^Tx$ and thus $(e^Tx)^2 = n$. Therefore, $x = e/\sqrt{n}$.

□

The significance of Lemma 3.12 will become clear in the next chapter where we study regular EDMs. The following theorem shows how to construct a new EDM of order $n + 1$ from an old EDM of order n and its Perron eigenvector.

Theorem 3.17 (Hayden et al. [105]) *Let D be an $n \times n$ EDM and let (λ, x) be the Perron eigenpair of D and assume that x is normalized. Further, let $Dw = e$ and let*

$$\alpha_l = \frac{\lambda}{e^Tx + \sqrt{\lambda e^Tw}} \text{ and } \alpha_u = \begin{cases} +\infty & \text{if } e^Tx = \sqrt{\lambda e^Tw}, \\ \frac{\lambda}{e^Tx - \sqrt{\lambda e^Tw}} & \text{otherwise.} \end{cases}$$

Then

$$D' = \begin{bmatrix} 0 & tx^T \\ tx & D \end{bmatrix}$$

is an EDM if and only if $\alpha_l \leq t \leq \alpha_u$.

Proof. By Theorem 3.3, D' is an EDM iff $t(xe^T + ex^T) - D \succeq 0$.
Let $D = \lambda xx^T - W\Lambda W^T$ be the spectral decomposition of D where $\Lambda \succeq 0$. Then

$$\begin{bmatrix} x^T \\ W^T \end{bmatrix} (t(xe^T + ex^T) - D)[x \ W] = \begin{bmatrix} 2te^T x - \lambda & t\,e^T W \\ t W^T e & \Lambda \end{bmatrix}. \qquad (3.27)$$

But, by Lemma 3.12, $e^T x = \sqrt{\lambda e^T w}$ iff $x = e/\sqrt{n}$ iff $W^T e = \mathbf{0}$. Therefore, if $x = e/\sqrt{n}$, i.e., if $W^T e = \mathbf{0}$, then D' is an EDM iff

$$t \geq \frac{\lambda}{2e^T x} = \frac{e^T De}{2n\sqrt{n}}.$$

But in this case, $e^T x = \sqrt{\lambda e^T w}$ and hence, $\alpha_l = \lambda/2e^T x$ and $\alpha_u = +\infty$. Therefore, the result follows in this case.

On the other hand, if $W^T e \neq \mathbf{0}$, then D' is an EDM iff

$$t > \frac{\lambda}{2e^T x} \quad \text{and} \quad \Lambda - \frac{t^2}{(2te^T x - \lambda)} W^T ee^T W \succeq \mathbf{0}.$$

Therefore, assume that $x \neq e/\sqrt{n}$ and $t > \frac{\lambda}{2e^T x}$. But since $e = Dw$ and $DW = -W\Lambda$, we have $W^T e = -\Lambda W^T w$. Hence,

$$\Lambda - \frac{t^2}{2te^T x - \lambda} W^T ee^T W = \Lambda^{1/2}(I - \frac{t^2}{2te^T x - \lambda} \Lambda^{1/2} W^T ww^T W \Lambda^{1/2})\Lambda^{1/2}.$$

Thus, by Lemmas 3.11 and 3.12, D' is an EDM iff $t > \lambda/2e^T x$ and

$$1 - \frac{t^2}{2te^T x - \lambda} w^T W\Lambda W^T w = 1 - \frac{t^2}{2te^T x - \lambda} (\frac{(e^T x)^2}{\lambda} - e^T w) \geq 0;$$

i.e.,

$$t > \frac{\lambda}{2e^T x} \quad \text{and} \quad ((e^T x)^2 - \lambda e^T w)t^2 - 2e^T x\lambda t + \lambda^2 \leq 0.$$

The roots of this quadratic equation are

$$\lambda \frac{e^T x \pm \sqrt{\lambda e^T w}}{(e^T x - \sqrt{\lambda e^T w})(e^T x + \sqrt{\lambda e^T w})} = \frac{\lambda}{e^T x \pm \sqrt{\lambda e^T w}}.$$

To complete the proof we need to show that $\alpha_u \geq \alpha_l > \lambda/2e^T x$. But this follows since $e^T x > \sqrt{\lambda e^T w}$.

$\square$

Two remarks are in order. First, $e^T w$ is well defined since $e \in \mathrm{col}(D)$, i.e., e is orthogonal to $\mathrm{null}(D)$. Thus, if $y \in \mathrm{null}(D)$, then $D(w+y) = e$. However, $e^T(w+y) = e^T w$. Second, if $e^T w = 0$, then $\alpha_l = \alpha_u = \lambda/e^T x$.

Example 3.8 *Consider the EDM* $D = \begin{bmatrix} 0 & 2 & 4 & 2 \\ 2 & 0 & 2 & 4 \\ 4 & 2 & 0 & 2 \\ 2 & 4 & 2 & 0 \end{bmatrix}$. *Then* $\lambda = 8$, $x = e/2$ *and* $w =$

$e/8$. *Thus* $e^T x = \sqrt{\lambda} e^T w = 2$. *Hence,* $\alpha_u = +\infty$ *and* D' *is an EDM for all* $t \geq \alpha_l = 2$.

Example 3.9 *Consider the EDM* $D = \begin{bmatrix} 0 & 0 & 1 \\ 0 & 0 & 1 \\ 1 & 1 & 0 \end{bmatrix}$. *Then* $\lambda = \sqrt{2}$, $x = \frac{1}{2}[1\ 1\ \sqrt{2}]^T$

and $w = \frac{1}{2}[1\ 1\ 2]^T$. *Thus*

$$\alpha_l = \frac{2}{\sqrt{2} + 1 + 2^{5/4}} \quad \text{and} \quad \alpha_u = \frac{2}{\sqrt{2} + 1 - 2^{5/4}}.$$

Example 3.10 *Consider the EDM* $D = \begin{bmatrix} 0 & 5 & 4 & 5 \\ 5 & 0 & 5 & 16 \\ 4 & 5 & 0 & 5 \\ 5 & 16 & 5 & 0 \end{bmatrix}$. *Then* $\lambda = 10 + 2\sqrt{34}$, $w =$

$\frac{1}{6}[-1\ 1\ -1\ 1]^T$ *and*

$$x = \frac{1}{\sqrt{136 + 12\sqrt{34}}} \begin{bmatrix} 5 \\ 3 + \sqrt{34} \\ 5 \\ 3 + \sqrt{34} \end{bmatrix}.$$

Thus $e^T w = 0$ *and* $e^T x = (8 + \sqrt{34})/\sqrt{34 + 3\sqrt{34}}$. *Therefore,*

$$\alpha_l = \alpha_u = \frac{10 + 2\sqrt{34}}{8 + \sqrt{34}} \sqrt{34 + 3\sqrt{34}}.$$

Next, we turn our attention to the problem of constructing an EDM of order $n_1 + n_2$ from two EDMs of orders n_1 and n_2. Let (λ, x) and (μ, y) be the Perron eigenpairs of EDMs D_1 and D_2, respectively. Then the off-diagonal blocks in the new EDM are txy^T and tyx^T, where t is a positive scalar. First, we discuss the case where $t \neq \sqrt{\lambda\mu}$ followed by the case where $t = \sqrt{\lambda\mu}$. In these two cases, we assume that at least one Perron eigenvector is not equal to e. After that, we discuss the case where both Perron eigenvectors are equal to e. The significance of an EDM having e as a Perron eigenvector will become clear in the next chapter.

Theorem 3.18 (Hayden et al. [105]) *Let* D_1 *and* D_2 *be two EDMs of orders* n_1 *and* n_2 *respectively. Let* (λ, x) *and* (μ, y) *be the Perron eigenpairs of* D_1 *and* D_2, *respectively. Assume that* x *and* y *are normalized and assume that either* $x \neq e_{n_1}/\sqrt{n_1}$ *or* $y \neq e_{n_2}/\sqrt{n_2}$. *Further, let* $D_1 w_1 = e_{n_1}$, $D_2 w_2 = e_{n_2}$, *and* $t^2 \neq \lambda\mu$. *Then*

$$D' = \begin{bmatrix} D_1 & txy^T \\ tyx^T & D_2 \end{bmatrix}$$

is an EDM if and only if the following three conditions hold:

1. $t^2 > \lambda \mu$,
2. $\left((e^T x)^2 - \lambda (e^T w_1 + e^T w_2)\right)\left((e^T y)^2 - \mu (e^T w_1 + e^T w_2)\right) \geq 0$,
3. $\alpha_l \leq t \leq \alpha_u$, *where*

$$\alpha_l = \frac{e^T x\, e^T y - \sqrt{\left((e^T x)^2 - \lambda (e^T w_1 + e^T w_2)\right)\left((e^T y)^2 - \mu (e^T w_1 + e^T w_2)\right)}}{\left(\frac{(e^T x)^2}{\lambda} + \frac{(e^T y)^2}{\mu} - e^T w_1 - e^T w_2\right)}.$$

and

$$\alpha_u = \frac{e^T x\, e^T y + \sqrt{\left((e^T x)^2 - \lambda (e^T w_1 + e^T w_2)\right)\left((e^T y)^2 - \mu (e^T w_1 + e^T w_2)\right)}}{\left(\frac{(e^T x)^2}{\lambda} + \frac{(e^T y)^2}{\mu} - e^T w_1 - e^T w_2\right)}.$$

Proof. The proof uses Theorem 3.11. Thus we show, first, that D' has exactly one positive eigenvalue iff $t > \sqrt{\lambda \mu}$. Let $D_1 = \lambda x x^T - W_1 \Lambda_1 W_1^T$ and $D_2 = \mu y y^T - W_2 \Lambda_2 W_2^T$ be the spectral decompositions of D_1 and D_2, respectively, where $\Lambda_1 \succeq \mathbf{0}$ and $\Lambda_2 \succeq \mathbf{0}$. Then

$$\begin{bmatrix} x^T & 0 \\ W_1^T & 0 \\ 0 & y^T \\ 0 & W_2^T \end{bmatrix} D' \begin{bmatrix} x & W_1 & 0 & 0 \\ 0 & 0 & y & W_2 \end{bmatrix} = \begin{bmatrix} \lambda & 0 & t & 0 \\ 0 & -\Lambda_1 & 0 & 0 \\ t & 0 & \mu & 0 \\ 0 & 0 & 0 & -\Lambda_2 \end{bmatrix}.$$

Now trace $\begin{bmatrix} \lambda & t \\ t & \mu \end{bmatrix} > 0$. Thus, D' has exactly one positive eigenvalue iff $\det \begin{bmatrix} \lambda & t \\ t & \mu \end{bmatrix} \leq 0$; i.e., iff $t \geq \sqrt{\lambda \mu}$. But since $t^2 \neq \lambda \mu$, assume that $t > \sqrt{\lambda \mu}$.

Next we find w such that $D'w = e$. Since we are interested in $e^T w$, it suffices to find one such w. For if $y \in \mathrm{null}(D')$, then $e^T (w + y) = e^T w$. To this end, we have to solve for $a \in \mathbb{R}^{n_1}$ and $b \in \mathbb{R}^{n_2}$ such that

$$\begin{bmatrix} D_1 & t x y^T \\ t y x^T & D_2 \end{bmatrix} \begin{bmatrix} a \\ b \end{bmatrix} = \begin{bmatrix} e_{n_1} \\ e_{n_2} \end{bmatrix}.$$

Thus, we have $D_1 a + t x y^T b = e_{n_1}$ or $D_1 a + t x y^T D_2 b / \mu = e_{n_1}$. Similarly, $D_2 b + t y x^T D_1 a / \lambda = e_{n_2}$. Hence,

$$D_1 a = e_{n_1} - \frac{t}{\mu} e^T y\, x + \frac{t^2}{\lambda \mu} x x^T D_1 a, \qquad (3.28)$$

or

$$(I_{n_1} - \frac{t^2}{\lambda \mu} x x^T) D_1 a = e_{n_1} - \frac{t}{\mu} e^T y\, x.$$

But

$$(I_{n_1} - \frac{t^2}{\lambda \mu} x x^T)^{-1} = I_{n_1} + \frac{t^2}{(\lambda \mu - t^2)} x x^T.$$

Therefore,

$$D_1 a = e_{n_1} + \frac{t}{(\lambda\mu - t^2)}(te^T x - \lambda e^T y)x.$$

Now $D_1^\dagger = xx^T/\lambda - W_1 \Lambda_1^\dagger W_1^T$. Thus $D_1^\dagger x = x/\lambda$ and $D_1^\dagger e = w_1$. Therefore,

$$a = w_1 + \frac{t}{(\lambda\mu - t^2)}(\frac{t}{\lambda}e^T x - e^T y)\, x.$$

Similarly,

$$b = w_2 + \frac{t}{(\lambda\mu - t^2)}(\frac{t}{\mu}e^T y - e^T x)\, y,$$

and thus

$$w = \begin{bmatrix} w_1 + \frac{t}{(\lambda\mu - t^2)}(\frac{t}{\lambda}e^T x - e^T y)\, x \\ w_2 + \frac{t}{(\lambda\mu - t^2)}(\frac{t}{\mu}e^T y - e^T x)\, y \end{bmatrix}$$

satisfies $D'w = e$. Next we require that $e^T w \geq 0$, or

$$e^T w_1 + e^T w_2 - \frac{t}{(t^2 - \lambda\mu)}(\frac{t}{\lambda}(e^T x)^2 + \frac{t}{\mu}(e^T y)^2 - 2e^T x\, e^T y) \geq 0.$$

Note that $t^2 \neq \lambda\mu$. Thus

$$(e^T w_1 + e^T w_2)(t^2 - \lambda\mu) - (\frac{t^2}{\lambda}(e^T x)^2 + \frac{t^2}{\mu}(e^T y)^2 - 2te^T x\, e^T y) \geq 0,$$

and hence

$$(\frac{(e^T x)^2}{\lambda} + \frac{(e^T y)^2}{\mu} - e^T w_1 - e^T w_2)t^2 - 2te^T x\, e^T y + \lambda\mu(e^T w_1 + e^T w_2) \leq 0. \quad (3.29)$$

Note that, by Lemma 3.12, the coefficient of t^2 in (3.29) is > 0 since $x \neq e/\sqrt{n_1}$ or $y \neq e/\sqrt{n_2}$. The discriminant of this quadratic equation is given by

$$(e^T x)^2 (e^T y)^2 - \lambda\mu(e^T w_1 + e^T w_2)(\frac{(e^T x)^2}{\lambda} + \frac{(e^T y)^2}{\mu} - e^T w_1 - e^T w_2)$$

which is equal to

$$(e^T x)^2 \left((e^T y)^2 - \mu(e^T w_1 + e^T w_2)\right) - \lambda(e^T w_1 + e^T w_2)\left((e^T y)^2 - \mu(e^T w_1 + e^T w_2)\right)$$

which in turn can be factorized as

$$\left((e^T x)^2 - \lambda(e^T w_1 + e^T w_2)\right)\left((e^T y)^2 - \mu(e^T w_1 + e^T w_2)\right).$$

Thus, Condition 2 of the theorem amounts to requiring this discriminant to be ≥ 0, in which case, Inequality (3.29) holds iff $\alpha_l \leq t \leq \alpha_u$. Notice that Inequality (3.29) does not hold for any t if this discriminant is negative.

□

Next we consider the case where $t = \sqrt{\lambda\mu}$.

Theorem 3.19 (Hayden et al. [105]) *Let D_1 and D_2 be two EDMs of orders n_1 and n_2, respectively. Let (λ, x) and (μ, y) be the Perron eigenpairs of D_1 and D_2, respectively. Assume that x and y are normalized. Further, let $D_1 w_1 = e_{n_1}$, $D_2 w_2 = e_{n_2}$ and $t^2 = \lambda\mu$. Then*

$$D' = \begin{bmatrix} D_1 & txy^T \\ tyx^T & D_2 \end{bmatrix}$$

is an EDM if and only if the following two conditions hold:

1. $\sqrt{\mu}\, e^T x = \sqrt{\lambda}\, e^T y,$
2. $\sqrt{\lambda\mu}\, (e^T w_1 + e^T w_2) \geq e^T x\, e^T y.$

Proof.　The proof is similar to that of Theorem 3.18. We saw in the proof of Theorem 3.18 that D' has exactly one positive eigenvalue if $t^2 \geq \lambda\mu$. In this case, Eq. (3.28) reduces to

$$D_1 a = e_{n_1} - \frac{t}{\mu} e^T y\, x + x x^T D_1 a,$$

or

$$(I - x x^T) D_1 a = e_{n_1} - \frac{t}{\mu} e^T y\, x. \tag{3.30}$$

Equation (3.30) has a solution iff its RHS lies in $x^\perp$, i.e., iff $x^T (e_{n_1} - \frac{t}{\mu} e^T y\, x) = e^T x - \frac{\sqrt{\lambda}}{\sqrt{\mu}} e^T y = 0$. Thus, assume that

$$\sqrt{\mu}\, e^T x = \sqrt{\lambda}\, e^T y.$$

Therefore, Eq. (3.30) reduces to

$$(I - x x^T) D_1 a = (I - x x^T) e.$$

Thus, $D_1 a = e + \alpha x$ for some scalar α. But since we are interested in only one solution of $D'w = e$, we set $\alpha = 0$. Hence,

$$a = D_1^\dagger e = w_1.$$

Now

$$D_2 b = e_{n_2} - \frac{t y x^T D_1 a}{\lambda} = e_{n_2} - \frac{t e^T x}{\lambda} y.$$

Thus

$$b = D_2^\dagger (e - \frac{t e^T x}{\lambda} y) = w_2 - \frac{t e^T x}{\lambda\mu} y = w_2 - \frac{e^T x}{\sqrt{\lambda\mu}} y.$$

Therefore, $e^T w = e^T a + e^T b = e^T w_1 + e^T w_2 - e^T x\, e^T y / \sqrt{\lambda\mu} \geq 0$ is equivalent to

$$\sqrt{\lambda\mu}\left(e^T w_1 + e^T w_2\right) \geq e^T x\, e^T y.$$

$$\square$$

The lower and upper limits α_l and α_u in Theorem 3.18 have simpler forms in the following three cases. The interpretation of these cases in terms of the different classes of EDMs is given in the next chapter.

Case 1: Assume that $e^T w_1 = e^T w_2 = 0$. Then $\alpha_l = 0$ and

$$\alpha_u = 2\lambda\mu \frac{e^T x\, e^T y}{\mu(e^T x)^2 + \lambda(e^T y)^2}.$$

Now for $a > 0$ and $x > 0$, $f(x) = ax + 1/(ax)$ attains its minimum value of 2 at $x = 1/a$. Thus for $x > 0$, $1/f(x)$ attains its maximum value of $1/2$ at $x = 1/a$. Hence,

$$\lambda\mu \frac{e^T x\, e^T y}{\mu(e^T x)^2 + \lambda(e^T y)^2} = \sqrt{\lambda\mu}\, \frac{1}{\sqrt{\frac{\mu}{\lambda}}\frac{e^T x}{e^T y} + \sqrt{\frac{\lambda}{\mu}}\frac{e^T y}{e^T x}} \leq \frac{1}{2}\sqrt{\lambda\mu}.$$

Therefore, $\alpha_u \leq \sqrt{\lambda\mu}$. Hence, by Theorem 3.18, D' is not an EDM for all $t > \sqrt{\lambda\mu}$ since Condition 1 of Theorem 3.18 requires that $t^2 > \lambda\mu$. Moreover, for $t = \sqrt{\lambda\mu}$, Theorem 3.19 implies that D' is not an EDM since Condition 2 does not hold. Consequently, D' is not an EDM for all t. Another way to see this is to let $c^T = [w_1^T \ w_2^T]$. Then obviously, $c \in e^\perp$. Moreover, $c^T D' c = 2tx^T w_1\, y^T w_2 = 2te^T x\, e^T y/(\lambda\mu) > 0$ for all $t > 0$. As a result, D' is not negative semidefinite on the subspace $e^\perp$ for all $t > 0$.

Case 2: Assume that $e^T w_1 = 0$ and $y = e/\sqrt{n_2}$. Then it follows from Lemma 3.12 that $(e^T y)^2 = \mu e^T w_2$. Thus $\alpha_l = \alpha_u = \lambda e^T y/e^T x$. As a result, D' is an EDM for $t = \alpha_l$ iff $\alpha_l \geq \sqrt{\lambda\mu}$.

Case 3: Assume that $D_1 = D_2$ and let $a = \lambda^2 \frac{e^T w_1}{(e^T x)^2 - \lambda e^T w_1}$. Then

$$[\alpha_l, \alpha_u] = \begin{cases} [a, \lambda] & \text{if } (e^T x)^2 > 2\lambda e^T w_1, \\ [\lambda, \lambda] & \text{if } (e^T x)^2 = 2\lambda e^T w_1, \\ [\lambda, a] & \text{if } (e^T x)^2 < 2\lambda e^T w_1. \end{cases}$$

As a result, if $(e^T x)^2 > 2\lambda e^T w_1$, then D' is not an EDM for all t since Condition 1 of Theorem 3.18 requires $t > \lambda$. On the other hand, if $(e^T x)^2 = 2\lambda e^T w_1$, then the conditions of Theorem 3.19 hold and thus D' is an EDM iff $t = \lambda$. Finally, if $(e^T x)^2 < 2\lambda e^T w_1$, then D' is an EDM for all $t : \lambda \leq t \leq a$.

Example 3.11 *To illustrate Case 2, consider the EDMs* $D_1 = \begin{bmatrix} 0 & 1 & 4 \\ 1 & 0 & 1 \\ 4 & 1 & 0 \end{bmatrix}$ *and* $D_2 = \begin{bmatrix} 0 & 1 \\ 1 & 0 \end{bmatrix}$. *Then* $\lambda = 2 + \sqrt{6}$, $e^T x = \sqrt{6 + 2\sqrt{6}}/2$ *and* $e^T w_1 = 0$; *and* $\mu = 1$, $e^T y = \sqrt{2}$

and $e^T w_2 = 2$. Therefore, α_l and α_u for $D' = \begin{bmatrix} D_1 & txy^T \\ tyx^T & D_2 \end{bmatrix}$ are

$$\alpha_l = \alpha_u = \lambda \frac{e^T y}{e^T x} = \frac{4 + 2\sqrt{6}}{\sqrt{3 + \sqrt{6}}}.$$

Note that $\sqrt{\lambda \mu} = \sqrt{2 + \sqrt{6}} < \alpha_u$. D' is an EDM iff $t = \alpha_l$.

Example 3.12 *To illustrate Case 3, consider the EDM $D_1 = \begin{bmatrix} 0 & 0 & 1 \\ 0 & 0 & 1 \\ 1 & 1 & 0 \end{bmatrix}$. Then $\lambda = \sqrt{2}$, $e^T x = (2 + \sqrt{2})/2$ and $e^T w_1 = 2$. Thus $(e^T x)^2 = 3/2 + \sqrt{2}$ and $2\lambda e^T w_1 = 4\sqrt{2}$. Therefore, $D' = \begin{bmatrix} D_1 & txx^T \\ txx^T & D_1 \end{bmatrix}$ is an EDM for all $t \in [\lambda, a]$, where $a = 4/(3/2 + \sqrt{2} - 2\sqrt{2}) = 8/(3 - 2\sqrt{2})$.*

Example 3.13 *Again to illustrate Case 3, consider the EDM $D_1 = \begin{bmatrix} 0 & 1 & 4 \\ 1 & 0 & 1 \\ 4 & 1 & 0 \end{bmatrix}$. Then $\lambda = 2 + \sqrt{6}$, $e^T x = \sqrt{6 + 2\sqrt{6}}/2$ and $e^T w_1 = 0$. Therefore, $(e^T x)^2 > 2\lambda e^T w_1$ and thus D' is not an EDM for all t. Note that $\alpha_l = a = 0$ and $\alpha_u = \lambda = 2 + \sqrt{6}$.*

So far, we have assumed that at least one Perron eigenvector is not equal to e. Next, we discuss the case where both Perron eigenvectors are equal to e.

Theorem 3.20 (Jaklič and Modic [117]) *Let D_1 and D_2 be two EDMs of orders n_1 and n_2, respectively. Let (λ, x) and (μ, y) be the Perron eigenpairs of D_1 and D_2, respectively. Assume that $x = e_{n_1}/\sqrt{n_1}$ and $y = e_{n_2}/\sqrt{n_2}$. Then*

$$D' = \begin{bmatrix} D_1 & txy^T \\ tyx^T & D_2 \end{bmatrix}$$

is an EDM iff

$$t \geq \frac{n_1 \mu + n_2 \lambda}{2\sqrt{n_1 n_2}}.$$

Proof. Let V be the block matrix defined in (3.15). Then

$$V^T D' V = \begin{bmatrix} V_1^T D_1 V_1 & 0 & 0 \\ 0 & V_2^T D_2 V_2 & 0 \\ 0 & 0 & (n_1 \mu + n_2 \lambda - 2t\sqrt{n_1 n_2})/n \end{bmatrix}.$$

Thus, $\mathcal{T}_V(D') \succeq 0$ iff $t \geq \frac{n_1 \mu + n_2 \lambda}{2\sqrt{n_1 n_2}}$.   □

Note that if $D_1 = D_2$ in Theorem 3.20, then D' is an EDM iff $t \geq \lambda$. On the other hand, in this case $(e^T x)^2 = n_1$ and $\lambda e^T w_1 = n_1$. Thus Condition 2 of Theorem 3.19 holds and hence Theorem 3.19 also implies that D' is an EDM for $t = \lambda$.

Example 3.14 *Let $D_1 = E_{n_1} - I_{n_1}$ and $D_2 = E_{n_2} - I_{n_2}$ be two EDMs of orders n_1 and n_2. Let $n = n_1 + n_2$. Then $\lambda = n_1 - 1$ and $\mu = n_2 - 1$. Moreover, $x = e_{n_1}/\sqrt{n_1}$ and $y = e_{n_2}/\sqrt{n_2}$. Thus, $D' = \begin{bmatrix} D_1 & txy^T \\ tyx^T & D_2 \end{bmatrix}$ is an EDM iff $t \geq \frac{2n_1n_2 - n}{2\sqrt{n_1n_2}}$.*

Example 3.15 *Consider the EDMs $D_1 = \begin{bmatrix} 0 & 1 \\ 1 & 0 \end{bmatrix}$ and $D_2 = \begin{bmatrix} 0 & 2 & 4 & 2 \\ 2 & 0 & 2 & 4 \\ 4 & 2 & 0 & 2 \\ 2 & 4 & 2 & 0 \end{bmatrix}$. Then $\lambda = 1$, $x = e_2/\sqrt{2}$, $w_1 = e$, $\mu = 8$, $y = e_4/2$, and $w_2 = e_2/8$. Thus, $D' = \begin{bmatrix} D_1 & txy^T \\ tyx^T & D_2 \end{bmatrix}$ is an EDM iff $t \geq \frac{n_1\mu + n_2\lambda}{2\sqrt{n_1n_2}} = 5/\sqrt{2}$.*

Example 3.16 *Consider the EDM $D_1 = \begin{bmatrix} 0 & 2 & 4 & 2 \\ 2 & 0 & 2 & 4 \\ 4 & 2 & 0 & 2 \\ 2 & 4 & 2 & 0 \end{bmatrix}$. Then $\lambda = 8$, $x = e_4/2$ and $w_1 = e_2/8$. Thus, $D' = \begin{bmatrix} D_1 & txx^T \\ txx^T & D_1 \end{bmatrix}$ is an EDM iff $t \geq \frac{n_1\lambda}{n_1} = \lambda = 8$.*

We conclude this section by showing how to use Kronecker product to construct new EDMs. Such construction will prove useful when studying Manhattan distances matrices on rectangular grids. Recall that E is the matrix of all 1's.

Theorem 3.21 ([12]) *Let D_1 be an $m \times m$ EDM of embedding dimension r_1 and let D_2 be an $n \times n$ EDM of embedding dimension r_2. Then*

$$D = E_m \otimes D_2 + D_1 \otimes E_n$$

is an EDM of embedding dimension $r = r_1 + r_2$.

Proof. Since $I_{mn} = I_m \otimes I_n$ and $E_{mn} = E_m \otimes E_n$, it follows that

$$\mathscr{T}(E_m \otimes D_2) = -\frac{1}{2}(I_m \otimes I_n - \frac{1}{nm}E_m \otimes E_n)(E_m \otimes D_2)(I_m \otimes I_n - \frac{1}{nm}E_m \otimes E_n)$$
$$= -\frac{1}{2}(E_m \otimes (D_2 - \frac{1}{n}(E_nD_2 + D_2E_n) + \frac{1}{n^2}E_nD_2E_n))$$
$$= E_m \otimes \mathscr{T}(D_2) \succeq 0.$$

Similarly, $\mathscr{T}(D_1 \otimes E_n) = \mathscr{T}(D_1) \otimes E_n \succeq 0$. Hence, $\mathscr{T}(D) = \mathscr{T}(E_m \otimes D_2) + \mathscr{T}(D_1 \otimes E_n) \succeq 0$ and thus D is an EDM.

Finally, $(E_m \otimes \mathscr{T}(D_2))(\mathscr{T}(D_1) \otimes E_n) = (\mathscr{T}(D_1) \otimes E_n))(E_m \otimes \mathscr{T}(D_2)) = 0$. Thus, it follows from Theorem 1.12 that $\mathrm{rank}(\mathscr{T}(D)) = \mathrm{rank}(\mathscr{T}(E_m \otimes D_2)) + \mathrm{rank}(\mathscr{T}(D_1 \otimes E_n))$. Hence, $r = r_1 + r_2$.

□

3.6 Some Necessary and Sufficient Inequalities for EDMs

Lower and upper bounds on the smallest eigenvalue of a real symmetric matrix A give rise to sufficient and necessary conditions for the positive semidefiniteness of A. These conditions, in turn, give rise to sufficient and necessary conditions for EDMs if $A = \mathcal{T}(D)$ or $\mathcal{T}_V(D)$. In this section, we present such conditions using two different approaches to bound the smallest eigenvalue of a real symmetric matrix.

3.6.1 The Trace Approach

In this approach, lower and upper bounds on the smallest eigenvalue of a real symmetric matrix A are given in terms of trace(A) and trace(A^2). These bounds are derived by solving a nonlinear optimization problem or by using Cauchy–Schwarz inequality.

Theorem 3.22 (Wolkowicz and Styan [196, 197]) *Let A be an $n \times n$ real symmetric matrix of nonzero eigenvalues $\lambda_1 \geq \cdots \geq \lambda_r$, $r \geq 2$, and let*

$$ m = \frac{trace(A)}{r}, \ and \ s^2 = \frac{trace(A^2)}{r} - \left(\frac{trace(A)}{r}\right)^2. $$

Then, the smallest nonzero eigenvalue λ_r satisfies

$$ m - s\sqrt{r-1} \leq \lambda_r \leq m - \frac{s}{\sqrt{r-1}}. $$

Proof.   Let $\lambda = (\lambda_i) \in \mathbb{R}^r$ be the vector consisting of the nonzero eigenvalues of A and let $J = I_r - e_r e_r^T / r$. Then $m = e^T \lambda / r$ and $s^2 = \lambda^T J \lambda / r$. Let $w \in \mathbb{R}^r$. Then, since $J^2 = J$, Cauchy–Schwarz inequality implies that

$$ |w^T J \lambda| \leq (w^T J w)^{1/2} (\lambda^T J \lambda)^{1/2}. $$

But $\lambda^T J \lambda = rs^2$. Thus,

$$ -s\sqrt{r}(w^T J w)^{1/2} \leq w^T J \lambda \leq s\sqrt{r}(w^T J w)^{1/2}. $$

Now let w be the rth standard unit vector. Then $w^T J w = (r-1)/r$ and $w^T J \lambda = \lambda_r - m$. Hence,

$$ -s\sqrt{r-1} \leq \lambda_r - m \leq s\sqrt{r-1}. $$

This establishes the lower bound. To establish the upper bound, note that $\sum_{i=1}^r (\lambda_i - \lambda_r) = rm - r\lambda_r$, and $(\lambda_i - \lambda_r)(\lambda_j - \lambda_r) \geq 0$ for all i and j since λ_r is the smallest eigenvalues of A. Consequently, $\sum_{i \neq j} (\lambda_i - \lambda_r)(\lambda_j - \lambda_r) \geq 0$. Thus

$$ r^2(m - \lambda_r)^2 = \left(\sum_{i=1}^r (\lambda_i - \lambda_r)\right)^2 \geq \sum_{i=1}^r (\lambda_i - \lambda_r)^2 = \sum_i \lambda_i^2 - 2rm\lambda_r + r\lambda_r^2. $$

But $\sum_{i=1}^{r} \lambda_i^2 = rs^2 + rm^2$. Thus, $\sum_{i=1}^{r} (\lambda_i - \lambda_r)^2 = rs^2 + r(m - \lambda_r)^2$. Hence,

$$(r-1)(m - \lambda_r)^2 \geq s^2.$$

Therefore, $m - \lambda_r \geq s/\sqrt{r-1}$ and this establishes the upper bound.

$\square$

Theorem 3.23 (Alfakih and Wolkowicz [19]) *Let $D \neq 0$ be an $n \times n$ nonnegative real symmetric matrix, $n \geq 3$, whose diagonal entries are all 0's.*

1. If

$$\frac{2}{n} e^T D^2 e - \frac{(n-3)}{n^2(n-2)}(e^T D e)^2 \geq trace(D^2), \qquad (3.31)$$

then D is an EDM.

2. If D is an $n \times n$ EDM, then

$$\frac{2}{n} e^T D^2 e \geq trace(D^2).$$

Proof. Let $\operatorname{rank}(-V^T D V) = r$, then obviously $r \leq n-1$. Let λ_r denote the smallest eigenvalue of $(-V^T D V)$. To prove Statement 1, it suffices to show that if (3.31) holds, then $\lambda_r \geq 0$. To this end, using Theorem 3.22, we have

$$m = \frac{\operatorname{trace}(-V^T D V)}{r} = \frac{\operatorname{trace}(-DJ)}{r} = \frac{e^T D e}{nr}$$

since $\operatorname{trace}(D) = 0$. Now,

$$s^2 = \frac{1}{r}\operatorname{trace}(D^2) - 2\frac{e^T D^2 e}{rn} + \frac{1}{rn^2}(e^T D e)^2 - \frac{1}{r^2 n^2}(e^T D e)^2,$$

$$= \frac{1}{r}\operatorname{trace}(D^2) - 2\frac{e^T D^2 e}{rn} + \frac{(r-1)}{r^2 n^2}(e^T D e)^2.$$

Moreover,

$$m^2 - (r-1)s^2 = -\frac{(r-1)}{r}\operatorname{trace}(D^2) + 2\frac{(r-1)e^T D^2 e}{rn} - \frac{(r-2)}{rn^2}(e^T D e)^2,$$

$$= \frac{r-1}{r}\left(-\operatorname{trace}(D^2) + 2\frac{e^T D^2 e}{n} - \frac{(r-2)}{(r-1)n^2}(e^T D e)^2\right).$$

Now $f(r) = (r-2)/(r-1)$ is an increasing function. Thus $f(r) \leq f(n-1)$. Consequently,

$$m^2 - (r-1)s^2 \geq \frac{r-1}{r}\left(-\operatorname{trace}(D^2) + 2\frac{e^T D^2 e}{n} - \frac{(n-3)}{(n-2)n^2}(e^T D e)^2\right) \geq 0.$$

Hence, $m^2 \geq (r-1)s^2$ or $m \geq s\sqrt{r-1}$ since $m > 0$. Therefore, $\lambda_r \geq 0$.

To prove Statement 2, assume that D is an EDM and assume, by way of contradiction, that $2e^T D^2 e/n < \text{trace}(D^2)$. Then

$$m^2 - \frac{s^2}{(r-1)} = \frac{1}{r(r-1)}(-\text{trace}(D^2) + 2\frac{e^T D^2 e}{n}).$$

Hence, $m^2 - s^2/(r-1) < 0$ or $m - s/\sqrt{r-1} < 0$ since $m > 0$. Therefore, $\lambda_r < 0$, a contradiction.

$\square$

Two remarks concerning the sufficient condition in Theorem 3.23 are in order. First, the assumption that D is nonnegative cannot be dropped since D appears quadratically. Thus, if D satisfies the sufficient condition, then so does $(-D)$. Second, if $n = 3$, then this sufficient condition becomes also necessary.

Corollary 3.4 *Let $D \neq 0$ be a 3×3 nonnegative real symmetric matrix whose diagonal entries are all 0's. Then D is an EDM if and only if*

$$\frac{2}{3}e^T D^2 e \geq trace(D^2).$$

Corollary 3.4 has an interesting interpretation in terms of the triangular inequality. Let D be a 3×3 EDM and let $d_{12} = a$, $d_{13} = b$ and $d_{23} = c$. Then
$$D^2 = \begin{bmatrix} a^2 + b^2 & bc & ac \\ bc & a^2 + c^2 & ab \\ ac & ab & b^2 + c^2 \end{bmatrix}. \text{ Thus, } \text{trace}(D^2) = 2(a^2 + b^2 + c^2) \text{ and}$$
$e^T D^2 e = \text{trace}(D^2) + 2(ab + ac + bc)$. Hence,

$$\frac{2}{3}e^T D^2 e \geq \text{trace}(D^2) \text{ iff } -(a^2 + b^2 + c^2) + 2(ab + ac + bc) \geq 0.$$

But
$$-(a^2 + b^2 + c^2) + 2(ab + ac + bc) = 4ab - (a + b - c)^2.$$

Thus, it follows from the proof of Theorem 3.4 that the condition of Corollary 3.4 is equivalent to the triangular inequality.

Theorem 3.23 did not make use of the rank of D. Let $\text{rank}(D) = k$ and assume that $k \leq n - 1$. Let $r = \text{rank}(-V^T DV)$. Then $r \leq k$. Note that we cannot assume that $r \leq k - 1$ since we have not established yet that D is an EDM. Also, note that $k \geq 2$ since $\text{trace}(D) = 0$ and $D \neq 0$. Therefore, the sufficient condition of Theorem 3.23 can be weakened. On the other hand, as the proof of Statement 2 of Theorem 3.23 shows, the necessary condition of Theorem 3.23 is independent of k.

Theorem 3.24 (Alfakih and Wolkowicz [19]) *Let $D \neq 0$ be an $n \times n$ nonnegative real symmetric matrix, $n \geq 3$, whose diagonal entries are all 0's. Assume that $\text{rank}(D) = k$ where $k \leq n - 1$. If*

$$\frac{2}{n}e^T D^2 e - \frac{(k-2)}{n^2(k-1)}(e^T De)^2 \geq trace(D^2),$$

then D is an EDM.

3.6.2 The Norm Approach

This approach, due to Bénasséni [40], is based on a result of Bauer and Fike [37]. Let $\alpha > 0$ and let $\Delta = \alpha(E - I)$, i.e., Δ is the EDM associated with the standard simplex. Then $\mathcal{T}_V(\Delta) = \alpha I_{n-1}/2$. Let D be a nonzero nonnegative matrix whose diagonal entries are all 0's. If $\mathcal{T}_V(D)$ is close to $\mathcal{T}_V(\Delta)$, then we expect D to be an EDM. Thus, an upper bound on the norm $||\mathcal{T}_V(D) - \mathcal{T}_V(\Delta)||$ gives rise to a sufficient condition for D to be an EDM.

Let A and B be two real symmetric matrices and let λ be an eigenvalue of A with corresponding eigenvector x. Then $(\lambda I - B)x = (A - B)x$. Assume that λ is not an eigenvalue of B, then $x = (\lambda I - B)^{-1}(A - B)x$. Thus, for any induced matrix norm we have $||x|| \leq ||(\lambda I - B)^{-1}|| \, ||A - B|| \, ||x||$ since induced matrix norms are submultiplicative. Hence, for the matrix norm induced by the Euclidean vector norm, we have

$$\frac{1}{||(\lambda I - B)^{-1}||_2} \leq ||A - B||_2 \leq ||A - B||_F, \tag{3.32}$$

where $||.||_F$ denotes the Frobenius norm.

Let D be a nonzero nonnegative matrix whose diagonal entries are all 0's. Further, let $A = \mathcal{T}_V(D)$, $B = \mathcal{T}_V(\Delta)$ and let λ be an eigenvalue of $\mathcal{T}_V(D)$ and assume that $\lambda \neq \alpha/2$. Then

$$||(\lambda I_{n-1} - \mathcal{T}_V(\Delta))^{-1}||_2 = ||(\lambda - \alpha/2)^{-1} I_{n-1}||_2 = \frac{1}{|\lambda - \alpha/2|}$$

since $||I_{n-1}||_2 = 1$. Therefore, it follows from (3.32) that

$$|\lambda - \alpha/2| \leq ||\mathcal{T}_V(D) - \mathcal{T}_V(\Delta)||_F. \tag{3.33}$$

Now since each eigenvalue of $\mathcal{T}_V(D)$ either satisfies (3.33) or is equal to $\alpha/2$, we conclude that all eigenvalues of $\mathcal{T}_V(D)$ lie in a disk centered at $\alpha/2$ and of radius $||\mathcal{T}_V(D) - \mathcal{T}_V(\Delta)||_F$. Therefore, if we can find $\alpha > 0$ such that $||\mathcal{T}_V(D) - \mathcal{T}_V(\Delta)||_F \leq \alpha/2$, then the eigenvalues of $\mathcal{T}_V(D)$ are all nonnegative and consequently, D is an EDM.

Let $f(\alpha) = \alpha^2/4 - ||\mathcal{T}_V(D) - \mathcal{T}_V(\Delta)||_F^2$. We need to find α^*, as a function of D, which maximizes $f(\alpha)$ and then find a condition on D such that $f(\alpha^*) \geq 0$. To this end,

$$4||\mathscr{T}_V(D) - \mathscr{T}_V(\Delta)||_F^2 = ||V^T(D - \alpha E + \alpha I)V||_F^2$$
$$= \text{trace}(J(D - \alpha E + \alpha I)J(D - \alpha E + \alpha I))$$
$$= \text{trace}((D - Dee^T/n + \alpha J)(D - Dee^T/n + \alpha J))$$
$$= \text{trace}(D^2) + \frac{(e^T De)^2}{n^2} - 2\frac{e^T D^2 e}{n} + \alpha^2(n-1) - 2\alpha\frac{e^T De}{n}.$$

Thus, $4f(\alpha)$ is maximized at $\alpha^* = e^T De/(n(n-2))$ and

$$4f(\alpha^*) = -\text{trace}(D^2) - \frac{(n-3)}{n^2(n-2)}(e^T De)^2 + 2\frac{e^T D^2 e}{n}.$$

Hence, the condition $f(\alpha^*) \geq 0$ implies the sufficient condition of Theorem 3.23.

Bénasséni derived a stronger sufficient condition by considering $||D - \Delta||_F$ instead of $||\mathscr{T}_V(D - \Delta)||_F$. To this end, $||\mathscr{T}_V(D - \Delta)||_2 = ||V^T(D - \Delta)V||_2 \leq ||V^T||_2||V||_2||D - \Delta||_2$. But $||V^T||_2 = ||V||_2 = 1$. Thus, $||\mathscr{T}_V(D - \Delta)||_2 \leq ||D - \Delta||_2 \leq ||D - \Delta||_F$. Therefore, if $||D - \Delta||_F \leq \alpha/2$, then $||\mathscr{T}_V(D - \Delta)||_F \leq \alpha/2$ and thus D is an EDM. Therefore,

$$4||D - \Delta||_F^2 = ||D - \alpha E + \alpha I||_F^2$$
$$= \text{trace}((D - \alpha E + \alpha I)(D - \alpha E + \alpha I))$$
$$= \text{trace}(D^2) + (n^2 - n)\alpha^2 - 2\alpha e^T De.$$

Let $g(\alpha) = \alpha^2/4 - ||D - \Delta||_F^2$. We need to find α^*, as a function of D, which maximizes $g(\alpha)$ and then find a condition on D such that $g(\alpha^*) \geq 0$. Therefore, $4g(\alpha)$ is maximized at $\alpha^* = e^T De/(n^2 - n - 1)$ and

$$4g(\alpha^*) = -\text{trace}(D^2) + \frac{(e^T De)^2}{n^2 - n - 1}.$$

Hence, the condition that $g(\alpha^*) \geq 0$ leads to the following stronger sufficient condition (weaker result) for EDMs.

Theorem 3.25 (Bénasséni [40]) *Let $D \neq 0$ be an $n \times n$ nonnegative real symmetric matrix whose diagonal entries are all 0's. If*

$$\frac{(e^T De)^2}{n^2 - n - 1} \geq \text{trace}(D^2),$$

then D is an EDM.

The sufficient condition of Theorem 3.25 can be interpreted in terms of the variance of the off-diagonal entries of D. The mean of D is $\bar{d} = \sum_{i,j} d_{ij}/(n(n-1)) = e^T De/(n(n-1))$. Thus

$$\text{var}(D) = \sum_{i,j} d_{ij}^2/(n^2 - n) - (\bar{d})^2 = \frac{1}{n^2 - n}\text{trace}(D^2) - \frac{1}{(n^2 - n)^2}(e^T De)^2.$$

Therefore, the sufficient condition of Theorem 3.25 [40] is equivalent to

$$\text{var}(D) \leq \frac{1}{n^2(n-1)^2(n^2-n-1)}(e^T De)^2.$$

Example 3.17 *To illustrate the sufficient condition of 3.23, consider the 5×5 matrix $D(t) = \begin{bmatrix} 0 & te^T \\ te & E-I \end{bmatrix}$. Then it follows from Theorem 3.3 that $D(t)$ is an EDM for all $t \geq 3/8$. This result can also be obtained by using Theorem 3.17. Indeed, in this case, $x = e/2$, $\lambda = 3$, $w = e/3$ and $t' = 2t$. Thus, $\alpha_l = 3/4$ and $\alpha_u = \infty$. Consequently, $D(t')$ is an EDM iff $t' \geq \alpha_l$.*

Now

$$\frac{2}{n}e^T(D(t))^2 e = \frac{8}{5}(5t^2 + 6t + 9),$$

$$\frac{(n-3)}{n^2(n-2)}(e^T D(t)e)^2 = \frac{32}{75}(4t^2 + 12t + 9),$$

$$\text{trace}(D^2) = 4(2t^2 + 3).$$

Thus, the sufficient condition in Theorem 3.23 holds iff $-32t^2 + 84t - 27 \geq 0$, i.e., iff

$$\frac{3}{8} \leq t \leq \frac{18}{8}.$$

Observe that $D(1)$ is the EDM of the standard simplex. Thus, as expected, the sufficient condition of Theorem 3.23 holds for values of t close enough to 1.

3.7 Schoenberg Transformations

This section addresses the following natural question. What real functions f, when applied entrywise, map EDMs to EDMs? A characterization of such functions was obtained by Schoenberg [169, 170] and thus, they are known as Schoenberg transformations. A good reference on Schoenberg transformations in data analysis is [38]. Recall that $f[D] = (f(d_{ij}))$ denotes the matrix obtained from D by applying f to D entrywise.

Theorem 3.26 (Schoenberg [168, 169, 170]) *Let $D = (d_{ij})$ be an EDM. Then $f[D]$ is an EDM if and only if*

$$f(d) = \int_0^\infty \frac{(1 - e^{-td})}{t} g(t)dt,$$

where $g(t)$ is nonnegative for $t \geq 0$ such that $\int_1^\infty g(t)dt/t$ exists.

Note that $f(0) = 0$ and $f'(d) = \int_0^\infty e^{-td}g(t)dt$. Hence, it readily follows that $f(d)$ satisfies

$$(-1)^{i-1} f^{(i)}(d) \geq 0 \text{ for all } d > 0 \text{ and for all } i \geq 1, \qquad (3.34)$$

where $f^{(i)}$ denotes the ith derivative of $f(d)$. In what follows, we present several examples of Schoenberg transformations.

Corollary 3.5 (Schoenberg [168]) *Let* $D = (d_{ij})$ *be an EDM. Then* $[D]^a$ *is an EDM for all* $a: 0 < a < 1$.

Proof. Let $g(t) = at^{-a}/\Gamma(1-a)$, where $\Gamma(1-a)$ is the well-known gamma function, i.e.,

$$\Gamma(1-a) = \int_0^\infty t^{-a} e^{-t} dt.$$

Observe that $\Gamma(1-a) \geq 0$ since $a < 1$. Thus

$$f(d) = \frac{a}{\Gamma(1-a)} \int_0^\infty \frac{(1 - e^{-td})}{t^{a+1}} dt.$$

Let $y = td$, then

$$\int_0^\infty \frac{(1 - e^{-td})}{t^{a+1}} dt = d^a \int_0^\infty \frac{(1 - e^{-y})}{y^{a+1}} dy.$$

Integrating by parts, we get

$$\int_0^\infty \frac{(1 - e^{-y})}{y^{a+1}} dy = \left[\frac{1 - e^{-y}}{ay^a}\right]_\infty^0 + \frac{1}{a} \int_0^\infty y^{-a} e^{-y} dy = \frac{\Gamma(1-a)}{a}.$$

Hence, $f(d) = d^a$ is a Schoenberg transformation.   □

It is worth noting that $f(d) = d^a$, $0 < a < 1$, satisfies (3.34) since $f^{(i)}(d) = a(a-1)\cdots(a-i+1)d^{a-i}$. Also, note that $f^{(1)}(d) = a/\Gamma(1-a) \int_0^\infty t^{-a} e^{-td} dt$.

Let $D = \begin{bmatrix} 0 & 1 & 4 \\ 1 & 0 & 1 \\ 4 & 1 & 0 \end{bmatrix}$. Then it is easy to verify that D is an EDM of embedding dimension 1 and $\sqrt{[D]}$ is an EDM of dimension 2. On the other hand, $[D]^2$ is not an EDM. Trivially, $D = (\sqrt{[D]})^2$. Therefore, for an arbitrary EDM D and for $a > 1$, $[D]^a$ may or may not be an EDM.

Corollary 3.6 *Let* $D = (d_{ij})$ *be an EDM and let* $a > 0$. *Then* $D' = E - \exp[-aD]$ *is an EDM.*

Proof. Let $g(t) = a\delta(t - a)$ where δ is the *Dirac delta function*. Then

$$f(d) = \int_0^\infty \frac{(1 - e^{-td})}{t} a\delta(t - a) dt = 1 - e^{-ad}.$$

Hence, $f(d) = 1 - e^{-ad}$ is a Schoenberg transformation.

                                                                                            □

Corollary 3.6 can also be proved directly [70] using Theorem 2.8. Indeed, assume that D is generated by points $p^2, \ldots, p^{n+1}$, and assume that the embedding dimension of D is r. Let p^1 be any point in $\mathbb{R}^r$ and let $d_i = ||p^1 - p^i||^2$ for $i = 2, \ldots, n+1$. Thus, by construction, $\begin{bmatrix} 0 & d^T \\ d & D \end{bmatrix}$ is an $(n+1) \times (n+1)$ EDM. Hence, it follows from Theorem 3.3 that $a(ed^T + de^T - D)$ is PSD. Moreover, by Theorem 2.8, $B = \exp[a(ed^T + de^T - D)]$ is PSD, where $b_{ij} = \exp(ad_i + ad_j - ad_{ij})$. Let $S = (s_{ij})$ be the diagonal matrix where $s_{ii} = \exp(-ad_i)$. Then $\exp[-aD] = SBS$ is PSD since $(SBS)_{ij} = s_{ii}b_{ij}s_{jj} = \exp(-ad_{ij})$. Moreover, $\text{diag}(\exp[-aD]) = e$. Therefore,

$$\mathcal{K}(\exp[-aD]) = 2E - 2\exp[-aD]$$

is an EDM. Next, two more Schoenberg transformations are given.

Example 3.18 *Let $g(t) = e^{-at}$, where $a > 0$. Then*

$$f(d) = \int_0^\infty (e^{-at} - e^{-t(a+d)}) \frac{dt}{t} = \ln(1 + \frac{d}{a}).$$

Thus, if D is an EDM, then so is $D' = (d'_{ij})$ where $d'_{ij} = \ln(1 + d/a)$.

Example 3.19 *Let $g(t) = te^{-t}$. Then*

$$f(d) = \int_0^\infty (1 - e^{-td})e^{-t}dt = 1 - \frac{1}{d+1} = \frac{d}{d+1}.$$

Thus, if $D = (d_{ij})$ is an EDM, then so is $D' = (d'_{ij})$ where $d'_{ij} = d_{ij}/(d_{ij}+1)$. Furthermore, $D'' = (d''_{ij})$ where $d''_{ij} = d^a_{ij}/(d^a_{ij}+1)$ is an EDM for $0 < a < 1$.

Finally, it should be pointed out that all Schoenberg transformations $f(d)$ considered above satisfy (3.34) and $f(0) = 0$.

3.8 Notes

Schoenberg [167] considered only the cases where $s = e/n$ and $e = e^i$, while Young and Householder [200] considered the case where $s = e^n$, the nth standard unit vector in $\mathbb{R}^n$. Gower [93] generalized Schoenberg and Young–Householder result to all s such that $e^T s = 1$. The case $x = e/\sqrt{n_1}$ and $y = e/\sqrt{n_2}$ in Theorem 3.18 was not considered in Hayden et al. [105]. It was first considered by Jaklič and Modic in [117].

Chapter 4
Classes of EDMs

Euclidean Distance Matrices fall into two classes: spherical and nonspherical. The first part of this chapter discusses various characterizations and several subclasses of spherical EDMs. Among the examples of spherical EDMs discussed are: regular EDMs, cell matrices, Manhattan distance matrices, Hamming distance matrices on the hypercube, distance matrices of trees and resistance distance matrices of electrical networks. The second part focuses on nonspherical EDMs and their characterization. As an interesting example of nonspherical EDMs, we discuss multispherical EDMs.

An EDM matrix D is said to be *spherical* if the generating points of D lie on a hypersphere. Otherwise, D is said to be *nonspherical*.

4.1 Spherical EDMs

Since EDMs are either spherical or nonspherical, any characterization of spherical EDMs is at the same time a characterization of nonspherical EDMs. This section presents six different characterizations of spherical EDMs. In the theorem that follows, we provide the first of these characterizations.

Theorem 4.1 (Tarazaga et al. [186]) *Let D be a nonzero EDM of embedding dimension r. Let P ($P^T e = 0$) be a configuration matrix of D and let $B = PP^T$ be the Gram matrix of D. Further, let J denote the orthogonal projection on $e^\perp$. Then D is spherical if and only if there exists $a \in \mathbb{R}^r$ such that*

$$Pa = \frac{1}{2}J diag(B),$$

in which case, the generating points of D lie on a hypersphere centered at a and of radius

$$\rho = (a^T a + \frac{1}{2n^2} e^T De)^{1/2}. \tag{4.1}$$

© Springer Nature Switzerland AG 2018
A.Y. Alfakih, *Euclidean Distance Matrices and Their Applications in Rigidity Theory*, https://doi.org/10.1007/978-3-319-97846-8_4

Proof. Assume that the generating points of D lie on a hypersphere centered at a and of radius $\rho = (a^T a + e^T De/2n^2)^{1/2}$. After a translation along a, the configuration matrix becomes $P - ea^T$. Hence, $\text{diag}((P - ea^T)(P^T - ae^T)) = \rho^2 e$. Thus, $\text{diag}(B) - 2Pa + a^T ae = (a^T a + e^T De/2n^2)e$. Therefore, $2Pa = \text{diag}(B) - (e^T De/2n^2)e$. Multiplying both sides by J, we get $2Pa = J\text{diag}(B)$ since $e^T P = \mathbf{0}$.

To prove the other direction, assume that there exists a such that $Pa = \frac{1}{2}J\text{diag}(B)$. Then, after a translation along a, the configuration matrix becomes $P - ea^T$. Thus, the Gram matrix becomes $B' = (P - ea^T)(P^T - ae^T)$. Thus, $B' = B - J\text{diag}(B)e^T/2 - e\text{diag}(B)^T J/2 + a^T aee^T$. Thus, $\text{diag}(B') = (I - J)\text{diag}(B) + a^T ae = (e^T \text{diag}(B)/n + a^T a)e$. But $e^T De = 2ne^T \text{diag}(B)$. Therefore, $\text{diag}(B') = (e^T De/2n^2 + a^T a)e = \rho^2 e$.

$\square$

Example 4.1 *Consider the EDM $D = \begin{bmatrix} 0 & 18 & 36 \\ 18 & 0 & 18 \\ 36 & 18 & 0 \end{bmatrix}$ with configuration matrix $P = \begin{bmatrix} -3 & -1 \\ 0 & 2 \\ 3 & -1 \end{bmatrix}$. Then $\text{diag}(B) = \begin{bmatrix} 10 \\ 4 \\ 10 \end{bmatrix}$ and hence $J\text{diag}(B)/2 = \begin{bmatrix} 1 \\ -2 \\ 1 \end{bmatrix}$. Thus, the equation $Pa = J\text{diag}(B)/2$ has solution $a = [0 \ -1]^T$. It is easy to verify that the generating points of D lie on hypersphere centered at a and of radius $\rho = (a^T a + e^T De/2n^2)^{1/2} = 3$.*

Example 4.2 *Consider the EDM $D = \begin{bmatrix} 0 & 1 & 4 & 17 \\ 1 & 0 & 1 & 16 \\ 4 & 1 & 0 & 17 \\ 17 & 16 & 17 & 0 \end{bmatrix}$ with configuration matrix $P = \begin{bmatrix} -1 & -1 \\ 0 & -1 \\ 1 & -1 \\ 0 & 3 \end{bmatrix}$. Then $\text{diag}(B) = \begin{bmatrix} 2 \\ 1 \\ 2 \\ 9 \end{bmatrix}$ and hence $J\text{diag}(B)/2 = \frac{1}{4}\begin{bmatrix} -3 \\ -5 \\ -3 \\ 11 \end{bmatrix}$. It is easy to verify that $J\text{diag}(B)/2$ is not in the column space of P and thus D is not spherical.*

An easy consequence of Theorem 4.1 is that all $n \times n$ EDMs of embedding dimension $n - 1$ are spherical.

Corollary 4.1 *Let D be an $n \times n$ EDM of embedding dimension $r = n - 1$. Then $\text{rank}(D) = n$ and D is spherical*

Proof. If $r = n - 1$, then $\text{rank}(D) = n$ since $\text{rank}(D) \geq r + 1$. Moreover, $\text{rank}(P) = n - 1$ and hence $\text{col}(P) = e^\perp$. The result follows since $J\text{diag}(B)$ lies in $e^\perp$.

$\square$

We turn, next, to the characterization of spherical EDMs of embedding dimension $\leq n - 2$. But first, we will need the following lemma.

Lemma 4.1 ([18]) *Let D be an $n \times n$ EDM of embedding dimension $r \leq n - 2$ and let Z be a Gale matrix of D. Then $\text{null}(D) = \text{gal}(D)$ if and only if there exists a scalar β such that $\beta ee^T - D \succeq \mathbf{0}$.*

Proof. Let P be a configuration matrix of D, $(P^T e = \mathbf{0})$, and let $B = PP^T$ be the Gram matrix of D. Then, it follows from the definition of $\mathcal{K}$ that $(-P^T DP) = 2(P^T P)^2$. Thus, $(-P^T DP)$ is PD since P has full column rank.

Lemma 3.10 implies that $DZ = e(\text{diag}(B))^T Z$ and hence $P^T DZ = Z^T DZ = \mathbf{0}$. Let $S = [P \ Z \ e]$. Then S is nonsingular and

$$S^T(\beta ee^T - D)S = \begin{bmatrix} -P^T DP & 0 & -P^T De \\ 0 & 0 & -Z^T De \\ -e^T DP & -e^T DZ & \beta n^2 - e^T De \end{bmatrix}.$$

Now assume that $\beta ee^T - D \succeq \mathbf{0}$, then $e^T DZ = n \ (\text{diag}(B))^T Z = \mathbf{0}$ and hence $(\text{diag}(B))^T Z = \mathbf{0}$. Therefore $DZ = 0$ and consequently $\text{gal}(D) \subseteq \text{null}(D)$. But $\text{null}(D) \subseteq \text{gal}(D)$ (Theorem 3.7). Therefore, $\text{gal}(D) = \text{null}(D)$.

On the other hand, assume that $\text{gal}(D) = \text{null}(D)$, i.e., $DZ = \mathbf{0}$. Now, by Schur complement, $\begin{bmatrix} -P^T DP & -P^T De \\ -e^T DP & \beta n^2 - e^T De \end{bmatrix}$ is PSD iff

$$n^2 \beta - e^T De - \frac{1}{2} e^T DP(P^T P)^{-2} P^T De \geq 0,$$

where we have substituted $(-P^T DP) = 2(P^T P)^2$. Therefore, $\beta ee^T - D$ is PSD for a sufficiently large β.

$\square$

Three additional characterizations of spherical EDMs are given in the following theorem.

Theorem 4.2 *Let D be an $n \times n$ EDM of embedding dimension $r \leq n - 2$. Then the following statements are equivalent:*

1. *D is spherical.*
2. *$\text{null}(D) = \text{gal}(D)$, i.e., $DZ = \mathbf{0}$, where Z is a Gale matrix of D.*
3. *$\text{rank}(D) = r + 1$.*
4. *There exists a scalar β such that $\beta ee^T - D \succeq \mathbf{0}$.*

Proof. The equivalence between Statements 2 and 4 follows from Lemma 4.1. Moreover, Statements 2 and 3 are equivalent since $\dim \text{gal}(D) = n - r - 1$ and since $\text{null}(D) \subseteq \text{gal}(D)$. Next, we prove the equivalence between Statements 1 and 2.

Assume that D is spherical. Therefore, by Theorem 4.1, there exists a such that $2Pa = J\text{diag}(B)$ and thus $2Z^T Pa = Z^T J\text{diag}(B) = Z^T \text{diag}(B) = \mathbf{0}$. Hence, it follows from Lemma 3.10 that $DZ = \mathbf{0}$ and thus $\text{gal}(D) \subseteq \text{null}(D)$. But $\text{null}(D) \subseteq \text{gal}(D)$ (Theorem 3.7). Therefore, $\text{gal}(D) = \text{null}(D)$ and hence, Statement 2 holds. On the other hand, assume that Statement 2 holds. Then $Z^T \text{diag}(B) = \mathbf{0}$ and thus $\text{diag}(B) = Pa' + \gamma e$ for some vector a' and scalar γ. Hence, $J\text{diag}(B) = Pa'$ and thus D is spherical.

$\square$

The equivalence between Statements 1 and 2 was proven by Alfakih and Wolkowicz in [18]. The equivalence between Statements 1 and 3 was proven by

Gower in [93]. Finally, the equivalence between Statements 1 and 4 was proven by Neumaier in [153] and was later independently proven by Tarazaga et al. in [186].

As the next theorem shows, the minimum value of β in Theorem 4.2 can be expressed in terms of the radius ρ.

Theorem 4.3 (Neumaier [153]) *Let D be an $n \times n$ spherical EDM of radius ρ and let β^* be the minimum scalar such that $\beta^* ee^T - D \succeq 0$. Then*

$$\beta^* = 2\rho^2. \tag{4.2}$$

Proof.   It follows from the proof of Lemma 4.1 above that

$$n^2 \beta^* = e^T De + \frac{1}{2} e^T DP(P^T P)^{-2} P^T De.$$

Moreover, $2Pa = J\mathrm{diag}(B)$ and hence $2a = (P^T P)^{-1} P^T \mathrm{diag}(B)$. On the other hand, $P^T De = nP^T \mathrm{diag}(B)$. Therefore, $2na = (P^T P)^{-1} P^T De$. Accordingly,

$$\beta^* = 2(\frac{1}{2n^2} e^T De + a^T a) = 2\rho^2.$$

□

Example 4.3 *Consider the EDM D of Example 4.1. Then it is easy to verify that $\mathrm{rank}(D) = 3 = r + 1$. Let $\beta = 18\beta'$, then using double-sided Gaussian elimination we have that $(\beta ee^T - D) = (\beta' ee^T - D/18) \succeq 0$ iff*

$$\begin{bmatrix} \beta' & \cdot & \cdot \\ \cdot & \frac{2\beta'-1}{\beta'} & \cdot \\ \cdot & \cdot & \frac{4(\beta'-1)}{2\beta'-1} \end{bmatrix} \succeq 0.$$

Thus, $(\beta ee^T - D) \succeq 0$ iff $\beta' \geq 1$; i.e., iff $\beta \geq 18 = 2\rho^2$.

The following theorem is an easy consequence of Theorem 4.3.

Theorem 4.4 (Kurata and Sakuma [124]) *Let D be an $n \times n$ spherical EDM of radius ρ. Let $\hat{D} = \sum_{i=1}^{k} \lambda_i Q^i D Q^{i^T}$, where $\sum_{i=1}^{k} \lambda_i = 1$ and where Q^i is a permutation matrix and $\lambda_i \geq 0$ for $i = 1, \ldots, k$. Then $\hat{D}$ is a spherical EDM of radius $\hat{\rho} \leq \rho$.*

Proof.   Clearly $\hat{D}$ is an EDM since it is a convex combination of EDMs. Now Theorem 4.3 implies that $2\rho^2 ee^T - D \succeq 0$. Thus, $Q^i(2\rho^2 ee^T - D)Q^{i^T} = 2\rho^2 ee^T - Q^i D Q^{i^T} \succeq 0$. Hence, $\sum_{i=1}^{k} \lambda_i (2\rho^2 ee^T - Q^i D Q^{i^T}) = 2\rho^2 ee^T - \sum_{i=1}^{k} \lambda_i Q^i D Q^{i^T} \succeq 0$. Therefore, $\hat{D}$ is a spherical EDM of radius $\hat{\rho}$ and by Theorem 4.3, $\hat{\rho} \leq \rho$.

□

We say that points p^i and p^j are *antipodal* if $d_{ki} + d_{kj} = d_{ij}$ for all $k = 1, \ldots, n$; that is, $D_{.i} + D_{.j} = d_{ij}e$, where $D_{.i}$ denotes the ith column of D. Then it is an immediate consequence of Theorem 4.3 that each entry d_{ij} is $\leq 2\beta^*$, with equality holding if and only if p^i and p^j are antipodal. This fact was observed by Neumaier [152] who gave a direct proof of it without appealing to Theorem 4.3.

Example 4.4 *Consider the EDM D in Example 4.1. Then $\rho = 3$ and hence $\beta^* = 18$. Moreover, $d_{13} = 36 = 2\beta^*$ and thus points p^1 and p^3 are antipodal. Notice that $D_{.1} + D_{.3} = 36e = d_{13}e$.*

Let $D^\dagger$ denote the Moore–Penrose inverse of an EDM D. Then it follows from Theorem 3.9 that e lies in $\text{col}(D)$ or $DD^\dagger e = e$. The solution of the system of equations $Dw = e$ is $w = D^\dagger e + (I - D^\dagger D)z$ where z is any vector in $\mathbb{R}^n$. Thus

$$e^T w = w^T D w = e^T D^\dagger e. \tag{4.3}$$

In addition to the above characterizations, spherical EDMs have two more characterizations in terms of w. The first of these characterizations is given next.

Theorem 4.5 (Gower [93, 92]) *Let D be a nonzero EDM and let $Dw = e$. Then D is spherical if and only if $e^T w > 0$, in which case, the generating points of D lie on a hypersphere of radius*

$$\rho = \left(\frac{1}{2e^T w}\right)^{1/2}.$$

Proof. $e^T w \geq 0$ since D is an EDM (Theorem 3.11). Assume that $e^T w > 0$ and let $B = -(I - ew^T/(e^T w))D(I - we^T/(e^T w))/2$. Then $-2B = D - ee^T/(e^T w)$. Therefore, $\text{diag}(B) = \frac{1}{2e^T w}e = \rho^2 e$ and hence D is spherical.

On the other hand, assume that $e^T w = 0$. Then by Theorem 3.10, there exists a configuration matrix P such that $P^T w = \mathbf{0}$. Therefore, $w \in \text{gal}(D)$. But $Dw = e$, thus $\text{null}(D) \neq \text{gal}(D)$ and hence D is nonspherical.

$\square$

Remark 4.1 *w in Theorem 4.5 is not unique. If $y \in \text{null}(D)$, then $D(w + y) = e$. However, $e \perp \text{null}(D)$ since $e \in \text{col}(D)$. Thus, ρ is well defined since $e^T(w + y) = e^T w$.*

Example 4.5 *Consider the EDM of Example 4.1 and let $w = \frac{1}{36}[1 \ 0 \ 1]^T$. Then $Dw = e$ and $e^T w = 1/18$. Note that $\rho^2 = 1/(2e^T w) = 9$.*

The second characterization of spherical EDMs in terms of w is given in the following theorem.

Theorem 4.6 *Let D be a nonzero EDM and let $Dw = e$. Further, let B be the Gram matrix of D such that $Bw = \mathbf{0}$. Then*

$$w^T \text{diag}(B) = \frac{1}{2} \text{ or } 1. \tag{4.4}$$

Moreover, D is spherical if and only if $w^T \text{diag}(B) = \frac{1}{2}$.

Proof.   The existence of a Gram matrix B such that $Bw = \mathbf{0}$ follows from Theorem 3.10. Also, it follows from Theorems 3.11 and 4.5 that $e^T w \geq 0$ and D is spherical iff $e^T w > 0$.

Observe that $e = Dw = w^T \operatorname{diag}(B) \, e + e^T w \operatorname{diag}(B)$ and thus $e^T w = w^T Dw = 2w^T \operatorname{diag}(B) \, e^T w$. Therefore,

$$(1 - w^T \operatorname{diag}(B)) \, e = e^T w \operatorname{diag}(B) \tag{4.5}$$

and

$$(1 - 2w^T \operatorname{diag}(B)) \, e^T w = 0. \tag{4.6}$$

Now if $e^T w > 0$, then Eq. (4.6) implies that $w^T \operatorname{diag}(B) = \frac{1}{2}$. Furthermore, since $B \neq \mathbf{0}$, Eq. (4.5) implies that $w^T \operatorname{diag}(B) = 1$ if and only if $e^T w = 0$. As a result, (4.4) holds since $e^T w \geq 0$. Consequently, $e^T w > 0$ iff $w^T \operatorname{diag}(B) = \frac{1}{2}$. $\qquad\square$

Remark 4.2 *Assume that D is a spherical EDM and $Bw = \mathbf{0}$. Then $\operatorname{diag}(B) = \rho^2 e$ (see the proof of Theorem 4.5). Thus, it follows from Theorem 4.5 that $w^T \operatorname{diag}(B) = \rho^2 e^T w = 1/2$.*

Example 4.6 *Consider the EDM D of Example 4.1, where $w = \frac{1}{36}[1 \ 0 \ 1]^T$. Thus, a configuration matrix P of D that satisfies $P^T w = \mathbf{0}$ is $P = \begin{bmatrix} -3 & 0 \\ 0 & 3 \\ 3 & 0 \end{bmatrix}$. Hence, $\operatorname{diag}(B) = 9e$ and therefore $w^T \operatorname{diag}(B) = 1/2$.*

On the other hand, if we use configuration matrix $P' = \begin{bmatrix} -3 & -1 \\ 0 & 2 \\ 3 & -1 \end{bmatrix}$, then $B'w \neq \mathbf{0}$.

In this case, we have $w^T \operatorname{diag}(B') = 5/9$. Consequently, it is imperative that the Gram matrix B in Theorem 4.6 satisfies $Bw = \mathbf{0}$.

Now consider the EDM D of Example 4.2, where $w = \frac{1}{2}[1 \ -2 \ 1 \ 0]^T$. Thus, a configuration matrix P of D that satisfies $P^T w = \mathbf{0}$ is $P = \begin{bmatrix} 1 & -1 \\ 0 & -1 \\ -1 & -1 \\ 0 & 3 \end{bmatrix}$. Hence, $\operatorname{diag}(B) = [2 \ 1 \ 2 \ 9]^T$ and thus $w^T \operatorname{diag}(B) = 1$.

Next, we collect the above sixth characterizations of spherical EDMs in the following theorem.

Theorem 4.7 *Let D be a nonzero $n \times n$ EDM of embedding dimension r. If $r = n - 1$, then D is spherical. Otherwise, if $r \leq n - 2$, then the following statements are equivalent:*

1. *D is spherical.*
2. *There exists $a \in \mathbb{R}^r$ such that $Pa = \frac{1}{2} J \operatorname{diag}(B)$, where B is the Gram matrix of D such that $B = -JDJ/2$; i.e., $Be = \mathbf{0}$.*
3. *$\operatorname{null}(D) = \operatorname{gal}(D)$; i.e., $DZ = \mathbf{0}$, where Z is a Gale matrix of D.*

4. $rank(D) = r+1$.
5. *There exists a scalar β such that $\beta ee^T - D \succeq 0$.*
6. *$e^T w > 0$, where $Dw = e$, in which case, the generating points of D lie on a hypersphere of radius*

$$\rho = (\frac{1}{2e^T w})^{1/2}.$$

7. *$w^T diag(B) = \frac{1}{2}$, where $Dw = e$ and $B = -(I - ew^T/(e^T w))D(I - we^T/(e^T w))/2$; i.e., $Bw = 0$.*

Two observations regarding Theorem 4.7 are discussed next. These observations, which are immediate consequences of parts 3 and 5, were made in [186, 184].

First, assume that the Gram matrix B satisfies $Be = 0$. Then, since $Z^T D = Z^T diag(B) e^T$, it follows from part 3 that D is spherical iff $Z^T diag(B) = 0$ iff $diag(B)$ lies in $col([P\ e])$. Note that Theorem 4.1 implies that D is spherical iff $Jdiag(B) \in col(P)$.

Second, suppose that D is an EDM such that $\beta E - D = A \succeq 0$ for some scalar β. Then $x^T Dx = \beta(e^T x)^2 - x^T Ax \le \beta(e^T x)^2$ for all x. Thus,

$$\sup\{\frac{x^T Dx}{(e^T x)^2} : x \notin e^\perp\} \le \beta. \tag{4.7}$$

Conversely, assume that (4.7) holds. Then $x^T(D - \beta E)x \le 0$ for all $x \notin e^\perp$. Moreover, if $x \in e^\perp$, then $x^T Dx \le 0$ since D is an EDM. As a result, $x^T(D - \beta E)x \le 0$ for all x and hence $\beta E - D$ is PSD. Therefore, it follows from part 5 that D is spherical iff $\sup\{\frac{x^T Dx}{(e^T x)^2} : x \notin e^\perp\} < \infty$. Finally, we should mention that a detailed investigation of vector $s = w/e^T w = 2\rho^2 w$ is given in [188].

Now assume that D is a nonsingular spherical EDM and let X be its projected Gram matrix. Then X is nonsingular. Consequently, the Moore–Penrose inverse of B, the Gram matrix of D, is given by $B^\dagger = VX^{-1}V^T$ and hence $B^\dagger B = VV^T = J$. Note that in case X is singular, i.e., if $rank(B) \le n-2$, then it is easy to verify that $B^\dagger = P(P^T P)^{-2}P^T$, where P is a configuration matrix of D. As the following theorem shows, D^{-1} can be expressed in terms of $B^\dagger$.

Theorem 4.8 (Styan and Subak-Sharpe [182]) *Let D be a nonsingular spherical EDM and let $B = -JDJ/2$ be its Gram matrix. Further, let $Dw = e$ and let ρ be the radius of the hypersphere containing the generating points of D. Then*

$$D^{-1} = -\frac{1}{2}B^\dagger + 2\rho^2 ww^T. \tag{4.8}$$

Proof. It follows from the definition of $\mathcal{K}$ that

$$B^\dagger D = B^\dagger diag(B) e^T - 2I + \frac{2}{n}ee^T. \tag{4.9}$$

Thus, multiplying (4.9) by D^{-1} yields

$$B^\dagger = -2D^{-1} + (B^\dagger \mathrm{diag}(B) + \frac{2}{n}e)w^T. \tag{4.10}$$

Moreover, $B^\dagger e = 0$ implies that

$$B^\dagger \mathrm{diag}(B) + \frac{2}{n}e = 4\rho^2 w. \tag{4.11}$$

Therefore, substituting (4.11) into (4.10) yields

$$B^\dagger = -2D^{-1} + 4\rho^2 ww^T.$$

□

The significance of Theorem 4.8 will be come clear when we discuss, below, the distance matrices of trees and the resistance distance matrices of electrical networks.

An immediate consequence of Theorem 4.8 is that if D is a nonsingular spherical EDM, then the inverse of the Cayley–Menger matrix M given in (3.26) is also given by

$$M^{-1} = 2\rho^2 \begin{bmatrix} -1 & w^T \\ w & -B^\dagger/(4\rho^2) \end{bmatrix}. \tag{4.12}$$

We should point out that Eq. (4.12) was also obtained in [79, 81, 182] and that Theorem 4.8 was generalized by Balaji and Bapat in [32].

At this point, making a connection with Jung's Theorem is in order.

Theorem 4.9 (Jung [119]) *Let S be a compact set in $\mathbb{R}^r$ and let d be the diameter of S; i.e., $d = \max\{\|p^i - p^j\| : p^i, p^j \in S\}$. Then S is contained in a ball of radius ρ, where*

$$\rho^2 \le d^2 \frac{r}{2(r+1)}.$$

Let D be a spherical $n \times n$ EDM of embedding dimension r and let d_{max} be the maximum entry of D. Then Jung's Theorem implies that

$$a^T a + \frac{e^T De}{2n^2} \le d_{max} \frac{\mathrm{rank}(D) - 1}{2\,\mathrm{rank}(D)}. \tag{4.13}$$

Furthermore, equality holds in (4.13) if $D = \Delta = E - I$ is the EDM of the standard simplex. This follows since in this case, $2\rho^2 = 1 - 1/n$, $d_{max} = 1$ and $\mathrm{rank}(D) = n$.

Before proceeding to discuss several subclasses of spherical EDMs, we show, next, how to construct a new spherical EDM from two old ones by using Kronecker product.

Theorem 4.10 ([12]) *Let D_1 and D_2 be two spherical EDMs of orders m and n, and of radii ρ_1 and ρ_2, respectively. Then*

$$D = E_m \otimes D_2 + D_1 \otimes E_n$$

is a spherical EDM generated by points that lie of a hypersphere of radius

$$\rho = (\rho_1^2 + \rho_2^2)^{1/2}.$$

Proof. Let $D_1 w_1 = e_m$ and $D_2 w_2 = e_n$ where $e_m^T w_1 > 0$ and $e_n^T w_2 > 0$. Then

$$
\begin{aligned}
D(w_1 \otimes w_2) &= (E_m \otimes D_2)(w_1 \otimes w_2) + (D_1 \otimes E_n)(w_1 \otimes w_2) \\
&= E_m w_1 \otimes e_n + e_m \otimes E_n w_2 \\
&= e_m^T w_1 \, e_m \otimes e_n + e_m \otimes e_n^T w_2 \, e_n \\
&= (e_m^T w_1 + e_n^T w_2) e_{mn}.
\end{aligned}
$$

Let $w = (w_1 \otimes w_2)/(e_m^T w_1 + e_n^T w_2)$. Then $Dw = e$ and

$$
e^T w = \frac{1}{(e_m^T w_1 + e_n^T w_2)} (e_m^T \otimes e_n^T)((w_1 \otimes w_2) = \frac{1}{(e_m^T w_1 + e_n^T w_2)} (e_m^T w_1 \, e_n^T w_2) > 0.
$$

Therefore, D is a spherical EDM. Furthermore,

$$
\rho^2 = \frac{1}{2 e^T w} = \frac{1}{2(e_m^T w_1 \, e_n^T w_2)} (e_m^T w_1 + e_n^T w_2) = \rho_2^2 + \rho_1^2.
$$

□

In the following subsections, we discuss several subclasses of spherical EDMs.

4.1.1 Regular EDMs

An important subclass of spherical EDMs is that of regular EDMs. A spherical EDM D is *regular* if the generating points of D lie on a hypersphere centered at the centroid of these points; i.e., if $a = \mathbf{0}$ in Eq. (4.1) (assuming that the centroid coincides with the origin). Consequently, the generating points of a regular EDM lie on a hypersphere of radius $\rho = (e^t De/2n^2)^{1/2}$. As a result, Inequality (4.13) in case of a regular EDM reduces to

$$
\frac{e^T De}{n^2} \le d_{\max} \frac{\text{rank}(D) - 1}{\text{rank}(D)}. \tag{4.14}
$$

An example of a regular EDM is Δ, the EDM of the standard simplex. Regular EDMs have properties that mirror those of adjacency matrices of regular graphs. A generalization of regular EDMs is given in [188]. We begin, first, with the following simple characterization of regular EDMs. We should point out here that, by Rayleigh–Ritz Theorem, the Perron eigenvalue $\lambda_1 \ge e^T De/n$ for any EDM D.

Theorem 4.11 (Hayden and Tarazaga [101]) *Let D be a nonzero $n \times n$ EDM, then D is regular if and only if $(e^T De/n, e)$ is the Perron eigenpair of D.*

Proof. Assume that D is regular then $\text{diag}(B) = \rho^2 e$. Hence, $D = 2\rho^2 ee^T - 2B$. Thus, $De = 2n\rho^2 e = (e^T De/n) e$; i.e., $(e^T De/n, e)$ is the Perron eigenpair of D.

To prove the reverse direction, assume that $De = (e^T De/n) e$ and let $e^T De/(2n^2) = \rho^2$. Then by the definition of $\mathscr{T}$, we have that $B = \mathscr{T}(D) = -(D - 2\rho^2 ee^T)/2$. Thus, $Be = \mathbf{0}$ and $\text{diag}(B) = \rho^2 e$ and hence, D is regular.

□

The following corollary is an immediate consequence of Theorems 3.12 and 4.11. It extends the Hoffman polynomial of graphs [110] to EDMs.

Corollary 4.2 *Let D be an $n \times n$ EDM and let $\lambda > -\alpha_1 > \cdots > -\alpha_k$ be the distinct eigenvalues of D. Then there exists a polynomial $f(D)$ such that $f(D) = E$ if and only if D is regular, in which case*

$$f(D) = n \frac{\prod_{i=1}^{k}(D + \alpha_i I)}{\prod_{i=1}^{k}(\frac{e^T De}{n} + \alpha_i)}. \tag{4.15}$$

f is called the Hoffman polynomial *of D.*

Proof. By Theorem 3.12, there exists a polynomial g such that $g(D) = \gamma x x^T$, where x is the Perron eigenvector of D. Assume that D is regular. Then $x = e$ and thus there exists $g(x) = \gamma E$. Hence, $f(x) = g(x)/\gamma$. Now to find the scalar γ, notice that $(D + \alpha_i I)e = (e^T De/n + \alpha_i)e$. Therefore, $g(D)e = \prod_{i=1}^{k}(e^T De/n + \alpha_i)e = n\gamma e$ and thus $\gamma = \prod_{i=1}^{k}(e^T De/n + \alpha_i)/n$.

On the other hand, assume that an EDM D satisfies (4.15) and let $De = u$. Then D commutes with E since $DE = Df(D) = f(D)D = ED$. Consequently, $DE = ue^T = ED = eu^T$. Thus $u = (u^T e/n) e$ and hence D is regular.

□

Example 4.7 *Consider the EDM $D = \begin{bmatrix} 0 & 2 & 4 & 2 \\ 2 & 0 & 2 & 4 \\ 4 & 2 & 0 & 2 \\ 2 & 4 & 2 & 0 \end{bmatrix}$ with configuration matrix $P = \begin{bmatrix} -1 & 0 \\ 0 & -1 \\ 1 & 0 \\ 0 & 1 \end{bmatrix}$. Then $De = 8e$. Obviously, the generating points of D lie on a hyper-sphere centered at the origin and of radius $\rho = e^T De/2n^2 = 1$. Moreover, $w = e/8$ and $\text{diag}(B) = e$. Thus $w^T \text{diag}(B) = 1/2$. Note that $\beta^* = 2\rho^2 = 2$ and hence, p^1 and p^3 are antipodal since $d_{13} = 2\beta^* = 4$. Likewise, p^2 and p^4 are antipodal.*

The eigenvalues of D are $8, 0, -4, -4$. Thus $k = 2$, $\alpha_1 = 0$, $\alpha_2 = 4$. Hence, $\gamma = \prod_{i=1}^{2}(\frac{e^T De}{n} + \alpha_i)/n = 24$. Therefore, the Hoffman polynomial of D is

$$f(x) = \frac{1}{24}x(x+4).$$

We saw earlier that a spherical EDM can be constructed from two spherical EDMs by using Kronecker product. The same result also applies to regular EDMs.

Theorem 4.12 *Let D_1 and D_2 be two regular EDMs of orders m and n, respectively. Then*

$$D = E_m \otimes D_2 + D_1 \otimes E_n$$

is a regular EDM.

Proof. D is an EDM by Theorem 4.10. Thus, it suffices to show that e is an eigenvector of D. To this end,

$$
\begin{aligned}
De &= (E_m \otimes D_2)(e_m \otimes e_n) + (D_1 \otimes E_n)(e_m \otimes e_n) \\
&= (m\, e_n^T D_2 e_n / n + n\, e_m^T D_1 e_m / m)(e_m \otimes e_n) \\
&= \frac{1}{mn} e^T De\, e.
\end{aligned}
$$

$\square$

Next, we turn to another subclass of spherical EDMs.

4.1.2 Cell Matrices

An $n \times n$ matrix $D = (d_{ij})$ is called a *cell matrix* if for $i \neq j$, we have

$$d_{ij} = c_i + c_j \text{ for some } c \geq \mathbf{0} \text{ in } \mathbb{R}^n.$$

Consequently, D is a cell matrix if $D = ec^T + ce^T - 2\,\mathrm{Diag}(c)$ for some $c \geq \mathbf{0}$. For example, $\Delta = E - I$, the EDM of the standard simplex, is a cell matrix corresponding to $c = e/2$. Cell matrices, which were introduced by Jaklič and Modic in [116], model a star graph; i.e., a tree with one root node and $n-1$ adjacent leaves.

It is readily seen that cell matrices are EDMs since the projected Gram matrix of a cell matrix D is $\mathscr{T}_V(D) = V^T \mathrm{Diag}(c) V \succeq \mathbf{0}$. Furthermore, $\mathscr{T}(D) = J\mathrm{Diag}(c)J$. Consequently, the Gram matrix of D is given by $B = \mathrm{Diag}(c) - ce^T/n - ec^T/n + e^T c\, ee^T/n^2$. Assume that c has $s \geq 2$ zero entries and wlog assume that $c_{n-s+1} = \cdots = c_n = 0$. Then, the following two facts are immediate consequence of the definition. First, $p^{n-s+1} = \cdots = p^n$ since $d_{ij} = 0$ for all $i, j = n-s+1, \ldots, n$. This fact is used, next, to determine the embedding dimension of a cell matrix. Second, the last s columns (hence rows) of D are identical since for all i, $d_{ij} = c_i$ (independent of j) for all $j = n-s+1, \ldots, n$.

Lemma 4.2 *Let c in $\mathbb{R}^n$ be $\geq \mathbf{0}$ and let D be the cell matrix corresponding to c. Let s denote the number of zero entries of c. Then the embedding dimension of D is given by*

$$r = \begin{cases} n-1 \text{ if } s = 0 \text{ or } s = 1, \\ n-s \text{ if } s \geq 2. \end{cases}$$

Proof.   Recall that V, as defined in (3.11), has full column rank and every $(n-1) \times (n-1)$ submatrix of V is nonsingular. Let $\mathrm{Diag}(\sqrt{c})Vx = \mathbf{0}$. If $s = 0$, then $Vx = \mathbf{0}$ and hence $x = \mathbf{0}$. Also, if $s = 1$, then again $x = \mathbf{0}$. Thus, if $s \leq 1$, null$(\mathrm{Diag}(\sqrt{c})V)$ is trivial and hence rank$(\mathrm{Diag}(\sqrt{c})V) = n-1$. Consequently, $r = \mathrm{rank}\ \mathscr{T}_V(D) = \mathrm{rank}(V^T \mathrm{Diag}(c)V) = \mathrm{rank}(\mathrm{Diag}(\sqrt{c})V) = n-1$.

Now assume that $s \geq 2$ and wlog assume that $c_{n-s+1} = \cdots = c_n = 0$. Then obviously, $p^{n-s+1} = \cdots = p^n$. Thus r, the embedding dimension of D, is equal to the embedding dimension of the EDM generated by $p^1, \ldots, p^{n-s+1}$; i.e., r is equal to the embedding dimension of the cell matrix corresponding to $\bar{c} = [c_1 \; \cdots \; c_{n-s+1}]^T$. Notice that $\bar{c}$ has one zero entry. Therefore, by the previous case, it follows that $r = n - s + 1 - 1 = n - s$.

$\square$

Therefore, Lemma 4.2 implies that, if c has $s \geq 2$ zero entries, say, $c_{n-s+1} = \cdots = c_n = 0$, then $p^1, \ldots, p^{n-s+1}$ are affinely independent and $p^{n-s+2} = \cdots = p^n = p^{n-s+1}$. Hence, $\mathrm{rank}(D) = n - s + 1$ since the $(n - s + 1)$ leading principal submatrix of D is spherical of embedding dimension $n - s$; and since the last s columns of D are identical. Moreover, it is easy to see that in this case, i.e., if $s \geq 2$, then

$$
Z = \begin{bmatrix} \mathbf{0} \\ I_{s-1} \\ -e_{s-1}^T \end{bmatrix}
$$

is a Gale matrix of D. As a result, cell matrices are spherical EDMs.

Theorem 4.13 (Jaklič and Modic [116]) *Cell matrices are spherical Euclidean distance matrices.*

The proof of Theorem 4.13 in [116] is based on Theorem 3.11 and part 6 of Theorem 4.7. However, this theorem is an immediate consequence of part 3 of Theorem 4.7 since it is easy to verify that $DZ = \mathbf{0}$ if $s \geq 2$. Note that if $s = 0$ or $s = 1$, then D is obviously spherical. Also, this theorem follows from part 4 of Theorem 4.7 since if $s = 0$ or 1, then r, the embedding dimension of D, is equal to $n - 1$. Otherwise, if $s \geq 2$, then $r = n - s$. Accordingly, the result follows since in this case $\mathrm{rank}(D) = n - s + 1 = r + 1$.

Example 4.8 *Let $c = [1 \; 2 \; 3 \; 0]^T$. Then, the cell matrix corresponding to c is*

$$
D = \begin{bmatrix} 0 & 3 & 4 & 1 \\ 3 & 0 & 5 & 2 \\ 4 & 5 & 0 & 3 \\ 1 & 2 & 3 & 0 \end{bmatrix}.
$$

The embedding dimension of D is $r = 3$ and the generating points of D are affinely independent.

Now let $c' = [c^T \; 0]^T$. Then, the cell matrix corresponding to c' is $D' = \begin{bmatrix} D & c \\ c^T & 0 \end{bmatrix}$. Moreover, the embedding dimension of D' is again $r = 3$ and in this case $p^4 = p^5$.

Next, we turn to a third subclass of spherical EDMs, namely, the Manhattan distance matrices on grids.

4.1.3 Manhattan Distance Matrices on Grids

In this subsection, we focus on rectangular grids of unit squares with m rows and n columns. First, we consider the special case when $m = 1$. Let $G_n = (g_{ij})$ be the $n \times n$ Manhattan distance matrix of a rectangular grid of 1 row and n columns. Then

$$g_{ij} = |i - j|.$$

Let $p^1, \ldots, p^n$ be the points in $\mathbb{R}^{n-1}$ such that the first $i - 1$ entries of p^i are 1's and the remaining $n - i$ entries are 0's. Thus, p^1 coincides with the origin and $p^n = e_{n-1}$. Moreover, $||p^i - p^j||^2 = |i - j|$ for all $i, j = 1, \ldots, n$. Hence, G_n is an EDM with embedding dimension $r = n - 1$ generated by $p^1, \ldots, p^n$. Let $a = e/2$, then $||p^i - a||^2 = (n - 1)/4$ for all $i = 1, \ldots, n$. As a result, the points $p^1, \ldots, p^n$ lie on a hypersphere centered at $a = e/2$ and of radius $\rho = \frac{1}{2}(n - 1)^{1/2}$ and hence G is a spherical EDM. Another way to show that G_n is spherical is to observe that $g_{i1} + g_{in} = n - 1$ for all $i = 1, \ldots, n$. Thus, if we let $w = [1 \ 0 \ \cdots \ 0 \ 1]^T/(n-1)$, then $G_n w = e$ and $e^T w = 2/(n-1) > 0$.

Now consider a rectangular grid of m rows and n columns and let $\hat{d}_{ij,kl}$ be the Manhattan distance between the grid points at (i, j) and (k, l). Then

$$\hat{d}_{ij,kl} = |i - k| + |j - l|.$$

To represent these distances as the entries of an $mn \times mn$ matrix, we replace the double indices ij and kl by single indices s and t, respectively. First, let $s = j + n(i - 1)$ for $i = 1, \ldots, m$ and $j = 1, \ldots, n$. This relation produces a lexicographic ordering, i.e,

$$11, 12, \ldots, 1n, 21, 22, \ldots, 2n, \ldots, m1, m2, \ldots, mn.$$

In terms of s, the indices i and j are given by $i = \lceil s/n \rceil$ and $j = s - n(\lceil s/n \rceil - 1)$. Similarly, let $t = l + n(k - 1)$ for $k = 1, \ldots, m$ and $l = 1, \ldots, n$. Thus, $k = \lceil t/n \rceil$ and $l = t - n(\lceil t/n \rceil - 1)$.

Consider $(A \otimes B)$ where A and B are any two matrices of orders m and n, respectively. Then

$$(A \otimes B)_{st} = a_{ik} b_{jl}. \tag{4.16}$$

It is important to keep in mind that s depends on the first indices i and j of the entries of A and B, while t depends on the second indices. For example, let $m = 2$ and $n = 3$. Then the lexicographic ordering is

$$11, 12, 13, 21, 22, 23.$$

and $a_{12}b_{11} = (A \otimes B)_{14}$. This follows since in this case, $ij = 11$ and $kl = 21$ and thus $s = 1$ and $t = 4$. Similarly, $a_{21}b_{13} = (A \otimes B)_{43}$ and $a_{22}b_{23} = (A \otimes B)_{56}$.

Therefore,

$$(E_m \otimes G_n)_{st} = (E_m)_{ik}(G_n)_{jl} = (G_n)_{jl} = |j - l|.$$

and

$$(G_m \otimes E_n)_{st} = (G_m)_{ik}(E_n)_{jl} = (G_m)_{ik} = |i - k|.$$

Thus, the $mn \times mn$ matrix $D = (\hat{d}_{ij,kl})$ is given by

$$D = E_m \otimes G_n + G_m \otimes E_n. \tag{4.17}$$

Consequently, Theorem 4.10 implies that D is a spherical EDM generated by points that lie on a hypersphere of radius $\rho = \frac{1}{2}(n + m - 2)^{1/2}$; and Theorem 4.3 implies that

$$\frac{1}{2}(n + m - 2)E - D \succeq 0. \tag{4.18}$$

It should be noted that (4.18) was first obtained by Mettlemann and Peng in [146].

Example 4.9 *Consider the rectangular grid with unit squares of two rows and three columns. Then*

$$G_2 = \begin{bmatrix} 0 & 1 \\ 1 & 0 \end{bmatrix} \text{ and } G_3 = \begin{bmatrix} 0 & 1 & 2 \\ 1 & 0 & 1 \\ 2 & 1 & 0 \end{bmatrix}.$$

Thus, $D = E_2 \otimes G_3 + G_2 \otimes E_3$ is given by

$$D = \begin{bmatrix} 0 & 1 & 2 & 0 & 1 & 2 \\ 1 & 0 & 1 & 1 & 0 & 1 \\ 2 & 1 & 0 & 2 & 1 & 0 \\ 0 & 1 & 2 & 0 & 1 & 2 \\ 1 & 0 & 1 & 1 & 0 & 1 \\ 2 & 1 & 0 & 2 & 1 & 0 \end{bmatrix} + \begin{bmatrix} 0 & 0 & 0 & 1 & 1 & 1 \\ 0 & 0 & 0 & 1 & 1 & 1 \\ 0 & 0 & 0 & 1 & 1 & 1 \\ 1 & 1 & 1 & 0 & 0 & 0 \\ 1 & 1 & 1 & 0 & 0 & 0 \\ 1 & 1 & 1 & 0 & 0 & 0 \end{bmatrix} = \begin{bmatrix} 0 & 1 & 2 & 1 & 2 & 3 \\ 1 & 0 & 1 & 2 & 1 & 2 \\ 2 & 1 & 0 & 3 & 2 & 1 \\ 1 & 2 & 3 & 0 & 1 & 2 \\ 2 & 1 & 2 & 1 & 0 & 1 \\ 3 & 2 & 1 & 2 & 1 & 0 \end{bmatrix}.$$

Let $w_2 = e_2$ and $w_3 = \frac{1}{2}[1\ 0\ 1]^T$. Then $G_2 w_2 = e_2$ and $G_3 w_3 = e_3$. Now let $w = w_2 \otimes w_3/(e_2^T w_2 + e_3^T w_3) = \frac{1}{6}[1\ 0\ 1\ 1\ 0\ 1]^T$. Then $Dw = e$ and $e^T w = 2/3$. Accordingly, the generating points of D lie on a hypersphere of radius $\rho = \sqrt{3}/2$.

The fourth subclass of spherical EDMs is that of Hamming distance matrices on the hypercube.

4.1.4 Hamming Distance Matrices on the Hypercube

Let Q_r denote the r-dimensional hypercube; i.e., the vertices of Q_r are all points in $\mathbb{R}^r$ whose entries are either 0 or 1. Let $p^1, \dots, p^{2^r}$ be the vertices of Q_r and let $D = (d_{ij})$ be the $2^r \times 2^r$ matrix such that d_{ij} is the Hamming distance between p^i and p^j. Thus

$$d_{ij} = \sum_{k=1}^{r} |p_k^i - p_k^j| = \sum_{k=1}^{r} (p_k^i - p_k^j)^2 = \|p^i - p^j\|^2,$$

where the second equality follows since $p_k^i - p_k^j$ is either 1 or 0. Therefore, D is an EDM of embedding dimension r. Let a be the centroid of the generating points of D. Then $a = e/2$ and hence, these points lie on a hypersphere centered at a and of radius $\rho = \frac{1}{2}r^{1/2}$. Consequently, D is a regular EDM. Notice that, for these matrices, the origin coincides with one of the generating points and not with their centroid.

Note that, for $r \geq 2$, $\det(D) = 0$ since D is of order 2^r and $\mathrm{rank}(D) = r + 1$. However, some nonzero minors of D have a simple form.

Theorem 4.14 (Graham and Winkler [96]) *Assume that the vertices $p^1, \ldots, p^{r+1}$ of the hypercube Q_r form a simplex. Then the determinant of the submatrix of D induced by these points is given by*

$$(-1)^r r 2^{r-1}.$$

Proof. Let D' denote the $(r+1) \times (r+1)$ submatrix of D induced by $p^1, \ldots, p^{r+1}$. Then, it follows from Theorem 3.15 that

$$\mathrm{V}^2(p^1, \ldots, p^{r+1}) = \frac{(-1)^{r+1}}{2^r((r)!)^2} \det\left(\begin{bmatrix} 0 & e^T \\ e & D' \end{bmatrix}\right).$$

But, $\mathrm{V}(p^1, \ldots, p^{r+1}) = 1/r!$ since the parallelepiped generated by these points is the unit hypercube. Therefore,

$$\det\left(\begin{bmatrix} 0 & e^T \\ e & D' \end{bmatrix}\right) = (-1)^{r+1} 2^r.$$

Now by Schur complement,

$$\det\left(\begin{bmatrix} 0 & e^T \\ e & D' \end{bmatrix}\right) = \det(D')\det(0 - e^T D'^{-1} e) = -(e^T D'^{-1} e)\det(D').$$

But since D is a regular EDM of radius $\rho = \sqrt{r}/2$, it follows that D' is a spherical EDM of the same radius. Thus, $e^T D'^{-1} e = e^T w = 1/(2\rho^2) = 2/r$ and hence $\det(D') = (-1)^r r 2^{r-1}$.

$\square$

Distance matrices of trees are the fifth subclass of spherical EDMs.

4.1.5 Distance Matrices of Trees

Let T be a tree on n nodes. The distance matrix of T is the $n \times n$ matrix $D = (d_{ij})$ where d_{ij} is the number of edges in the path between node i and node j. For example, $d_{ij} = 1$ for every edge $\{i, j\}$ of T. By definition, $d_{ii} = 0$. As will be shown in this subsection, distance matrices of trees are spherical EDMs [32, 33]. Moreover, these matrices form a subset of the resistance distance matrices of electrical networks [121] to be discussed the following subsection.

Example 4.10 *Let T be the tree depicted in Fig. 4.1. Then the distance matrix of T is*

$$D = \begin{bmatrix} 0 & 1 & 2 & 2 & 3 \\ 1 & 0 & 1 & 1 & 2 \\ 2 & 1 & 0 & 2 & 3 \\ 2 & 1 & 2 & 0 & 1 \\ 3 & 2 & 3 & 1 & 0 \end{bmatrix}.$$

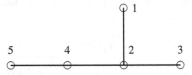

Fig. 4.1 The tree of Example 4.10

Distance matrices of trees have nice properties. For instance, the determinant and the inverse of these matrices have simple forms. More precisely, as shown by the following remarkable theorem, the determinant of a distance matrix of a tree T has a surprisingly simple form which is independent of the structure of T.

Theorem 4.15 (Graham and Pollak [95]) *Let D be the distance matrix of a tree on n nodes. Then*

$$\det(D) = (-1)^{n-1}(n-1)2^{n-2}.$$

Proof. Wlog assume that n is a leaf node adjacent to node $n-1$. Then it is easy to see that $d_{in} = d_{in-1} + 1$ for $i = 1, \ldots, n-1$. Therefore, by subtracting the $(n-1)$th column $((n-1)$th row) of D from the nth column (nth row), the (n,n)th entry of D becomes (-2) and all other entries in the nth column and the nth row become 1's. Now let T_{n-1} be the tree obtained by deleting node n and edge $\{n, n-1\}$. Let node i be a leaf of T_{n-1} and let node j be adjacent to i. Then by subtracting the jth column (jth row) of D from the ith column (ith row), the (i,i)th entry of D becomes (-2), the (i,n)th and (n,i)th entries become 0's and all other entries in the ith row and the ith column become 1's. By repeating this process, assuming that the last remaining node is node 1, we arrive at a *bordered diagonal matrix* whose determinant is easy to compute. More precisely, we get that

$$\det(D) = \det(\begin{bmatrix} 0 & 1 & \cdots & 1 \\ 1 & -2 & \cdots & 0 \\ \vdots & 0 & \ddots & 0 \\ 1 & 0 & \cdots & -2 \end{bmatrix}) = \det(\begin{bmatrix} \frac{n-1}{2} & 0 & \cdots & 0 \\ 1 & -2 & \cdots & 0 \\ \vdots & 0 & \ddots & 0 \\ 1 & 0 & \cdots & -2 \end{bmatrix}),$$

where the last determinant is obtained by adding (row $2 + \cdots +$ row n)/2 to row 1. Consequently, $\det(D) = (-2)^{n-1}(n-1)/2$.

□

An immediate consequence of Theorem 4.15 is that distance matrices of trees are nonsingular elliptic matrices.

Theorem 4.16 (Graham and Pollak [95]) *Let D be the distance matrix of a tree on n nodes. Then D has exactly one positive and $n-1$ negative eigenvalues.*

Proof. The proof is by induction on n. The assertion is obviously true for $n=2$ since the tree consists of one edge and thus the eigenvalues of D are clearly ± 1. Thus, assume that the assertion is true for $n=k$ and consider the $(k+1)\times(k+1)$ matrix $D = \begin{bmatrix} \bar{D} & d \\ d^T & 0 \end{bmatrix}$, where $\bar{D}$ is of order k and $d \in \mathbb{R}^k$. Therefore, by Cauchy interlacing theorem, matrix D has one positive eigenvalue, $k-1$ negative eigenvalues and one eigenvalue which can be either positive or negative. However, by Theorem 4.15, this last eigenvalue must be negative since $\det(D)$ has sign $(-1)^k$. Therefore, D has exactly k negative eigenvalues.

□

Similar to the determinant, the inverse of the distance matrix of a tree has a simple form as shown by the following theorem.

Theorem 4.17 (Graham and Lovász [94]) *Let D be the distance matrix of a tree T on n nodes. Let L denote the Laplacian of T and* deg *denote the vector of the degrees of the nodes of T. Then*

$$D^{-1} = -\frac{1}{2}L + \frac{1}{2(n-1)}(2e - \deg)(2e - \deg)^T.$$

Two remarks are in order here. First, Theorem 4.17 is a special case of Theorem 4.8. Second, suppose that node i is a leaf of T and let T' be the tree obtained from T by deleting node i and the edge incident with it. Let D' be the distance matrix of T'. Hence, the (i,i)-cofactor of D is equal to $\det(D') = (-1)^{n-2}(n-2)2^{n-3}$. Consequently, the (i,i)th entry of D^{-1} is

$$\frac{(-1)^{n-2}(n-2)2^{n-3}}{(-1)^{n-1}(n-1)2^{n-2}} = -\frac{n-2}{2(n-1)}$$

which is independent of i. This agrees, as should be the case, with the implication of Theorem 4.17 that the (i,i)th entry of D^{-1} is

$$-\frac{1}{2} + \frac{1}{2(n-1)} = \frac{2-n}{2(n-1)}.$$

We should point out that alternative proofs of Theorems 4.15 and 4.17 are given in [32, 33].

Example 4.11 *Consider the matrix D of Example 4.10. Then*

$$D^{-1} = \frac{1}{8} \begin{bmatrix} -3 & 3 & 1 & 0 & 1 \\ 3 & -11 & 3 & 4 & -1 \\ 1 & 3 & -3 & 0 & 1 \\ 0 & 4 & 0 & -8 & 4 \\ 1 & -1 & 1 & 4 & -3 \end{bmatrix}.$$

Observe that $D_{11}^{-1} = D_{33}^{-1} = D_{55}^{-1} = -3/8$ since nodes 1, 3, and 5 are leaves of T.

In the theorem that follows, we establish that the distance matrix of a tree is a spherical EDM and we determine its radius.

Theorem 4.18 *Let D be the distance matrix of a tree on n nodes. Then D is a spherical EDM of radius*

$$\rho = \frac{(n-1)^{1/2}}{2}.$$

Proof. Recall that D is elliptic. Let $Dw = e$. Then $w = D^{-1}e = (2e - \deg)/(n-1)$ since $Le = \mathbf{0}$ and since $e^T \deg = 2(n-1)$. Consequently, $e^T w = e^T D^{-1} e = 2/(n-1)$. The result follows from Theorem 4.7.

$\square$

As we mentioned earlier, distance matrices of trees form a subset of resistance distance matrices of electrical networks which we discuss next.

4.1.6 Resistance Distance Matrices of Electrical Networks

Let us regard a simple connected graph G as an electrical network where each edge of G is a unit resistor [73, 175]. Identify two nodes of G as a source node s and a sink node t and connect s and t to the terminals of a battery. Let the voltage across s and t be $v_s - v_t$ and the current flowing into s and out of t be i_{st}. Then the *effective resistance* between s and t, denoted by ω_{st}, is defined as

$$\omega_{st} = \frac{v_s - v_t}{i_{st}}.$$

As a result, graph G is equivalent to one edge $\{s,t\}$ with resistance ω_{st}. Let the resistances of two edges of G be ω_1 and ω_2. It is well known that these two edges can be replaced by a single edge of resistance $\omega_1 + \omega_2$ if they are in series, and of resistance $(\omega_1^{-1} + \omega_2^{-1})^{-1}$ if they are in parallel. Consequently, for series-parallel graphs, ω_{st} can be calculated by iteratively using these two rules.

Example 4.12 *Consider the electrical network of unit resistors of Fig. 4.2, where node 1 is identified as the source node s and node 4 is identified as the sink node t. It is easy to see that the effective resistance across s,t is $\omega_{st} = (2^{-1} + 2^{-1} + 1^{-1})^{-1} = 1/2$.*

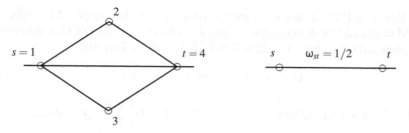

Fig. 4.2 The electrical network of unit resistors of Example 4.12. Node 1 is identified as the source node s, while node 4 is identified as the sink node t

As will be shown in this subsection, the matrix of all pair-wise effective resistors of G is a spherical EDM [121, 182]. To this end, assume that the magnitude of the current flowing into s and out of t is 1. Thus, since each edge of G is a unit resistor, *Kirchhoff current law*, based on the conservation of electric charge, implies that

$$\sum_{k:\{k,j\}\in E(G)} (v_j - v_k) = \delta_{js} - \delta_{jt} \text{ for all } j = 1,\ldots,n \qquad (4.19)$$

where δ_{ij} is the Kronecker delta. Let $i^{ext} = e^s - e^t$, where e^s and e^t are the sth and the tth standard unit vectors in $\mathbb{R}^n$; i.e.,

$$i_j^{ext} = \begin{cases} 1 \text{ if } j = s \\ -1 \text{ if } j = t \\ 0 \text{ otherwise.} \end{cases}$$

Also, let v in $\mathbb{R}^n$ be the vector consisting of the voltages on the nodes of G. Then (4.19) can be written in matrix form as

$$Lv = i^{ext}, \qquad (4.20)$$

where L is the Laplacian of G. Hence, $v = L^\dagger i^{ext}$, where $L^\dagger$ is the Moore–Penrose inverse of L. Observe that $v + \alpha e$ satisfies (4.20) for any scalar α since $Le = 0$. Hence v is not unique. This should come as no surprise since voltages are not measured in absolute but in relative terms. Therefore, the effective resistance [121, 182] between s and t is given by

$$\omega_{st} = v_s - v_t = (e^s - e^t)^T L^\dagger (e^s - e^t) = L_{ss}^\dagger + L_{tt}^\dagger - 2L_{st}^\dagger.$$

As a result, the matrix of pair-wise effective resistances of G is given by

$$\Omega = \mathscr{K}(L^\dagger), \qquad (4.21)$$

where $\mathscr{K}$ is as defined in (3.3). Moreover, since L is PSD of rank $n - 1$ (graph G is connected) and $Le = 0$, it follows that $L = V\Phi V^T$ for some $(n - 1) \times (n - 1)$ PD symmetric matrix Φ, where V is as defined in (3.11). Consequently, $L^\dagger = V\Phi^{-1}V^T$

and thus $L^\dagger$ is PSD of rank $n-1$ and satisfies $L^\dagger e = 0$. As a result, Ω is a spherical EDM of embedding dimension $n-1$ and $L^\dagger$ is the Gram matrix of Ω. Furthermore, the projected Gram matrix of Ω is $X = V^T L^\dagger V = \Phi^{-1}$. Therefore,

$$\Omega = \mathscr{K}_V(X), \text{ where } X = (V^T L V)^{-1}. \tag{4.22}$$

Let $C = L + E/n$ and let $Q = [V \ e/\sqrt{n}]$. Then $C = Q \begin{bmatrix} \Phi & 0 \\ 0 & 1 \end{bmatrix} Q^T$ and hence $C^{-1} = Q \begin{bmatrix} X & 0 \\ 0 & 1 \end{bmatrix} Q^T = L^\dagger + E/n$. Therefore,

$$L^\dagger = (L + \frac{1}{n}E)^{-1} - \frac{1}{n}E. \tag{4.23}$$

Let the generating points of Ω lie on a hypersphere of center a and radius ρ. Next, we calculate a and ρ. To this end, the configuration matrix of Ω is $P = VX^{1/2}$. Part 2 of Theorem 4.7 implies that $2Pa = J\text{diag}(L^\dagger)$ or $2VX^{1/2}a = J\text{diag}(L^\dagger)$. Thus,

$$2a = X^{-1/2}V^T\text{diag}(L^\dagger)$$

and hence,

$$4a^T a = (\text{diag}(L^\dagger))^T L \, \text{diag}(L^\dagger).$$

On the other hand, $e^T De = 2n \, \text{trace}(L^\dagger)$. Consequently,

$$\rho^2 = a^T a + e^T De/(2n^2) = \frac{1}{4}(\text{diag}(L^\dagger))^T L \, \text{diag}(L^\dagger) + \frac{1}{n}\text{trace}(L^\dagger). \tag{4.24}$$

(4.24) can be alternatively obtained as follows. Premultiplying Eq. (4.11) by $\text{diag}(B)^T$ yields

$$(\text{diag}(B))^T B^\dagger \text{diag}(B) + \frac{2}{n}e^T\text{diag}(B) = 4\rho^2 w^T \text{diag}(B).$$

But since $n = w^T De = n \, w^T\text{diag}(B) + e^T\text{diag}(B) \, e^T w$, it follows that $w^T\text{diag}(B) = 1 - e^T\text{diag}(B)/2n\rho^2$. Thus, (4.24) follows by setting $B = L^\dagger$.

Example 4.13 *The Laplacian and its Moore–Penrose inverse of the graph of Example 4.12 are*

$$L = \begin{bmatrix} 3 & -1 & -1 & -1 \\ -1 & 2 & 0 & -1 \\ -1 & 0 & 2 & -1 \\ -1 & -1 & -1 & 3 \end{bmatrix} \text{ and } L^\dagger = \frac{1}{16}\begin{bmatrix} 3 & -1 & -1 & -1 \\ -1 & 5 & -3 & -1 \\ -1 & -3 & 5 & -1 \\ -1 & -1 & -1 & 3 \end{bmatrix}.$$

Consequently, the matrix of resistance distances is

$$\Omega = \mathscr{K}(L^{\dagger}) = \frac{1}{8} \begin{bmatrix} 0 & 5 & 5 & 4 \\ 5 & 0 & 8 & 5 \\ 5 & 8 & 0 & 5 \\ 4 & 5 & 5 & 0 \end{bmatrix}$$

and $\rho^2 = 17/64$.

Example 4.14 *Consider the electrical network corresponding to the complete graph K_n. The Laplacian and its Moore–Penrose inverse are given by $L = nI - E$ and $L^{\dagger} = J/n$. Therefore, the matrix of resistance distances is $\Omega = 2(E - I)/n$.*

Finally, showing that distance matrices of trees is a subset of resistance distance matrices of electrical networks [121] is straightforward. For assume that G is a tree, say T. Then the path between any two nodes s and t of T is unique. Consequently, the effective resistance between s and t is equal to the distance between s and t.

4.2 Nonspherical EDMs

Evidently, many characterizations of spherical EDMs give rise to characterizations of nonspherical ones. For example, if D is an EDM of embedding dimension r, then it follows at once from Theorems 4.7 and 3.8 that D is nonspherical iff rank$(D) = r + 2$. Also, an immediate consequence of Theorem 4.5 is that an EDM D is nonspherical if and only if $e^T w = 0$ where $Dw = e$. As we remarked earlier, such w is not unique, for if $y \in \text{null}(D)$, then $D(w + y) = e$ and $e^T(w + y) = 0$ since $e \in \text{col}(D)$ and since $\text{col}(D)$ is orthogonal to $\text{null}(D)$. However, there is a unique η such that $D\eta = e$ and $\eta \perp (e \oplus \text{null}(D))$. Such unique η plays an important role in determining the eigenvalues of nonspherical EDMs as well as in the characterizations of their null and column spaces.

Theorem 4.19 *Let D be an $n \times n$ nonspherical EDM of embedding dimension r. Then $\text{gal}(D) = \text{null}(D) \oplus \text{span}(\eta)$, where η is the unique vector in $\mathbb{R}^n$ such that*

$$D\eta = e, \quad \eta \perp (e \oplus \text{null}(D)). \tag{4.25}$$

Proof. Let Z be a Gale matrix of D and as always, let $\bar{r} = n - r - 1$. Then Lemma 3.10 implies that $DZ = e\xi^T$, where $\xi = (\xi_i) = Z^T \text{diag}(B) \in \mathbb{R}^{\bar{r}}$. If $r = n - 2$, i.e., if $\bar{r} = 1$, then D is nonsingular since rank$(D) = r + 2 = n$ and Z is $n \times 1$. Therefore, $DZ = \xi_1 e$ and hence $\eta = Z/\xi_1 = D^{-1}e$. Consequently, in this case, $\text{gal}(D) = \text{span}(\eta)$ and $\text{null}(D)$ is trivial. Note that $e^T \eta = 0$.

Now assume that $r \leq n - 3$ and wlog assume that $\xi_1 \neq 0$. Let $w = Z_{.1}/\xi_1$ where $Z_{.1}$ is the first column of Z. Then $Dw = e$. Define the $\bar{r} \times \bar{r}$ nonsingular upper triangular matrix

$$S = \begin{bmatrix} \xi_1^{-1} & -\xi_2 & \cdots & -\xi_{\bar{r}} \\ 0 & \xi_1 & \cdots & 0 \\ 0 & \ddots & \xi_1 & 0 \\ 0 & \cdots & 0 & \xi_1 \end{bmatrix}.$$

Then $\xi^T S = [1 \ 0]$ and thus $e\xi^T S = [e \ 0]$. On the other hand, $ZS = [w \ \bar{Z}]$ is a Gale matrix where $\bar{Z}$ is $n \times (\bar{r} - 1)$. Consequently, $DZS = D[w \ \bar{Z}] = [e \ 0]$. Therefore, $\text{col}(\bar{Z}) \subseteq \text{null}(D)$. But $\dim \text{null}(D) = n - r - 2 = \bar{r} - 1$, thus $\text{col}(\bar{Z}) = \text{null}(D)$. Let $Q = (I - \bar{Z}(\bar{Z}^T \bar{Z})^{-1} \bar{Z}^T)$ be the orthogonal projection onto $\text{null}(\bar{Z}^T)$ and let $\eta = Qw$. Then $\eta \in \text{null}(\bar{Z}^T)$ and hence, $\eta \perp \text{null}(D)$. Moreover, $D\eta = Dw = e$ (since $D\bar{Z} = \mathbf{0}$) and $e^T \eta = e^T w = 0$. Therefore,

$$\text{gal}(D) = \text{col}(ZS) = \text{col}([\eta \ \bar{Z}]) = \text{span}(\eta) \oplus \text{null}(D).$$

To show that η is unique, assume that $D\eta' = e$ where $\eta' \perp \text{null}(D)$. Thus, $\eta - \eta'$ lies in $\text{null}(D)$. But, $\eta - \eta'$ is $\perp \text{null}(D)$. Hence, $\eta - \eta' = \mathbf{0}$.

$\square$

Remark 4.3 *If* $\text{gal}(D) = \text{null}(D) \oplus \text{span}(x)$, *then* $x \perp e$ *since* $e \perp \text{gal}(D)$. *However,* x *may or may not be* $\perp \text{null}(D)$. *In Theorem 4.19,* $\text{gal}(D) = \text{null}(D) \oplus \text{span}(w) = \text{null}(D) \oplus \text{span}(\eta)$. *Both* w *and* η *are* $\perp e$, *but only* η *is* $\perp \text{null}(D)$.

Recall that $DD^\dagger$ is the orthogonal projection on $\text{col}(D)$ and that $DD^\dagger = D^\dagger D$ since D is symmetric. Also, recall that $Dw = e$ implies that $w = D^\dagger e + (I - D^\dagger D)z$ where z is an arbitrary vector. We saw above that η is the orthogonal projection of w onto $\text{null}(\bar{Z}^T)$ and that $\text{col}(\bar{Z}) = \text{null}(D)$. Thus η is, in fact, the orthogonal projection of w onto $\text{col}(D)$. Consequently,

$$\eta = DD^\dagger w = D^\dagger e.$$

Therefore, for nonspherical EDMs, we have

$$e^T \eta = \eta^T D\eta = e^T D^\dagger e = 0.$$

Example 4.15 *Consider the nonspherical EDM* $D = \begin{bmatrix} 0 & 1 & 4 & 9 & 16 \\ 1 & 0 & 1 & 4 & 9 \\ 4 & 1 & 0 & 1 & 4 \\ 9 & 4 & 1 & 0 & 1 \\ 16 & 9 & 4 & 1 & 0 \end{bmatrix}$ *with config-*

uration matrix $P = \begin{bmatrix} -2 \\ -1 \\ 0 \\ 1 \\ 2 \end{bmatrix}$. *Then* $\text{null}(D) = \text{col}(\begin{bmatrix} -1 & -3 \\ 3 & 8 \\ -3 & -6 \\ 1 & 0 \\ 0 & 1 \end{bmatrix})$ *and* $\eta = \frac{1}{14} \begin{bmatrix} 2 \\ -1 \\ -2 \\ -1 \\ 2 \end{bmatrix}$.

Moreover, $\text{diag}(B) = [4 \ 1 \ 0 \ 1 \ 4]^T$ *and hence,* $\eta^T \text{diag}(B) = 1$.

As an immediate consequence of the above characterizations of spherical EDMs, we have the following characterizations of nonspherical EDMs.

Theorem 4.20 *Let D be an $n \times n$ EDM of embedding dimension $r \leq n - 2$ and let Z be a Gale matrix of D. Then the following statements are equivalent:*

1. *D is nonspherical.*
2. $e^T \eta = 0$ *where* $D\eta = e$.
3. $DZ \neq \mathbf{0}$.
4. $rank(D) = r + 2$.
5. $\eta^T diag(B) = 1$, *where* $D\eta = e$ *and* $B = -JDJ/2$.

An interesting subclass of nonspherical EDMs is that of multispherical EDMs. We discuss this subclass next.

4.2.1 Multispherical EDMs

A nonspherical EDM is *multispherical* if its generating points lie on two or more concentric hyperspheres. More precisely, let D be an $n \times n$ nonspherical EDM and let $n_1, \ldots, n_k$ be positive integers such that $n_1 + \cdots + n_k = n$. Then D is said to be *k-multispherical* if there exists a sequence of k, $2 \leq k \leq n - 1$, distinct hyperspheres, each centered at the origin, such that the ith hypersphere contains n_i points. A vector $x \in \mathbb{R}^n$ is said to have a *k-block structure*, $2 \leq k \leq n - 1$, if the entries of x assume exactly k distinct values. For example, e^i, the ith standard unit vector, has a 2-block structure. Therefore, since the hyperspheres are centered at the origin, it immediately follows that D is k-multispherical if and only if $diag(B)$, where B is the Gram matrix; i.e., $B_{ij} = (p^i)^T p^j$, has a k-block structure. It is worth emphasizing here that the rank of B may not be equal to the embedding dimension of D since B may not be derived as $B = \mathscr{T}(D)$ (see Example 4.16 below).

Multispherical EDMs are characterized in the following two theorems.

Theorem 4.21 (Hayden et al. [104]) *Let D be an $n \times n$ EDM. Then the following two statements are equivalent:*

(i) *D is k-multispherical.*
(ii) *There exists $v \in \mathbb{R}^n$ such that $e^T v > 0$ and Dv has a k-block structure.*

Proof. Assume that Statement (ii) holds and let the Gram matrix of D be $B = -(I - ev^T/e^T v)D(I - ve^T/e^T v)/2$. Thus $Bv = \mathbf{0}$ and consequently

$$Dv = v^T diag(B) e + e^T v \, diag(B). \tag{4.26}$$

Thus, $diag(B)$ has a k-block structure and hence Statement (i) holds.

Conversely, assume that Statement (i) holds and wlog assume that $p^1, \ldots, p^{n_1}$ lie on the first hypersphere, $p^{n_1+1}, \ldots, p^{n_1+n_2}$ lie on the second hypersphere and so on. Let $B = (b_{ij} = (p^i)^T p^j)$ be the corresponding Gram matrix. Then $diag(B)$ has k-block structure. Now let x be any nonzero vector in $null(B)$. If $e^T x > 0$, set $v = x$ and thus, as in (4.26), Dv has a k-block structure. On the other hand, if $e^T x = 0$

for every $x \in \text{null}(B)$, then $\text{null}(B) \subseteq e^\perp$, or equivalently, $e \in \text{col}(B)$. Therefore, let $Bv = e$. Then $v^T Bv = e^T v > 0$ and

$$Dv = (v^T \text{diag}(B) - 2) \, e + e^T v \, \text{diag}(B).$$

Therefore, Dv has a k-block structure.

□

As a result, if D is k-multispherical, then there exists a system of coordinates, fixed by v, such that $D = \mathcal{T}(B)$, where $\text{diag}(B)$ has a k-block structure.

Example 4.16 *Let D be the EDM generated by $p^1 = \begin{bmatrix} -1 \\ 1 \end{bmatrix}$, $p^2 = \begin{bmatrix} 1 \\ 1 \end{bmatrix}$ and $p^3 = \begin{bmatrix} 0 \\ 1 \end{bmatrix}$. Then obviously, D is 2-multispherical. The null space of the corresponding Gram matrix B is the span of $x = [1 \ 1 \ -2]^T$. Notice that B is not derived as $\mathcal{T}(D)$ since its rank is 2, while the embedding dimension of D is 1. Thus, $e^T x = 0$ and hence $e \in \text{col}(B)$. Therefore, $Bv = e$, where $v = [0 \ 0 \ 1]^T$. Then the new Gram matrix is $B' = -(I - ev^T)D(I - ve^T)/2 = [1 \ -1 \ 0]^T [1 \ -1 \ 0]$. Obviously, $\text{rank}(B') = 1$ and $\text{diag}(B') = [1 \ 1 \ 0]^T$.*

Theorem 4.22 (Kurata and Matsuura [123]) *Let D be an $n \times n$ EDM. Then the following two statements are equivalent:*

(i) *D is k-multispherical, where $p^1, \ldots, p^{n_1}$ lie on the first hypersphere, $p^{n_1+1}, \ldots, p^{n_1+n_2}$ lie on the second hypersphere and so on.*

(ii) *There exist scalars $\beta_1, \ldots, \beta_k$ such that*

$$\begin{bmatrix} 2\beta_1 E_{n_1} & (\beta_1 + \beta_2)E_{n_1,n_2} & \cdots & (\beta_1 + \beta_k)E_{n_1,n_k} \\ (\beta_1 + \beta_2)E_{n_2,n_1} & 2\beta_2 E_{n_2} & \cdots & (\beta_2 + \beta_k)E_{n_2,n_k} \\ \cdots & \cdots & \ddots & \cdots \\ (\beta_1 + \beta_k)E_{n_k,n_1} & (\beta_2 + \beta_k)E_{n_k,n_2} & \cdots & 2\beta_k E_{n_k} \end{bmatrix} - D \succeq 0, \quad (4.27)$$

where E_{n_i} and E_{n_i,n_j} are the matrices of all 1's of orders $n_i \times n_i$ and $n_i \times n_j$ respectively.

Proof. Assume that Statement (i) holds. Then, by the previous theorem, there exists v such that $e^T v = 1$ and Dv has a k-block structure. Let $B = -(I - ev^T)D(I - ve^T)/2$. Then, $\text{diag}(B) = Dv - v^T Dve/2 = [\beta_1 e_{n_1}^T \ \cdots \ \beta_k e_{n_k}^T]^T$ and

$$D = \mathcal{K}(B) = \begin{bmatrix} \beta_1 e_{n_1} \\ \vdots \\ \beta_k e_{n_k} \end{bmatrix} [e_{n_1}^T \cdots e_{n_k}^T] + \begin{bmatrix} e_{n_1} \\ \vdots \\ e_{n_k} \end{bmatrix} [\beta_1 e_{n_1}^T \cdots \beta_k e_{n_k}^T] - 2B. \quad (4.28)$$

Thus Statement (ii) holds.

Conversely, assume that Statement (ii) holds and let the left-hand side of (4.27) be equal to $2B'$. Then $\text{diag}(B') = [\beta_1 e_{n_1}^T \ \cdots \ \beta_k e_{n_k}^T]^T$ and thus (4.27) can be written as $\mathrm{diag}(B')e^T + e(\mathrm{diag}(B'))^T - D = 2B'$. Hence, $\mathscr{K}(B') = D$ and thus B' is a Gram matrix of D. Therefore, D is k-multispherical and hence Statement (i) holds.

$\square$

Observe that if $k = 1$, then Theorem 4.22 reduces to part 5 of Theorem 4.7. Moreover, as was the case for a spherical EDM, the β_i's in Theorem 4.22 are related to the radii of the concentric hyperspheres. We should point out that Kurata and Tarazaga [125] obtained other characterizations of multispherical EDMs. Also, Tarazaga et al. [188] discussed the case where the centroid of the points in each of the concentric hypersphere coincides with the origin. Finally, Hayden et al. [104] presented a mixed-integer linear programming algorithm for finding the minimum number of concentric hyperspheres that contain the generating points of a given EDM.

We conclude this chapter by remarking that another interesting subclass of nonspherical EDMs, namely nonspherical centrally symmetric EDMs is considered in Chap. 6, where we study the eigenvalues of EDMs.

Chapter 5
The Geometry of EDMs

The geometric properties of EDMs are inherited from those of PSD matrices. Let $\mathscr{D}^n$ denote the set of EDMs of order n. This chapter focuses on the geometry of $\mathscr{D}^n$. In particular, we study the facial structure of $\mathscr{D}^n$ and its polar, and we highlight the similarities between $\mathscr{D}^n$ and the positive semidefinite cone $\mathscr{S}^n_+$.

5.1 The Basic Geometry of $\mathscr{D}^n$

Recall that $\mathscr{D}^n$ is the image of $\mathscr{S}^{n-1}_+$ under the linear transformation $\mathscr{K}_V$. As a result, the geometric properties of $\mathscr{D}^n$ are closely connected with those of $\mathscr{S}^n_+$. In particular, $\mathscr{D}^n$ is a pointed closed convex cone whose interior consists of all EDMs of embedding dimension $n-1$. Consequently, the interior of $\mathscr{D}^n$ is made up of spherical EDMs, while the boundary of $\mathscr{D}^n$ is made up of both spherical and nonspherical EDMs. Moreover, the following theorem is an immediate consequence of part 4 of Theorem 4.2.

Theorem 5.1 (Tarazaga [184]) *The set of spherical EDMs is convex.*

Proof. Let D_1 and D_2 be two spherical matrices. Then the two matrices $\beta_1 E - D_1$ and $\beta_2 E - D_2$ are PSD for some scalars β_1 and β_2. Hence, for any $\lambda : 0 \le \lambda \le 1$, it follows that $(\lambda \beta_1 + (1-\lambda)\beta_2)E - (\lambda D_1 + (1-\lambda)D_2)$ is PSD. Consequently, $\lambda D_1 + (1-\lambda)D_2$ is a spherical EDM.

$\square$

Now, $\mathscr{D}^n$ is the closure of the set of spherical EDMs. To see this, observe that for any EDM D and for any $\alpha > 0$, $D' = D + \alpha(E - I)$ is an EDM of embedding dimension $n-1$ since $\mathscr{T}_V(D') = \mathscr{T}_V(D) + (\alpha/2)I$ is PD. Hence, for any EDM D and for any $\varepsilon > 0$, there exists a spherical EDM D' such that $||D' - D|| \le \varepsilon$. In other words, every nonspherical EDM is the limit of a sequence of spherical EDMs.

© Springer Nature Switzerland AG 2018
A.Y. Alfakih, *Euclidean Distance Matrices and Their Applications in Rigidity Theory*, https://doi.org/10.1007/978-3-319-97846-8_5

Recall that cell matrices, i.e., matrices of the form $D = ce^T + ec^T - 2\text{Diag}(c)$ for some $c \geq 0$, are spherical EDMs. Let $T : \mathbb{R}^n \to \mathscr{S}^n$ where $T(x) = xe^T + ex^T - 2\text{Diag}(x)$. Then, obviously, the set of cell matrices is the image of the nonnegative orthant, $\mathbb{R}^n_+$, under T. A cone $K \in \mathscr{V}$ is said to be *polyhedral* if it is the conic hull of a finite number of vectors in $\mathscr{V}$. Clearly, $\mathbb{R}^n_+$ is a polyhedral cone since every $c \in \mathbb{R}^n_+$ can be written as $c = \sum_{i=1}^n c_i e^i$, where $c_i \geq 0$ and e^i is the ith standard unit vector in $\mathbb{R}^n$. Accordingly, every cell matrix can be written as $D = T(c) = \sum_{i=1}^n c_i T(e^i)$. Consequently, cell matrices form a polyhedral convex cone in $\mathscr{S}^n$. This result was obtained by Tarazaga and Kurata in [187, 126] where the geometric properties of cell matrices are discussed.

Also recall that $T_S(\hat{x})$, the tangent cone of convex set S at $\hat{x}$, is the closure of $F_S(\hat{x})$, the cone of feasible directions of S at $\hat{x}$; i.e.,

$$F_S(\hat{x}) = \{\beta(x - \hat{x}) : \text{ for all } x \in S \text{ and for all } \beta \geq 0\}.$$

The set $\mathscr{E}^n = \{A \in \mathscr{S}^n_+ : \text{diag}(A) = e\}$, i.e., the set of correlation matrices, is called the *elliptope*. Clearly, E, the matrix of all 1's, lies in $\mathscr{E}^n$. Moreover, the cone of feasible direction of the elliptope at E is given by

$$F_{\mathscr{E}^n}(E) = \{D' = \beta(A - E) : \text{ for all } A \in \mathscr{E}^n \text{ and for all } \beta \geq 0\}. \tag{5.1}$$

Let $D' \in F_{\mathscr{E}^n}(E)$. Then obviously $\text{diag}(D') = \mathbf{0}$ and $\mathscr{T}(-D') \succeq \mathbf{0}$. Consequently, $(-D')$ is an EDM, and more precisely, $(-D')$ is a spherical EDM (Theorem 4.7). On the other hand, if D is a spherical EDM, then $(-D)$ lies in $F_{\mathscr{E}^n}(E)$. As a result, $-F_{\mathscr{E}^n}(E)$ is exactly the set of spherical EDMs. Therefore, we have proved the following theorem.

Theorem 5.2 (Deza and Laurent [70]) *The EDM cone, $\mathscr{D}^n$, is the negative of the tangent cone of the elliptope at E.*

Next, we characterize the polar of $\mathscr{D}^n$ and we investigate more geometric connections between $\mathscr{D}^n$ and the elliptope $\mathscr{E}^n$.

5.2 The Polar of $\mathscr{D}^n$

We use two different approaches to find the polar of $\mathscr{D}^n$. The first one is direct and uses the adjoint of $\mathscr{K}_V$, while the second one is indirect and uses the geometric structure of the elliptope. We begin, first, with the direct approach.

Theorem 5.3 *The polar of $\mathscr{D}^n$ is given by*

$$(\mathscr{D}^n)^\circ = \{D' : D' = A + \text{Diag}(y), \text{ where } Ae = \mathbf{0}, A \succeq \mathbf{0} \text{ and } y \in \mathbb{R}^n\}.$$

Proof. Let $K = \{D' : D' = A + \text{Diag}(y), \text{ where } Ae = \mathbf{0}, A \succeq \mathbf{0} \text{ and } y \in \mathbb{R}^n\}$. Let $D' \in K$ and let $D \in \mathscr{D}^n$. Note that $D'e = y$. Then, since $\text{diag}(D) = \mathbf{0}$, it follows that

$$\text{trace}(D'D) = \text{trace}(AD) + \text{trace}(\text{Diag}(y)D) = \text{trace}(A\mathscr{K}_V(X)) = \text{trace}(\mathscr{K}_V^*(A)X)$$

where $X \in \mathscr{S}^{n-1}_+$. But, Lemma 3.5 implies that

$$\text{trace}(\mathscr{K}_V^*(A)X) = 2\,\text{trace}(V^T(\text{Diag}(Ae) - A)VX) = -2\,\text{trace}(V^T AVX) \le 0.$$

Hence, $K \subseteq (\mathscr{D}^n)^\circ$.

To prove the other direction, let $D' \in (\mathscr{D}^n)^\circ$ and let $D \in \mathscr{D}^n$. Then

$$\text{trace}(D'D) = \text{trace}(D'\mathscr{K}_V(X)) = \text{trace}(\mathscr{K}_V^*(D')X) \le 0.$$

Therefore, $-\mathscr{K}_V^*(D') \succeq 0$ and hence $V^T(D' - \text{Diag}(D'e))V \succeq 0$. Let $A = D' - \text{Diag}(D'e)$. Then $Ae = 0$ and hence $VV^T AVV^T = JAJ = A \succeq 0$. Therefore, $D' = A + \text{Diag}(D'e)$. Moreover, $D' \in K$ (set $D'e = y$). Hence $(\mathscr{D}^n)^\circ \subseteq K$.

$\square$

Next, we turn to the indirect approach for finding $(\mathscr{D}^n)^\circ$. It is an immediate consequence of Theorems 5.2 and 1.42 that the polar of $\mathscr{D}^n$ is the negative of $N_{\mathscr{E}^n}(E)$, the normal cone of the elliptope at E. Therefore, the characterization of $(\mathscr{D}^n)^\circ$ follows from the following characterization of $N_{\mathscr{E}^n}(E)$.

Theorem 5.4 (Laurent and Poljak [131]) *The normal cone of the elliptope at E is given by*

$$N_{\mathscr{E}^n}(E) = \{C : C = -A + \text{Diag}(y), \text{ where } Ae = 0, A \succeq 0 \text{ and } y \in \mathbb{R}^n\}.$$

Proof. Let $K = \{C : C = -A + \text{Diag}(y), \text{ where } Ae = 0, A \succeq 0 \text{ and } y \in \mathbb{R}^n\}$. Let $C \in K$ and let $Y \in \mathscr{E}^n$. Then

$$\begin{aligned}
\text{trace}(CE) - \text{trace}(CY) &= \text{trace}(\text{Diag}(y)E) + \text{trace}(AY) - \text{trace}(\text{Diag}(y)Y) \\
&= e^T y + \text{trace}(AY) - e^T y \\
&= \text{trace}(AY) \ge 0.
\end{aligned}$$

Therefore, $C \in N_{\mathscr{E}^n}(E)$ and hence $K \subseteq N_{\mathscr{E}^n}(E)$.

To prove the reverse direction, let $C \in N_{\mathscr{E}^n}(E)$ and consider the following pair of dual SDP problems:

$$\begin{array}{llll}
\text{(P) max} & \text{trace}(CY) & \text{(D) min} & e^T y \\
& \text{subject to } \text{diag}(Y) = e & & \text{subject to } \text{Diag}(y) \succeq C \\
& \phantom{\text{subject to }} Y \succeq 0 & &
\end{array}$$

Hence, $Y = E$ is an optimal solution of (P). Consequently, by SDP strong duality, there exists y such that $\text{Diag}(y) - C = A \succeq 0$ and $e^T y = \text{trace}(CE)$. Therefore, $C = \text{Diag}(y) - A$. Moreover, $e^T y = e^T Ce = e^T y - e^T Ae$ and hence, $Ae = 0$. Therefore, $C \in K$ and thus $N_{\mathscr{E}^n}(E) \subseteq K$.

$\square$

5.3 The Facial Structure of $\mathscr{D}^n$

The facial structure of the EDM cone is inherited from that of the positive semidefinite cone $\mathscr{S}_+^n$ [103]. Now by setting $T = \mathscr{K}_V$ and $\mathscr{T}_V|_{\mathscr{S}_h^n}$ in Theorem 1.30, we obtain the following:

Theorem 5.5 *Let D_1 and D_2 be in $\mathscr{D}^n$. Then $D_2 \in face(D_1, \mathscr{D}^n)$ if and only if $\mathscr{T}_V(D_2) \in face(\mathscr{T}_V(D_1), \mathscr{S}_+^{n-1})$.*

Therefore, using Theorem 2.16, we have

Theorem 5.6 (Tarazaga [184] and Alfakih [5]) *Let $D_1 \in \mathscr{D}^n$. Then the minimal face of $\mathscr{D}^n$ containing D_1 is given by*

$$face(D_1, \mathscr{D}^n) = \{D \in \mathscr{D}^n : null(\mathscr{T}_V(D_1)) \subseteq null(\mathscr{T}_V(D))\}.$$

Next, the minimal face of $\mathscr{D}^n$ containing D_1 and its relative interior is characterized in terms of the Gale space of D_1. This should come as no surprise since the Gale space of D is closely connected with the null space of its projected Gram matrix.

Theorem 5.7 (Tarazaga [184] and Alfakih [5]) *Let $D_1 \in \mathscr{D}^n$. Then the minimal face of $\mathscr{D}^n$ containing D_1 is given by*

$$face(D_1, \mathscr{D}^n) = \{D \in \mathscr{D}^n : gal(D_1) \subseteq gal(D)\}.$$

Moreover, the relative interior of $face(D_1, \mathscr{D}^n)$ is given by

$$relint(face(D_1, \mathscr{D}^n)) = \{D \in \mathscr{D}^n : gal(D_1) = gal(D)\}.$$

Proof.   Let U_1 be the matrix whose columns form an orthonormal basis of $null(\mathscr{T}_V(D_1))$. Then $col(VU_1)$ is a basis of $gal(D_1)$ (Lemma 3.8). Let P be a configuration matrix of D such that $P^T e = \mathbf{0}$. Then $\mathscr{T}_V(D)U_1 = \mathbf{0}$ iff $P^T VU_1 = \mathbf{0}$ iff $col(VU_1) \subseteq gal(D)$ iff $gal(D_1) \subseteq gal(D)$. To complete the proof, note that $D \in relint(face(D_1, \mathscr{D}^n))$ iff $D \in face(D_1, \mathscr{D}^n)$ and $D_1 \in face(D, \mathscr{D}^n)$.   □

It should be pointed out that Gale space is constant over the relative interior of a face of $\mathscr{D}^n$. Also, a characterization of the minimal faces of $\mathscr{D}^n$ in terms of the column space is given in Hayden et al. [103] and Tarazaga et al. [186].

Example 5.1 *Consider the two EDMs*

$$D_1 = \begin{bmatrix} 0 & 1 & 2 & 1 \\ 1 & 0 & 1 & 2 \\ 2 & 1 & 0 & 1 \\ 1 & 2 & 1 & 0 \end{bmatrix} \text{ and } D_2 = \begin{bmatrix} 0 & 1 & 4 & 1 \\ 1 & 0 & 1 & 0 \\ 4 & 1 & 0 & 1 \\ 1 & 0 & 1 & 0 \end{bmatrix}$$

with configuration and Gale matrices as follows:

$$P_1 = \frac{1}{2} \begin{bmatrix} -1 & 1 \\ 1 & 1 \\ 1 & -1 \\ -1 & -1 \end{bmatrix}, \ P_2 = \begin{bmatrix} -1 \\ 0 \\ 1 \\ 0 \end{bmatrix} \ and \ Z_1 = \begin{bmatrix} 1 \\ -1 \\ 1 \\ -1 \end{bmatrix}, \ Z_2 = \begin{bmatrix} 1 & 0 \\ -2 & 1 \\ 1 & 0 \\ 0 & -1 \end{bmatrix}.$$

Then $Z_1 = Z_2 e_2$ and thus $gal(D_1) \subset gal(D_2)$. Consequently, D_2 lies in $face(D_1, \mathscr{D}^n)$.

Similar to the positive semidefinite cone, the EDM cone is facially exposed. That is, for every face F of $\mathscr{D}^n$, there exists a hyperplane H such that $F = \mathscr{D}^n \cap H$.

Theorem 5.8 (Tarazaga [184]) *Let F be a proper face of $\mathscr{D}^n$. Let $D_1 \in relint(F)$ and let Z_1 be a Gale matrix of D_1. Further, let $H = \{D \in \mathscr{D}^n : trace(Z_1^T D Z_1) = 0\}$. Then $F = \mathscr{D}^n \cap H$.*

Proof. Note that $F = face(D_1, \mathscr{D}^n)$. Also, note that H is a supporting hyperplane to $\mathscr{D}^n$ since for all $D = \mathscr{K}_V(X) \in \mathscr{D}^n$, we have

$$trace(DZ_1 Z_1^T) = trace(X \mathscr{K}_V^*(Z_1 Z_1^T)) = -2 \, trace(Z_1^T V X V^T Z_1) \le 0. \qquad (5.2)$$

Let $D \in F$ and let Z be a Gale matrix of D. Then, $Z_1 = ZA$ for some matrix A. Moreover, $DZ_1 = DZA = e\, \xi^T A$ (Lemma 3.10). Thus $Z_1^T DZ_1 = \mathbf{0}$. Consequently, $D \in H$ and thus $F \subseteq (\mathscr{D}^n \cap H)$.

To prove the reverse inclusion, let $D \in (\mathscr{D}^n \cap H)$. Then (5.2) implies that $trace(Z_1^T DZ_1) = -2 \, trace(Z_1^T \mathscr{T}(D) Z_1) = 0$ and hence $Z_1^T \mathscr{T}(D) Z_1 = \mathbf{0}$. Therefore, $\mathscr{T}(D) Z_1 = \mathbf{0}$ and thus $gal(D_1) \subseteq gal(D)$. As a result, $D \in F$ and hence $(\mathscr{D}^n \cap H) \subseteq F$.

$\square$

Let G be a connected graph on n nodes and m edges and let π be the linear transformation that maps an $n \times n$ symmetric matrix A to the vector $a \in \mathbb{R}^m$ consisting of the entries of A indexed by the edges of G. That is, $\pi : \mathscr{S}^n \to \mathbb{R}^m$ such that

$$(\pi(A))_{ij} = a_{ij} \ if \ \{i, j\} \in E(G). \qquad (5.3)$$

Consequently, the adjoint of π is given by

$$(\pi^*(a))_{ij} = \begin{cases} a_{ij} & if \ \{i, j\} \in E(G), \\ 0 & otherwise. \end{cases} \qquad (5.4)$$

As the next theorem shows, $\pi(\mathscr{D}^n)$ is closed. $\pi(\mathscr{D}^n)$ is called the *coordinate shadow* of $\mathscr{D}^n$.

Theorem 5.9 (Drusvyatskiy et al. [75]) *The coordinate shadow of the cone of EDMs is closed, i.e., $\pi(\mathscr{D}^n)$ is closed.*

Proof. Let D be an EDM such that $\pi(D) = \mathbf{0}$. Since G is a connected graph on n nodes, it follows that $p^1 = \cdots = p^n$ and thus $D = \mathbf{0}$. The result follows from [160, Theorem 9.1, page 73].

$\square$

Note that $\pi(\mathscr{S}_+^n)$ is not necessarily closed. For example, let $\pi(\begin{bmatrix} a & b \\ b & c \end{bmatrix}) = \begin{bmatrix} a \\ b \end{bmatrix})$.
Then obviously $\pi(\mathscr{S}_+^2) = (0,0) \cup \{x \in \mathbb{R}^2 : x_1 > 0\}$ is not closed.

Next, we specialize Theorem 2.23 to EDMs. Let D be a given EDM and let G be a given simple connected graph. Further, let

$$\mathscr{X}(\mathscr{F}) = \{X \in \mathscr{S}_+^{n-1} : \pi(\mathscr{K}_V(X)) = \pi(D)\};$$

i.e., $\mathscr{X}(\mathscr{F})^1$ is the set of projected Gram matrices of all $n \times n$ EDMs which agree with D in the entries defined by G. It is easy to see that the adjoint of $\pi(\mathscr{K}_V)$ is given by

$$(\pi(\mathscr{K}_V))^*(\omega) = \mathscr{K}_V^*(\pi^*(\omega)) = 2V^T(\text{Diag}(\pi^*(\omega)e) - \pi^*(\omega))V. \qquad (5.5)$$

Thus, if we define $\Omega = \text{Diag}(\pi^*(\omega)e) - \pi^*(\omega)$, then

$$\Omega_{ij} = \begin{cases} -\omega_{ij} & \text{if } \{i,j\} \in E(G), \\ 0 & \text{if } i \neq j, \{i,j\} \notin E(G), \\ \sum_{k=1}^n \omega_{ik} & \text{if } i = j. \end{cases} \qquad (5.6)$$

Ω is called a *stress matrix*. Let $\Omega' = V^T \Omega V$. Then $\mathscr{K}_V^*(\pi^*(\omega)) = 2V^T \Omega V = 2\Omega'$. Observe that $\Omega e = 0$. Thus $\Omega = V \Omega' V^T$. Consequently, Ω is PSD of rank k iff Ω' is PSD of rank k. As a result, Theorem 2.23 reduces to

Theorem 5.10 (Drusvyatskiy et al. [75]) *Let D be an EDM of embedding dimension r, $r \leq n - 2$. Let $d = \pi(D)$ and let $\mathscr{X}(\mathscr{F}) = \{X \in \mathscr{S}_+^{n-1} : \pi(\mathscr{K}_V(X)) = d\}$. Assume that $\text{rank}(X) \leq r$ for all $X \in \mathscr{X}(\mathscr{F})$. Then the following statements are equivalent:*

(i) *The singularity degree of $\mathscr{X}(\mathscr{F})$ is 1.*
(ii) ω *exposes face$(d, \pi(\mathscr{D}^n))$.*
(iii) Ω' *exposes face$(\mathscr{X}(\mathscr{F}), \mathscr{S}_+^{n-1})$.*
(iv) $\Omega \succeq 0$, $d^T \omega = 0$ *and* $\text{rank}(\Omega) = n - r - 1$.

The reader should recall that unlike the faces of $\mathscr{D}^n$, the faces of $\pi(\mathscr{D}^n)$ may or may not be exposed. We should also point out that Theorem 5.10 is implicit in Gortler and Thurston [90].

5.4 Notes

Theorem 5.7 was first obtained by Tarazaga in [184] without explicitly stating it in terms of Gale spaces. More specifically, Tarazaga defines a space LGS in terms of null(D) if D is spherical, and in terms of null(D) and the span of η if D is nonspherical (see Theorem 4.19). A closer look at LGS reveals that it is identical to Gale space.

---

[1] This set will be discussed in great detail in Chap. 8, where the rationale for this notion will become clear.

Chapter 6
The Eigenvalues of EDMs

The focus of this chapter is on the eigenvalues of EDMs. In the first part, we present a characterization of the column space of an EDM D. This characterization is then used to express the eigenvalues of D in terms of the eigenvalues of its Gram matrix $B = \mathcal{T}(D) = -JDJ/2$. In case of regular and nonspherical centrally symmetric EDMs, the same result can also be obtained by using the notion of equitable partition. In the second part, we discuss some other topics related to eigenvalues such as: a method for constructing nonisomorphic cospectral EDMs; the connection between EDMs, graphs, and combinatorial designs; EDMs with exactly two or three distinct eigenvalues and the EDM inverse eigenvalue problem.

It should be pointed out that the eigenvalues of Gram matrix B are precisely the eigenvalues of the projected Gram matrix X with one extra zero eigenvalue; i.e., the characteristic polynomials of B and X satisfy $\chi_B(\mu) = \mu \, \chi_X(\mu)$. This is easy to see since

$$\begin{bmatrix} e^T/\sqrt{n} \\ V^T \end{bmatrix} B \begin{bmatrix} e/\sqrt{n} & V \end{bmatrix} = \begin{bmatrix} 0 & 0 \\ 0 & X \end{bmatrix}.$$

In the next section, we show how to exploit a characterization of the column space of D to express the eigenvalues of D in terms of those of B.

6.1 The Eigenvalues via the Column Space of D

Let D be an EDM of embedding dimension r and let $B = -JDJ/2$ be its Gram matrix. Let $B = W\Lambda W^T$ be the spectral decomposition of B, where Λ is the $r \times r$ diagonal matrix consisting of the positive eigenvalues of B, and W is the $n \times r$ matrix whose columns are the orthonormal eigenvectors of B corresponding to these positive eigenvalues. It is convenient in this chapter to let the configuration matrix be

$$P = W\Lambda^{1/2}.$$

Thus, $P^T D P = P^T \mathcal{K}(B) P = -2P^T B P = -2\Lambda^{1/2} W^T B W \Lambda^{1/2} = -2\Lambda^2$.

© Springer Nature Switzerland AG 2018
A.Y. Alfakih, *Euclidean Distance Matrices and Their Applications in Rigidity Theory*, https://doi.org/10.1007/978-3-319-97846-8_6

Example 6.1 *Let* $D = \begin{bmatrix} 0 & 2 & 6 & 2 \\ 2 & 0 & 6 & 2 \\ 6 & 6 & 0 & 6 \\ 2 & 2 & 6 & 0 \end{bmatrix}$. *Then* $B = \mathcal{T}(D) = \begin{bmatrix} 1 & 0 & -1 & 0 \\ 0 & 1 & -1 & 0 \\ -1 & -1 & 3 & -1 \\ 0 & 0 & -1 & 1 \end{bmatrix}$ *has* $\Lambda =$

Diag$(1,1,4)$. *Hence,* $P = \begin{bmatrix} -0.0674 & 0.8137 & 0.5774 \\ -0.6710 & -0.4652 & 0.5774 \\ 0.0000 & 0.0000 & -1.7321 \\ 0.7384 & -0.3485 & 0.5774 \end{bmatrix}$.

The cases of spherical and nonspherical EDMs are treated separately. We begin first with spherical EDMs.

6.1.1 The Eigenvalues of Spherical EDMs

Let D be a nonzero spherical EDM and let Z be a Gale matrix of D. Then, Theorem 4.2 implies that $\mathrm{null}(D) = \mathrm{gal}(D) = \mathrm{col}(Z)$ and hence the columns of $[P \; e]$ form a basis of $\mathrm{col}(D)$. Let $Q = [W \; e/\sqrt{n}] = [P\Lambda^{-1/2} \; e/\sqrt{n}]$. Then, the nonzero eigenvalues of D are equal to the eigenvalues of $Q^T D Q$. But

$$Q^T D Q = \begin{bmatrix} -2\Lambda & \frac{1}{\sqrt{n}}\Lambda^{-1/2} P^T D e \\ \frac{1}{\sqrt{n}} e^T D P \Lambda^{-1/2} & \frac{1}{n} e^T D e \end{bmatrix}. \tag{6.1}$$

Notice that $Q^T D Q$ as given in (6.1) is a bordered diagonal matrix. Also, notice that the nonzero eigenvalues of D are interlaced by the nonzero eigenvalues of $(-2B)$. This was also observed, in the case of regular EDMs, by Hayden and Tarazaga in [101]. Therefore, the characteristic polynomial of $Q^T D Q$ is

$$\chi_{Q^T D Q}(\mu) = \prod_{i=1}^{r} (a_i - \mu) \left[(\frac{1}{n} e^T D e - \mu) - \sum_{i=1}^{r} \frac{b_i^2}{(a_i - \mu)} \right].$$

Here, $a_i = -2\lambda_i$, where $\lambda_1, \ldots, \lambda_r$ are the positive eigenvalues of B, i.e.,

$$a = -2 \, \mathrm{diag}(\Lambda), \tag{6.2}$$

and

$$b = \frac{1}{\sqrt{n}} \Lambda^{-1/2} P^T D e. \tag{6.3}$$

Hence, the characteristic polynomial [4] of D is given by

$$\chi_D(\mu) = (-\mu)^{n-r-1} \left[(\frac{1}{n} e^T D e - \mu) \prod_{i=1}^{r} (a_i - \mu) - \sum_{i=1}^{r} b_i^2 \prod_{j=1, j \neq i}^{r} (a_j - \mu) \right]. \tag{6.4}$$

Notice that if $r = 1$, then

$$\chi_D(\mu) = (-\mu)^{n-2}\left[(\frac{1}{n}e^T De - \mu)(a_1 - \mu) - b_1^2\right].$$

Example 6.2 *Consider the EDM D of Example 6.1.*

$$Q^T DQ = \begin{bmatrix} -2 & & & 0 \\ & -2 & & 0 \\ & & -8 & -2\sqrt{3} \\ 0 & 0 & -2\sqrt{3} & 12 \end{bmatrix}.$$

Thus, $a_1 = -2\lambda_1 = -2$, $a_2 = -2\lambda_2 = -2$ and $a_3 = -2\lambda_3 = -8$. Also, $b_1 = b_2 = 0$, $b_3 = -2\sqrt{3}$ and $e^T De/4 = 12$. Therefore,

$$\chi_D(\mu) = (12 - \mu)(-2 - \mu)^2(-8 - \mu) - 12(-2 - \mu)^2$$
$$= (\mu + 2)^2(\mu^2 - 4\mu - 108).$$

Note that the coefficient of μ^3 is zero since trace$(D) = 0$. Thus, the eigenvalues of D are $\mu_1 = \mu_2 = -2$, $\mu_3 = 2 - 4\sqrt{7}$, and $\mu_4 = 2 + 4\sqrt{7}$.

Three remarks are in order here. First, the coefficient of μ^{n-1} in $\chi_D(\lambda)$ must be zero since trace$(D) = 0$. This is indeed the case since the coefficient of μ^r in the expression $(\frac{1}{n}e^T De - \mu)\prod_{i=1}^{r}(a_i - \mu)$ is $e^T De/n + \sum_{i=1}^{r} a_i = e^T \mathcal{K}(B)e/n - 2\,\text{trace}(B) = 0$. This follows since $e^T \mathcal{K}(B)e/n = 2e^T \text{diag}(B) = 2\,\text{trace}(B)$. Note that the highest possible power of μ in the expression $\sum_{i=1}^{r} b_i^2 \prod_{j=1, j\neq i}^{r}(a_j - \mu)$ is $r - 1$ and thus this expression does not contribute to the coefficient of μ^{n-1} in $\chi_D(\mu)$.

Second, suppose that $b_{i_0} = 0$ for some $1 \leq i_0 \leq r$. Then $\mu_{i_0} = -2\lambda_{i_0}$. To see this, observe that in this case, $(a_{i_0} - \mu)$ is a factor of the expression $\sum_{i=1}^{r} b_i^2 \prod_{j=1, j\neq i}^{r}(a_j - \mu)$. In Example 6.2 we saw that $b_1 = b_2 = 0$ and $\mu_1 = \mu_2 = -2$.

Third, suppose that $b_i \neq 0$ for all $i = 1, \ldots, r$ and suppose that λ_{i_0} is an eigenvalue of B with multiplicity k. Then $\mu_{i_0} = -2\lambda_{i_0}$ is an eigenvalue of D with multiplicity $k - 1$. This follows since in this case, $(a_{i_0} - \mu)^{k-1}$ is a factor of the expression $\sum_{i=1}^{r} b_i^2 \prod_{j=1, j\neq i}^{r}(a_j - \mu)$.

The characteristic polynomial $\chi_D(\mu)$ in (6.4) gives rise to a new characterization of regular EDMs. To this end, we need the following lemma.

Lemma 6.1 *Let D be a nonzero $n \times n$ spherical EDM of embedding dimension r. Let P be a configuration matrix of D, $P^T e = 0$, and let B be its Gram matrix. Then the following statements are equivalent:*

(i) *The r negative eigenvalues of D are precisely the nonzero eigenvalues of $(-2B)$.*

(ii) *The Perron eigenvalue of D is equal to $e^T De/n$.*

(iii) *$P^T De = 0$.*

Proof. Suppose that Statement (i) holds. Then, since trace$(D) = 0$, it follows that the positive eigenvalue of D is equal to $-\sum_{i=1}^{r} a_i = 2\,\text{trace}(B) = e^T De/n$ and thus Statement (ii) holds.

Now assume that Statement (ii) holds. Then $\chi_D(e^T De/n) = 0$ and therefore, $\sum_{i=1}^r b_i^2 \prod_{j=1, j\neq i}^r (a_j - e^T De/n) = 0$. But $a_i < 0$ for all $i = 1, \ldots, r$. Thus, for all $i = 1, \ldots, r$, we have $\prod_{j=1, j\neq i}^r (a_j - e^T De/n) = (-1)^{r-1} c_i$ where $c_i > 0$. Consequently, $b_1 = \cdots = b_r = 0$ and hence $P^T De = \mathbf{0}$; i.e., Statement (iii) holds.

To complete the proof, observe that Statement (iii) trivially implies Statements (i) and (ii).

$\square$

Theorem 6.1 (Hayden and Tarazaga [101] and Alfakih [4]) *Let D be a nonzero $n \times n$ EDM of embedding dimension r and let $B = -JDJ/2$ be its Gram matrix. Then D is regular if and only if the negative eigenvalues of D are precisely the nonzero eigenvalues of $(-2B)$; i.e.,iff the characteristic polynomial of D is*

$$\chi_D(\mu) = (-\mu)^{n-r-1} (\frac{1}{n} e^T De - \mu) \prod_{i=1}^r (a_i - \mu). \tag{6.5}$$

Proof. Assume that D is regular. Then, obviously, D is spherical and its Perron eigenvalue is $e^T De/n$. Thus, the result follows from Lemma 6.1.

To prove the other direction, assume that the negative eigenvalues of D are precisely the nonzero eigenvalues of $(-2B)$. Thus, $\text{rank}(D) = r + 1$ and hence D is spherical. Therefore, by Lemma 6.1, $e^T De/n$ is an eigenvalue of D. Consequently, it follows from Corollary 1.1 that e is the Perron eigenvector of D. Hence, D is regular. $\square$

The characteristic polynomial $\chi_D(\mu)$ in (6.5) is alternatively obtained below by using the notion of equitable partition. Next, we turn to nonspherical EDMs.

6.1.2 The Eigenvalues of Nonspherical EDMs

Let D be an $n \times n$ nonspherical EDM of embedding dimension r, then $\text{rank}(D) = r + 2$. Moreover, by Theorem 4.19, $\text{gal}(D) = \text{null}(D) \oplus \text{span}(\eta)$, where η is the unique vector in $\mathbb{R}^n$ such that $D\eta = e$, $\eta \perp e$ and $\eta \perp \text{null}(D)$. Thus, $\eta \in \text{col}(D)$ and hence the columns of $[P \; e \; \eta]$ form a basis for $\text{col}(D)$.

Similar to the spherical case, let $P = W\Lambda^{1/2}$ and let

$$Q = [P\Lambda^{-1/2} \quad e/\sqrt{n} \quad \eta/(\eta^T \eta)^{1/2}].$$

As a result, the nonzero eigenvalues of D are the eigenvalues of $Q^T DQ$. But

$$Q^T DQ = \begin{bmatrix} -2\Lambda & \frac{1}{\sqrt{n}}\Lambda^{-1/2} P^T De & \mathbf{0} \\ \frac{1}{\sqrt{n}} e^T DP\Lambda^{-1/2} & e^T De/n & \sqrt{n}/\sqrt{\eta^T \eta} \\ \mathbf{0} & \sqrt{n}/\sqrt{\eta^T \eta} & 0 \end{bmatrix}. \tag{6.6}$$

Then

$$\chi_{Q^T D Q}(\mu) = \prod_{i=1}^{r}(a_i - \mu) \det\left(\begin{bmatrix} e^T De/n - \mu - \sum_{i=1}^{r} \frac{b_i^2}{(a_i-\mu)} & \sqrt{n}/\sqrt{\eta^T \eta} \\ \sqrt{n}/\sqrt{\eta^T \eta} & -\mu \end{bmatrix}\right)$$

$$= \prod_{i=1}^{r}(a_i - \mu)\left(\mu^2 - \mu e^T De/n + \mu \sum_{i=1}^{r} \frac{b_i^2}{(a_i - \mu)} - n/\eta^T \eta\right)$$

$$= \left(\mu^2 - \mu \frac{e^T De}{n} - \frac{n}{\eta^T \eta}\right)\prod_{i=1}^{r}(a_i - \mu) + \mu \sum_{i=1}^{r} b_i^2 \prod_{j=1,j\neq i}^{r}(a_j - \mu),$$

where a and b are as defined in (6.2) and (6.3), respectively. Therefore, the characteristic polynomial of D is given by [4]

$$\chi_D(\mu) = (-\mu)^{n-r-2}\chi_{Q^T D Q}(\mu). \tag{6.7}$$

Notice that if $r = 1$, then

$$\chi_D(\mu) = (-\mu)^{n-3}\left((\mu^2 - \mu\frac{e^T De}{n} - \frac{n}{\eta^T \eta})(a_1 - \mu) + \mu b_1^2\right).$$

Similar to the spherical case, the coefficient of μ^{n-1} should be 0 since trace$(D) = 0$. Indeed, this is the case here since the coefficients of μ^r and μ^{r-1} in the expression $\prod_{i=1}^{r}(a_i - \mu)$ are $(-1)^r$ and $(-1)^{r-1}\sum_{i=1}^{r}a_i$, respectively. Thus, the coefficient of μ^{n-1} is $(-1)^{r+1}e^T De/n + (-1)^{r-1}\sum_{i=1}^{r}a_i = 0$. Also, similar to the spherical case, if $b_{i_0} = 0$ for some $1 \leq i_0 \leq r$, then a_{i_0} is an eigenvalue of D.

Example 6.3 *Consider the nonspherical EDM* $D = \begin{bmatrix} 0 & 1 & 4 & 2 \\ 1 & 0 & 1 & 1 \\ 4 & 1 & 0 & 2 \\ 2 & 1 & 2 & 0 \end{bmatrix}$ *with configuration*

matrix $P = \begin{bmatrix} 1/4 & 1 \\ 1/4 & 0 \\ 1/4 & -1 \\ -3/4 & 0 \end{bmatrix}$. *Then* $\eta - \frac{1}{2}[1 \ -2 \ 1 \ 0]^T$, *and* $\Lambda = \begin{bmatrix} 3/4 & \\ & 2 \end{bmatrix}$. *Moreover,*

$b = \frac{1}{\sqrt{n}}\Lambda^{-1/2}P^T De = [\frac{1}{2\sqrt{3}} \ 0]^T$. *Thus,*

$$\chi_D(\mu) = (\mu^2 - \frac{11}{2}\mu - \frac{8}{3})(-3/2 - \mu)(-4 - \mu) + \mu\frac{1}{12}(-4 - \mu)$$

$$= (-4 - \mu)(-\mu^3 + 4\mu^2 + 11\mu + 4).$$

Hence, the eigenvalues of D *are* -4 *(note that* $b_2 = 0$*),* -1.5159, -0.4428, *and* 5.9587.

Recall that regular EDMs have the property that $P^T De = 0$. A subclass of nonspherical EDMs is defined, next, that also satisfies this property. A nonspherical EDM is said to be *centrally symmetric* [4] if its configuration matrix can be written, after a possible relabeling of the generating points, as

$$P = \begin{bmatrix} P_1 \\ -P_1 \\ P_3 \end{bmatrix}, \tag{6.8}$$

where P_3 is either vacuous or the zero matrix. As a result, if D is a nonspherical centrally symmetric EDM, then it is easy to verify that $P^T De = 0$. Thus, the characteristic polynomial of D, in this case, reduces to

$$\chi_D(\mu) = (-\mu)^{n-r-2} \left(\mu^2 - \mu \frac{e^T De}{n} - \frac{n}{\eta^T \eta} \right) \prod_{i=1}^{r}(a_i - \mu). \tag{6.9}$$

As a result, we have the following theorem.

Theorem 6.2 (Alfakih [4]) *Let D be an $n \times n$ nonspherical centrally symmetric EDM with embedding dimension r and let $B = -JDJ/2$. Then r of the negative eigenvalues of D are equal to the nonzero eigenvalues of $(-2B)$, the $(r+1)$th negative eigenvalue of D is equal to $\frac{e^T De}{2n} - \sqrt{\frac{(e^T De)^2}{4n^2} + \frac{n}{\eta^T \eta}}$, and the Perron eigenvalue of D is equal to $\frac{e^T De}{2n} + \sqrt{\frac{(e^T De)^2}{4n^2} + \frac{n}{\eta^T \eta}}$.*

The characteristic polynomial $\chi_D(\mu)$ in (6.9) is alternatively obtained below using the notion of equitable partitions.

Example 6.4 *Consider the EDM $D = \begin{bmatrix} 0 & 4 & 2 & 4 & 8 \\ 4 & 0 & 2 & 8 & 4 \\ 2 & 2 & 0 & 2 & 2 \\ 4 & 8 & 2 & 0 & 4 \\ 8 & 4 & 2 & 4 & 0 \end{bmatrix}$ with configuration matrix $P =$*

$\begin{bmatrix} -1 & -1 \\ 1 & -1 \\ 0 & 0 \\ -1 & 1 \\ 1 & 1 \end{bmatrix}$. The positive eigenvalues of the corresponding Gram matrix B are 4 with multiplicity 2. Note that $\eta = \frac{1}{8}[1 \ 1 \ -4 \ 1 \ 1]^T$ and thus $5/\eta^T \eta = 16$. Also, $e^T De/5 = 16$. Now by relabeling the generating points, or by observing that $P^T De = 0$, we conclude that D is nonspherical centrally symmetric. Therefore,

$$\chi_D(\mu) = -\mu(\mu^2 - 16\mu - 16)(-8 - \mu)^2.$$

Hence, the nonzero eigenvalues of D are -8, -8, and $8 \pm 4\sqrt{5}$.

6.2 The Eigenvalues via Equitable Partitions

As we remarked earlier, the characteristic polynomials of regular and nonspherical centrally symmetric EDMs in the previous section can also be derived using equitable partitions. This section closely follows [8]. The notion of equitable partitions in algebraic graph theory was introduced by Sachs [163] and is related to auto-morphism groups of graphs and distance regular graphs [87]. Equitable partitions were used by Schwenk [172] to find the eigenvalues of graphs and by Hayden et al. [104], under the name *block structure*, to investigate multispherical EDMs. As we show next, the notion of equitable partitions easily extends to EDMs with many results mirroring those of graphs.

Let $N = \{1,\ldots,n\}$ and let $p^i, i \in N$, be the generating points of an $n \times n$ EDM D. An *m-partition* π of D is a sequence $\pi = (N_1, N_2, \ldots, N_m)$ of nonempty disjoint subsets of N whose union is N. The subsets $N_1, \ldots, N_m$ are called the *cells* of the partition. The n-partition where each cell is a singleton is called the *discrete partition*, while the 1-partition with only one cell is called the *single-cell partition.*

An *m-partition* π of D is said to be *equitable* if for all $i, j = 1, \ldots, m$ (case $i = j$ included), there exist nonnegative scalars α_{ij} such that for each $k \in N_i$, the sum of the squared Euclidean distances between p^k and all $p^l, l \in N_j$, is equal to α_{ij}; i.e.,

$$\forall i, j = 1, \ldots, m; \text{ and } \forall k \in N_i, \sum_{l \in N_j} d_{kl} = \alpha_{ij}. \tag{6.10}$$

Let $n_i = |N_i|$ and let $D_{[N_i,N_j]}$ denote the submatrix of D whose rows and columns are indexed by N_i and N_j, respectively. Then (6.10) is equivalent to

$$\forall i, j = 1, \ldots, m; D_{[N_i,N_j]} e_{n_j} = \alpha_{ij} e_{n_i}. \tag{6.11}$$

It immediately follows from (6.11) that the discrete partition of D is always equitable with $\alpha_{ij} = d_{ij}$. On the other hand, the single-cell partition of D is equitable if and only if D is regular, in which case, $\alpha_{11} = e^T De/n$.

Example 6.5 *Let D be the nonspherical centrally symmetric EDM considered in Example 6.4. Then $\pi_1 = (N_1 = \{1,2\}, N_2 = \{3\}, N_3 = \{4,5\})$ is equitable since*
$$D_{[N_1,N_1]} = D_{[N_3,N_3]} = \begin{bmatrix} 0 & 4 \\ 4 & 0 \end{bmatrix}, D_{[N_2,N_2]} = [0], D_{[N_1,N_2]} = D_{[N_3,N_2]} = \begin{bmatrix} 2 \\ 2 \end{bmatrix}, D_{[N_2,N_3]} =$$
$$D_{[N_2,N_1]} = [2\ 2], D_{[N_1,N_3]} = \begin{bmatrix} 4 & 8 \\ 8 & 4 \end{bmatrix}. \textit{ In this case, the α_{ij}'s can be collected in the}$$
following matrix
$$\begin{bmatrix} 4 & 2 & 12 \\ 4 & 0 & 4 \\ 12 & 2 & 4 \end{bmatrix}.$$

Note that the matrix of the α_{ij}'s is not symmetric. On the other hand, the partition
$$\pi_2 = (N_1 = \{1,2,4,5\}, N_2 = \{3\}) \textit{ is also equitable since } D_{[N_1,N_1]} = \begin{bmatrix} 0 & 4 & 4 & 8 \\ 4 & 0 & 8 & 4 \\ 4 & 8 & 0 & 4 \\ 8 & 4 & 4 & 0 \end{bmatrix},$$

$$D_{[N_1,N_2]} = \begin{bmatrix} 2 \\ 2 \\ 2 \\ 2 \end{bmatrix}, D_{[N_2,N_1]} = [2\ 2\ 2\ 2], \text{ and } D_{[N_2,N_2]} = [0]. \text{ The } \alpha_{ij}\text{'s can be collected}$$

in the following matrix $\begin{bmatrix} 16 & 2 \\ 8 & 0 \end{bmatrix}$.

For a partition π, define the $n \times m$ matrix $C_\pi = (c_{ij})$ where

$$c_{ij} = \begin{cases} \frac{1}{\sqrt{n_j}} & \text{if } i \in N_j, \\ 0 & \text{otherwise.} \end{cases} \tag{6.12}$$

C_π is called the *normalized characteristic matrix* [88] of π since its jth column is equal to $n_j^{-1/2}$ times the characteristic vector of N_j, and since $C_\pi^T C_\pi = I_m$.

Note that each row of C_π has exactly one nonzero entry since each i in N belongs to exactly one cell of the partition. For example, C_{π_1} and C_{π_2} of the partitions π_1 and π_2 of Example 6.5 are given by

$$C_{\pi_1} = \begin{bmatrix} e_2/\sqrt{2} & 0 & 0 \\ 0 & 1 & 0 \\ 0 & 0 & e_2/\sqrt{2} \end{bmatrix} \text{ and } C_{\pi_2} = \begin{bmatrix} e_2/2 & 0 \\ 0 & 1 \\ e_2/2 & 0 \end{bmatrix}.$$

The following lemma is the EDM equivalent of Godsil and McKay lemma for graphs [88].

Lemma 6.2 (Godsil and McKay [88]) *Let π be an m-partition of an $n \times n$ EDM D. Then π is equitable if and only if there exists an $m \times m$ matrix $S = (s_{ij})$ such that*

$$DC_\pi = C_\pi S, \tag{6.13}$$

in which case, $S = C_\pi^T DC_\pi$, i.e., $s_{ij} = (n_i/n_j)^{1/2} \alpha_{ij}$.

Proof. Assume that π is equitable and let $s_{ij} = (n_i/n_j)^{1/2} \alpha_{ij}$. Then, for all $k \in N_i$ and $j = 1, \ldots, m$, we have

$$(DC_\pi)_{kj} = \sum_{l=1}^n d_{kl} c_{lj} = \sum_{l \in N_j} d_{kl} \frac{1}{\sqrt{n_j}} = \frac{1}{\sqrt{n_j}} \alpha_{ij},$$

and

$$(C_\pi S)_{kj} = \sum_{l=1}^n c_{kl} s_{lj} = \frac{1}{\sqrt{n_i}} s_{ij} = \frac{1}{\sqrt{n_i}} \frac{\sqrt{n_i}}{\sqrt{n_j}} \alpha_{ij} = \frac{1}{\sqrt{n_j}} \alpha_{ij}.$$

Hence, $DC_\pi = C_\pi S$.

To prove the other direction, assume that $DC_\pi = C_\pi S$. Then for all $k \in N_i$ and $j = 1, \ldots, m$, we have

$$(DC_\pi)_{kj} = \sum_{l \in N_j} d_{kl} \frac{1}{\sqrt{n_j}} = (C_\pi S)_{kj} = \frac{1}{\sqrt{n_i}} s_{ij}.$$

Hence, $\sum_{l \in N_j} d_{kl} = \left(\frac{n_j}{n_i}\right)^{1/2} s_{ij} = \alpha_{ij}$ is independent of k. Consequently, the partition π is equitable and $s_{ij} = \left(\frac{n_i}{n_j}\right)^{1/2} \alpha_{ij}$.

□

It is worth pointing out here that S is symmetric since $S = C_\pi^T D C_\pi$. Another way to see this is to note that $D_{[N_i,N_j]} = (D_{[N_j,N_i]})^T$. Thus $D_{[N_i,N_j]} e_{n_j} = \alpha_{ij} e_{n_i}$. Therefore,

$$e_{n_i}^T D_{[N_i,N_j]} e_{n_j} = n_i \, \alpha_{ij} = e_{n_j}^T D_{[N_j,N_i]} e_{n_i} = n_j \, \alpha_{ji}.$$

Hence,

$$\alpha_{ij} = \frac{n_j}{n_i} \alpha_{ji},$$

and thus $s_{ij} = s_{ji}$. For example, matrices S_1 and S_2 of partitions π_1 and π_2 of Example 6.5 arc given by

$$S_1 = \begin{bmatrix} 4 & 2\sqrt{2} & 12 \\ 2\sqrt{2} & 0 & 2\sqrt{2} \\ 12 & 2\sqrt{2} & 4 \end{bmatrix} \text{ and } S_2 = \begin{bmatrix} 16 & 4 \\ 4 & 0 \end{bmatrix}.$$

Let π be an equitable partition of D where $m \le n-1$. Recall that $C_\pi^T C_\pi = I_m$. Let $\bar{C}_\pi$ be the $n \times (n-m)$ matrix such that $[C_\pi \ \bar{C}_\pi]$ is an $n \times n$ orthogonal matrix. For instance, $\bar{C}_{\pi_1}$ of Example 6.5 is given by $\bar{C}_{\pi_1} = \frac{1}{2} \begin{bmatrix} -1 & 1 \\ 1 & -1 \\ 0 & 0 \\ 1 & 1 \\ -1 & -1 \end{bmatrix}.$

Theorem 6.3 (Alfakih [8]) *Let π be an equitable m-partition of an $n \times n$ EDM D, where $m \le n-1$. Then*

$$\chi_D(\mu) = \chi_S(\mu)\, \chi_{\bar{S}}(\mu), \tag{6.14}$$

where $\bar{S} = \bar{C}_\pi^T D \bar{C}_\pi$ and S is as defined in (6.13).

Proof. Equation (6.13) implies that $C_\pi^T D = S C_\pi^T$. Recall that S is symmetric. Thus, it follows from the definition of $\bar{C}_\pi$ that $C_\pi^T D \bar{C}_\pi = S C_\pi^T \bar{C}_\pi = 0$. The result follows since

$$\chi_D(\mu) = \det\left[\begin{bmatrix} C_\pi^T \\ \bar{C}_\pi^T \end{bmatrix} D [C_\pi \ \bar{C}_\pi] - \mu I_n \right] = \det \begin{bmatrix} C_\pi^T D C_\pi - \mu I_m & 0 \\ 0 & \bar{C}_\pi^T D \bar{C}_\pi - \mu I_{n-m} \end{bmatrix}.$$

□

Note that in the case of discrete partition, i.e., if $m = n$, we have $S = D$ since $C_\pi = I$. Thus $\chi_D(\mu) = \chi_S(\mu)$ follows trivially. An analogous result for graphs, namely that the characteristic polynomial of S divides the characteristic polynomial of a graph was obtained by Mowshowitz [150] and by Schwenk [172]. The following theorem shows that if μ_i is an eigenvalue of $\bar{S}$, then $-\mu_i/2$ is an eigenvalue of $\mathcal{T}(D)$.

Theorem 6.4 ([8]) *Let π be an equitable m-partition of an $n \times n$ EDM D, where $m \leq n - 1$, and let $B = -JDJ/2$ be the Gram matrix of D. Then*

$$\chi_{\bar{S}}(\mu) \text{ divides } \chi_{(-2B)}(\mu), \tag{6.15}$$

where $\bar{S}$ is as defined in Theorem 6.3.

Proof. First, observe that if x is a vector in $\mathbb{R}^n$ that is constant on each cell N_i, then x lies in the column space of C_π. For instance, De lies in the column space of C_π since if $k \in N_i$, then $(De)_k = \sum_{j=1}^m \alpha_{ij}$, which is independent of k. Moreover, e is also in the column space of C_π. Therefore, $\bar{C}_\pi^T De = 0$ and $\bar{C}_\pi^T e = 0$. Thus,

$$\bar{C}_\pi^T B \bar{C}_\pi = -\frac{1}{2}\bar{C}_\pi^T D \bar{C}_\pi = -\frac{1}{2}\bar{S}.$$

Moreover, it follows from (6.13) that $\bar{C}_\pi^T D C_\pi = \bar{C}_\pi^T C_\pi S = 0$. Thus,

$$\bar{C}_\pi^T B C_\pi = -\frac{1}{2}\bar{C}_\pi^T D C_\pi = 0.$$

The result follows since

$$\chi_{(-2B)}(\mu) = \det\left[-2\begin{bmatrix} C_\pi^T \\ \bar{C}_\pi^T \end{bmatrix} B [C_\pi \ \bar{C}_\pi] - \mu I_n\right]$$

$$= \det\begin{bmatrix} -2C_\pi^T B C_\pi - \mu I_m & 0 \\ 0 & -2\bar{C}_\pi^T B \bar{C}_\pi - \mu I_{n-m} \end{bmatrix}$$

$$= \chi_{(-2C_\pi^T B C_\pi)}(\mu)\, \chi_{\bar{S}}(\mu).$$

$\square$

For instance, in Example 6.5, $\bar{S} = \bar{C}_{\pi_1}^T D \bar{C}_{\pi_1} = \begin{bmatrix} 0 & 0 \\ 0 & -8 \end{bmatrix}$. Thus $\chi_{\bar{S}}(\mu) = \mu(-8 - \mu)$. On the other hand, $\chi_{(-2B)}(\mu) = \mu^3(-8-\mu)^2$.

In the following two subsections, we use Theorem 6.3 to derive the characteristic polynomials of regular and nonspherical centrally symmetric EDMs.

6.2.1 The Eigenvalues of Regular EDMs

As was remarked earlier, the single-cell partition π is equitable for regular EDMs. Thus, in this case $m = 1$ and $C_\pi = e/\sqrt{n}$. Hence, $\bar{C}_\pi = V$ where V is as defined in (3.11). Moreover, $S = e^T De/n$ and $\bar{S} = V^T DV = -2V^T BV$, where $B = -JDJ/2$. Therefore,

$$\chi_S(\mu) = (\mu - e^T De/n) \text{ and } \chi_{\bar{S}}(\mu) = (-\mu)^{n-r-1}\prod_{i=1}^r (a_i - \mu),$$

where $\lambda_1, \ldots, \lambda_r$ are the positive eigenvalues of B and $a_i = -2\lambda_i$ for $i = 1, \ldots, r$. Thus, Theorem 6.3 provides an alternative derivation to the characteristic polynomial of a regular EDM given in (6.5).

6.2.2 The Eigenvalues of Nonspherical Centrally Symmetric EDMs

Let D be a $2n \times 2n$ nonspherical centrally symmetric EDM of embedding dimension r with configuration matrix $P = \begin{bmatrix} P_1 \\ -P_1 \end{bmatrix}$, where P_1 is $n \times r$. Observe that $P^T e_{2n} = \mathbf{0}$ and $B = PP^T = -JDJ/2$. Let $B_1 = P_1 P_1^T$ and note that $P_1^T e_n$ is not necessarily 0. Then

$$D = \begin{bmatrix} D_1 & A_1 \\ A_1 & D_1 \end{bmatrix},$$

where $D_1 = \mathscr{K}(B_1)$ and

$$A_1 = e_n(\mathrm{diag}(B_1))^T + \mathrm{diag}(B_1)e_n^T + 2B_1.$$

The partition π of D corresponding to

$$C_\pi = \frac{1}{\sqrt{2}} \begin{bmatrix} I_n \\ I_n \end{bmatrix} \quad \text{and} \quad \bar{C}_\pi = \frac{1}{\sqrt{2}} \begin{bmatrix} I_n \\ -I_n \end{bmatrix}$$

is obviously equitable, where

$$S = C_\pi^T D C_\pi = A_1 + D_1 = 2(e_n(\mathrm{diag}(B_1))^T + \mathrm{diag}(B_1)e_n^T) \qquad (6.16)$$

and

$$\bar{S} = \bar{C}_\pi^T D \bar{C}_\pi = D_1 - A_1 = -4B_1.$$

Now

$$B = \begin{bmatrix} B_1 & -B_1 \\ -B_1 & B_1 \end{bmatrix} = \begin{bmatrix} 1 & -1 \\ -1 & 1 \end{bmatrix} \otimes B_1.$$

Thus, the nonzero eigenvalues of B are equal to twice the nonzero eigenvalues of B_1. Hence, the nonzero eigenvalues of $\bar{S}$ are equal to -2 times the nonzero eigenvalues of B. Hence,

$$\chi_{\bar{S}}(\mu) = \chi_{(-2B)}(\mu) = (-\mu)^{n-r} \prod_{i=1}^{r} (a_i - \mu).$$

Therefore, to find $\chi_D(\mu)$, it remains to determine $\chi_S(\mu)$. To this end, note that

$$P^T D e_{2n} = [P_1^T \quad -P_1^T] \begin{bmatrix} (D_1 + A_1)e_n \\ (D_1 + A_1)e_n \end{bmatrix} = \mathbf{0}.$$

Hence, De_{2n} lies in the null space of P^T, i.e., De_{2n} lies in $\mathrm{span}(e_{2n}) \oplus \mathrm{null}(D) \oplus \mathrm{span}(\eta)$. But, since De_{2n} lies in $\mathrm{col}(D)$, it follows that $De_{2n} = \alpha e_{2n} + \beta \eta$ for some scalars α and β. Now $e_{2n}^T De_{2n} = \alpha 2n$ since D is nonspherical and thus $e_{2n}^T \eta = 0$. Furthermore, $\eta^T De_{2n} = 2n = \beta \eta^T \eta$. Therefore,

$$De_{2n} = \begin{bmatrix} (D_1 + A_1)e_n \\ (D_1 + A_1)e_n \end{bmatrix} = \frac{1}{2n} e_{2n}^T De_{2n} \begin{bmatrix} e_n \\ e_n \end{bmatrix} + \frac{2n}{\eta^T \eta} \begin{bmatrix} \xi \\ \xi \end{bmatrix},$$

where $\eta^T = [\xi^T \ \xi^T]$. Consequently,

$$Se_n = (D_1 + A_1)e_n = \frac{1}{2n} e_{2n}^T De_{2n} \, e_n + \frac{2n}{\eta^T \eta} \xi.$$

Note that (6.16) implies that $V^T SV = \mathbf{0}$. Also, note that $e_n^T \xi = 0$ since $e_{2n}^T \eta = 0$. Therefore,

$$\begin{aligned}
\chi_S(\mu) &= \det \left[\begin{bmatrix} \frac{e_n^T}{\sqrt{n}} \\ V^T \end{bmatrix} S \begin{bmatrix} \frac{e_n}{\sqrt{n}} & V \end{bmatrix} - \mu I_n \right] \\
&= \det \begin{bmatrix} \frac{1}{2n} e_{2n}^T De_{2n} - \mu & \frac{2\sqrt{n}}{\eta^T \eta} \xi^T V \\ \frac{2\sqrt{n}}{\eta^T \eta} V^T \xi & -\mu I_{n-1} \end{bmatrix} \\
&= (-\mu)^{n-2} \left(\mu^2 - \mu \frac{1}{2n} e_{2n}^T De_{2n} - \frac{4n}{(\eta^T \eta)^2} \xi^T V V^T \xi \right).
\end{aligned}$$

Now $\xi^T V V^T \xi = \xi^T \xi = \eta^T \eta / 2$. Therefore, Theorem 6.3 provides an alternative derivation of the characteristic polynomial of nonspherical centrally symmetric EDMs given in Theorem 6.2.

Example 6.6 *Consider the EDM* $D = \begin{bmatrix} 0 & 5 & 16 & 5 \\ 5 & 0 & 5 & 4 \\ 16 & 5 & 0 & 5 \\ 5 & 4 & 5 & 0 \end{bmatrix}$, *where D is nonspherical cen-*

trally symmetric with configuration matrix $P = \begin{bmatrix} -2 & 0 \\ 0 & -1 \\ 2 & 0 \\ 0 & 1 \end{bmatrix}$. *The partition π where*

$C_\pi = \frac{1}{\sqrt{2}} \begin{bmatrix} 1 & 0 \\ 0 & 1 \\ 1 & 0 \\ 0 & 1 \end{bmatrix}$ *and* $\bar{C}_\pi = \frac{1}{\sqrt{2}} \begin{bmatrix} 1 & 0 \\ 0 & 1 \\ -1 & 0 \\ 0 & -1 \end{bmatrix}$ *is equitable. In this case,* $S = \begin{bmatrix} 16 & 10 \\ 10 & 4 \end{bmatrix}$

and $\bar{S} = \begin{bmatrix} -16 & 0 \\ 0 & -4 \end{bmatrix}$. *Therefore,* $\chi_S(\mu) = \mu^2 - 20\mu - 36$, $\chi_{\bar{S}}(\mu) = (\mu + 16)(\mu + 4)$ *and* $\chi_{(-2B)}(\mu) = \mu^2(\mu + 16)(\mu + 4)$. *Note that* $\eta = \frac{1}{6}[1 \ -1 \ 1 \ -1]^T$ *and thus* $n/\eta^T \eta = 36$ *and* $e^T De/n = 20$. *Also, note that*

$$De = \begin{bmatrix} 26 \\ 14 \\ 26 \\ 14 \end{bmatrix} = 20 \begin{bmatrix} 1 \\ 1 \\ 1 \\ 1 \end{bmatrix} + 6 \begin{bmatrix} 1 \\ -1 \\ 1 \\ -1 \end{bmatrix} = \frac{1}{n} e^T De\, e + \frac{n}{\eta^T \eta}\, \eta.$$

6.3 Constructing Cospectral EDMs

Two $n \times n$ EDMs D_1 and D_2 are said to be *isomorphic* if there exists a permutation matrix Q such that $D_2 = QD_1Q^T$, and they are said to be *cospectral* if $\chi_{D_1}(\mu) = \chi_{D_2}(\mu)$. Obviously, isomorphic EDMs are cospectral, but the converse is not true. In this section, we show how to construct cospectral nonisomorphic EDMs.

Let D_1 be an $n \times n$ regular EDM of embedding dimension r generated by points $p^1, \ldots, p^n$ in $\mathbb{R}^r$. Let $\gamma > 1$ and assume that D_1 is not centrally symmetric. Let D^- and D^+ be the two $2n \times 2n$ EDMs generated as follows. D^- is generated by the points $p^1, \ldots, p^n, -\gamma p^1, \ldots, -\gamma p^n$; and D^+ is generated by the points $p^1, \ldots, p^n, \gamma p^1, \ldots, \gamma p^n$. Thus, the configuration matrices of D^- and D^+ are

$$P^- = \begin{bmatrix} P_1 \\ -\gamma P_1 \end{bmatrix} \text{ and } P^+ = \begin{bmatrix} P_1 \\ \gamma P_1 \end{bmatrix}$$

respectively, where P_1 is the configuration matrix of D_1. Let $B_1 = -JD_1J/2$. Then $\operatorname{diag}(B_1) = \rho^2 e = \frac{e^T D_1 e}{2n^2} e$ since D_1 is regular. Consequently,

$$D^- = \begin{bmatrix} D_1 & A_1 \\ A_1 & \gamma^2 D_1 \end{bmatrix}, \text{ and } D^+ = \begin{bmatrix} D_1 & A_2 \\ A_2 & \gamma^2 D_1 \end{bmatrix},$$

where

$$A_1 = \frac{1}{2n^2}(\gamma^2 + 1)e^T D_1 e\, E + 2\gamma B_1,$$

and

$$A_2 = \frac{1}{2n^2}(\gamma^2 + 1)e^T D_1 e\, E - 2\gamma B_1.$$

Moreover, $A_1 e = A_2 e = (\gamma^2 + 1)\frac{e^T D_1 e}{2n} e.$

It is worth pointing out that D^- and D^+ are nonspherical by construction. Furthermore, $P^- D^- e_{2n} = P^+ D^+ e_{2n} = 0$ since D_1 is regular. The implication of this fact will be highlighted below.

The 2-partition π of both D^- and D^+ where

$$C_\pi = \frac{1}{\sqrt{n}} \begin{bmatrix} e & 0 \\ 0 & e \end{bmatrix} \text{ and } \bar{C}_\pi = \begin{bmatrix} V & 0 \\ 0 & V \end{bmatrix}$$

is equitable since

$$D^- C_\pi = \frac{e^T D_1 e}{n\sqrt{n}} \begin{bmatrix} e & \frac{\gamma^2+1}{2} e \\ \frac{\gamma^2+1}{2} e & \gamma^2 e \end{bmatrix} = C_\pi S^-$$

and

$$D^+ C_\pi = \frac{e^T D_1 e}{n\sqrt{n}} \begin{bmatrix} e & \frac{\gamma^2+1}{2} e \\ \frac{\gamma^2+1}{2} e & \gamma^2 e \end{bmatrix} = C_\pi S^+,$$

where

$$S^- = S^+ = \frac{e^T D_1 e}{n} \begin{bmatrix} 1 & \frac{\gamma^2+1}{2} \\ \frac{\gamma^2+1}{2} & \gamma^2 \end{bmatrix}.$$

Therefore,

$$\chi_{S^-}(\mu) = \chi_{S^+}(\mu) = \mu^2 - \mu(\gamma^2+1)\frac{e^T D_1 e}{n} - \frac{(\gamma^2-1)^2}{4}\frac{(e^T D_1 e)^2}{n^2}. \qquad (6.17)$$

Consequently,

$$\chi_{S^-}(\mu) = \chi_{S^+}(\mu) = (\mu_1 - \mu)(\mu_2 - \mu),$$

where

$$\mu_1 = (1 + \gamma^2 + \sqrt{2\gamma^4+2})\frac{e^T D_1 e}{2n} \text{ and } \mu_2 = (1 + \gamma^2 - \sqrt{2\gamma^4+2})\frac{e^T D_1 e}{2n}. \qquad (6.18)$$

Note that $\mu_2 < 0$ since $\gamma > 1$. Also, note that

$$\frac{e_{2n}^T D^- e_{2n}}{2n} = \frac{e_{2n}^T D^+ e_{2n}}{2n} = (\gamma^2+1)\frac{e^T D_1 e}{n}$$

and

$$\eta^- = \eta^+ = \frac{n}{e^T D_1 e} \frac{2}{\gamma^2-1} \begin{bmatrix} -e \\ e \end{bmatrix} \text{ and thus } \frac{2n}{\eta^{-T} \eta^-} = \frac{(\gamma^2-1)^2}{4} \frac{(e^T D_1 e)^2}{n^2}.$$

As a result, (6.17) should come as no surprise since $P^- D^- e_{2n} = P^+ D^+ e_{2n} = \mathbf{0}$.

Now $V^T A_1 V = 2\gamma V^T B_1 V = -\gamma V^T D_1 V$ and similarly $V^T A_2 V = \gamma V^T D_1 V$. Recall that $V^T D_1 V = -2X_1$, where X_1 is the projected Gram matrix of D_1. Therefore,

$$\bar{S}^- = \bar{C}_\pi^T D^- \bar{C}_\pi = \begin{bmatrix} V^T D_1 V & -\gamma V^T D_1 V \\ -\gamma V^T D_1 V & \gamma^2 V^T D_1 V \end{bmatrix} = \begin{bmatrix} 1 & -\gamma \\ -\gamma & \gamma^2 \end{bmatrix} \otimes (-2X_1),$$

and

$$\bar{S}^+ = \bar{C}_\pi^T D^+ \bar{C}_\pi = \begin{bmatrix} V^T D_1 V & \gamma V^T D_1 V \\ \gamma V^T D_1 V & \gamma^2 V^T D_1 V \end{bmatrix} = \begin{bmatrix} 1 & \gamma \\ \gamma & \gamma^2 \end{bmatrix} \otimes (-2X_1).$$

Let

$$B^- = P^- P^{-T} = \begin{bmatrix} 1 & -\gamma \\ -\gamma & \gamma^2 \end{bmatrix} \otimes B_1 \text{ and } B^+ = P^+ P^{+T} = \begin{bmatrix} 1 & \gamma \\ \gamma & \gamma^2 \end{bmatrix} \otimes B_1.$$

It is easy to see that the matrices $\begin{bmatrix} 1 & -\gamma \\ -\gamma & \gamma^2 \end{bmatrix}$ and $\begin{bmatrix} 1 & \gamma \\ \gamma & \gamma^2 \end{bmatrix}$ have the same eigenvalues, namely $\gamma^2 + 1$ and 0. Consequently, B^- and B^+ are cospectral as well as $\bar{S}^-$ and $\bar{S}^+$.

Recall that X_1 and B_1 have the same nonzero eigenvalues. Let $\lambda_1, \ldots, \lambda_r$ be the nonzero eigenvalue of B^- and let $\lambda_1', \ldots, \lambda_r'$ be the nonzero eigenvalue of B_1. Then

$$\lambda_i = (\gamma^2 + 1)\lambda_i' \quad \text{for } i = 1, \ldots, r.$$

Moreover,

$$\chi_{\bar{S}^-}(\mu) = \chi_{\bar{S}^+}(\mu) = (-\mu)^{2n-r-2} \prod_{i=1}^{r} ((\gamma^2 + 1)a_i' - \mu), \tag{6.19}$$

where $a_i' = -2\lambda_i'$ for $i = 1, \ldots, r$, or

$$\chi_{\bar{S}^-}(\mu) = \chi_{\bar{S}^+}(\mu) = (-\mu)^{2n-r-2} \prod_{i=1}^{r} (a_i - \mu), \tag{6.20}$$

where $a_i = -2\lambda_i$ for $i = 1, \ldots, r$. Again (6.20) should come as no surprise since $P^- D^- e_{2n} = P^+ D^+ e_{2n} = 0$. As a result, by Theorem 6.3, the characteristic polynomial of D^- and D^+ is given by

$$\chi_{D^-}(\mu) = \chi_{D^+}(\mu) = (-\mu)^{2n-r-2}(\mu_1 - \mu)(\mu_2 - \mu) \prod_{i-1}^{r} ((\gamma^2 + 1)a_i' - \mu),$$

where μ_1 and μ_2 are as defined in (6.18). Since μ_1, μ_2, and a_i''s are nonzero, it follows that $\text{rank}(D^-) = \text{rank}(D^+) = r + 2$ as expected since D^- and D^+ are nonspherical with embedding dimension r. Moreover, D^- and D^+ are not isomorphic since D_1 is not centrally symmetric. Clearly, this assumption on D_1 cannot be dropped.

An example of two nonisomorphic cospectral EDMs is given next.

Example 6.7 *Consider the regular EDM $D_1 = E - I$. Then $\mathscr{T}(D_1) = J/2$ and $e^T D_1 e = n^2 - n$. Thus, $A_1 = \frac{(\gamma^2+1)(n-1)}{2n} ee^T + \gamma J$ and $A_2 = \frac{(\gamma^2+1)(n-1)}{2n} ee^T - \gamma J$. Therefore, for $\gamma = 2$ and $n = 3$, we have $A_1 = 2I_3 + E_3$ and $A_2 = 7E_3/3 - 2I$. Hence,*

$$D^- = \begin{bmatrix} E_3 - I_3 & 2I_3 + E_3 \\ 2I_3 + E_3 & 4(E_3 - I_3) \end{bmatrix} \text{ and } D^+ = \begin{bmatrix} E_3 - I_3 & -2I_3 + 7E_3/3 \\ -2I_3 + 7E_3/3 & 4(E_3 - I_3) \end{bmatrix}$$

are two nonisomorphic EDMs. Moreover, the nonzero eigenvalues of B_1 are $\lambda_1' = \lambda_2' = 1/2$. Hence, $a_1' = a_2' = -1$. Therefore,

$$\chi_{D^-}(\mu) = \chi_{D^+}(\mu) = (-\mu)^2 (5 + \sqrt{34} - \mu)(5 - \sqrt{34} - \mu)(-5 - \mu)^2.$$

6.4 EDMs, Graphs, and Combinatorial Designs

We investigate in this section the connections among EDMs, adjacency matrices of graphs and combinatorial designs. We begin by showing that the smallest eigenvalue of the adjacency matrix of a graph, or of an EDM, can be used to construct a new EDM.

Theorem 6.5 (Neumaier [152]) *Let A be the $n \times n$ adjacency matrix of a graph G and assume that its minimum eigenvalue $\lambda_n = -\alpha < 0$. Then $D' = \alpha(E - I) - A$ is a spherical EDM.*

Theorem 6.6 ([153]) *Let D an $n \times n$ EDM and assume that its minimum eigenvalue $\lambda_n = -\alpha < 0$. Then $D' = \alpha(E - I) - D$ is a spherical EDM.*

These two theorems are simple corollaries to Theorem 4.2 since in both cases $\alpha E - D'$ is PSD. Note that the nonnegativity of d'_{ij} for all i, j follows since, for example in the EDM case, the nonnegativity of 2×2 principal minors of $\alpha E - D' = \alpha I + D$ imply that $\alpha \geq d_{ij}$. The reader should recall that, if D is an EDM, then $D'' = \varepsilon(E - I) + D$ is a spherical EDM for any $\varepsilon > 0$.

The number of distinct off-diagonal entries of an EDM D is called the *degree* of D. Notice that the construction given in Theorem 6.5 yields EDMs of degree 2. Recall that $f[D] = (f(d_{ij}))$ denotes the matrix obtained from D by applying function f to D entrywise. Thus, for a nonnegative integer k, $[D]^k$ denotes the matrix whose (i, j)th entry is d_{ij}^k. Here, $0^0 = 1$ by definition and thus $[D]^0 = E$ and $[D]^1 = D$. An $n \times n$ EDM D is said to have *strength* t if for every nonnegative integers k and l where $k + l \leq t$, there exists a polynomial $f_{kl}(x)$ of degree $\leq \min\{k, l\}$ such that

$$([D]^k[D]^l)_{ij} = f_{kl}(d_{ij}). \tag{6.21}$$

The strength of D is a measure of its inner regularity. Notice that if D has strength t, then D has strength t' for all $t' \leq t$; and $[D]^k E = [D]^k[D]^0 = \text{const } E$ for all $k \leq t$ since $f_{k0}(x)$ must have degree 0. Let $f_{00}(x) = n$, i.e., $f_{00}(x)$ is the constant polynomial n. Then for any EDM D, we have $[D]^0[D]^0 = EE = nE$. Hence, every EDM D has strength zero. The theorem that follows characterizes EDMs of strength one.

Theorem 6.7 (Neumaier [152]) *Let D be an $n \times n$ EDM. Then D has strength 1 if and only if D is regular.*

Proof. An EDM D is regular iff $DE = \lambda E$, where $\lambda = e^T De/n$. Therefore, D is regular iff $[D]^1[D]^0 = DE = \lambda E$. The result follows by setting $f_{10}(x) = \lambda$. $\qquad\square$

For instance, the EDM of the simplex EDM $D = \gamma(E - I)$, where $\gamma > 0$, has strength 1 since D is regular; i.e., $DE = \lambda E$, where $\lambda = \gamma(n - 1)$.

A graph is said to be *regular* if all its nodes have the same degree. In particular, G is k-regular if each of its nodes has degree k. Consequently, graph G is k-regular iff its adjacency matrix A satisfies $AE = kE$.

We saw earlier how to construct a spherical EDM D from the adjacency matrix A of a graph G. The following theorem shows that such D is regular iff G is a regular.

Theorem 6.8 (Neumaier [152]) *Let A be the $n \times n$ adjacency matrix of graph G. Let λ_n, the minimum eigenvalue of A, be equal to $-\alpha < 0$ and of multiplicity s. Further, let $D = \alpha(E - I) - A$. Then graph G is regular if and only if D is a regular EDM, in which case the embedding dimension of D is equal to $n - 1 - s$.*

Proof.   D is regular iff $DE = \alpha(n - 1)E - AE = \lambda E$ iff $AE = kE$, where $k = \alpha(n - 1) - \lambda$ and $\lambda = e^T De/n$. This proves the first part. Now the projected Gram matrix of D is $X = \mathscr{T}_V(D) = (\alpha I_{n-1} + V^T AV)/2$. Thus, the embedding dimension of D is equal to the rank of $\alpha I_{n-1} + V^T AV$. Assume that G is k-regular and let $A = kee^T/n + V\Lambda V^T$ be the spectral decomposition of A, where the diagonal entries of $\Lambda = V^T AV$ are precisely the $n - 1$ eigenvalues of A other than k. Hence, the rank of $\alpha I_{n-1} + V^T AV$ is equal to $n - 1 - s$.

□

Notice that if $G = K_n$, then $A = E - I$ and thus $\alpha = 1$ and $s = n - 1$. Therefore, in this case, $D = \mathbf{0}$. A characterization of EDMs of strength two is given next.

Theorem 6.9 (Neumaier [152]) *Let D be an $n \times n$ real symmetric matrix with zero diagonal. Then D is an EDM of strength 2 if and only if*

$$DE = \lambda E, \quad D^2 + \alpha D = \frac{\lambda(\lambda + \alpha)}{n} E, \tag{6.22}$$

for some positive scalars λ and α, where λ/α is an integer.

Remark 6.1 *Suppose that D is a regular EDM with three distinct eigenvalues $\lambda > 0 > -\alpha$. Then it follows from Corollary 4.2 that D satisfies*

$$DE = \lambda E, \quad f(D) = n\frac{D(D + \alpha I)}{\lambda(\lambda + \alpha)} = E, \tag{6.23}$$

where $\lambda = e^T De/n$ and f is the Hoffman polynomial of D. Consequently, Condition 6.23 is identical to Condition 6.22.

Proof of Theorem 6.9. Assume that (6.22) holds and let $B' = \lambda E/n - D$. Then

$$B'^2 = -\lambda^2 E/n + D^2 = -\alpha D + \lambda \alpha E/n = \alpha B'.$$

Therefore, the eigenvalues of B' are either 0 or α. Hence, B' is PSD since $\alpha > 0$. Observe that $B' = 2\mathscr{T}(D) = -JDJ$. Therefore, D is a regular EDM and thus has strength 1.

Now, $([D]^1[D]^1)_{ij} = (D^2)_{ij} = f_{11}(d_{ij})$, where $f_{11}(x) = -\alpha x + \lambda(\lambda + \alpha)/n$. Also, $([D]^2[D]^0)_{ij} = ([D]^2 E)_{ij} = \sum_k d_{ik}^2$. Moreover, (6.22) implies that $(D^2)_{ii} = \lambda(\lambda + \alpha)/n$, a constant. But $(D^2)_{ii} = \sum_k d_{ik}^2$. Hence, $[D]^2[D]^0 = \lambda(\lambda + \alpha)/n\, E$ and thus $f_{20}(d_{ij}) = \lambda(\lambda + \alpha)/n$. Therefore, D has strength 2.

To prove the other direction, assume that D is an EDM with strength 2. Then D has strength 1 and thus $DE = \lambda E$ where $\lambda > 0$. Moreover, there exists a polynomial $f_{11}(x) = -\alpha x + \beta$ such that $(D^2)_{ij} = -\alpha d_{ij} + \beta$. Thus $D^2 = -\alpha D + \beta E$. But on the one hand, $D^2 E = -\alpha DE + n\beta E = (-\alpha\lambda + n\beta)E$, and on the other hand, $D^2E = \lambda^2 E$. Thus $\beta = \lambda(\lambda + \alpha)/n$. To complete the proof we need to show that $\alpha > 0$ and $\lambda/\alpha = k$, where k is a positive integer. To this end, let $B' = 2\mathcal{T}(D) = \lambda E/n - D$. Therefore, $B'^2 = \alpha B'$. Thus, the eigenvalues of B' are either 0's or α's and hence $\alpha > 0$. Moreover, let k be the multiplicity of the eigenvalue α. Then trace$(B') = \lambda = k\alpha$ and thus $\lambda/\alpha = k$.

□

As we saw in the proof of Theorem 6.9, the Gram matrix $B = \mathcal{T}(D)$ of any EDM that satisfies (6.22) has eigenvalues 0 or $\alpha/2$. Therefore, by Theorem 6.1, the negative eigenvalues of such EDMs are all equal to $(-\alpha)$. Moreover, if k denotes the multiplicity of the eigenvalue $(-\alpha)$, then $k = \lambda/\alpha$ since trace$(D) = \lambda - k\alpha = 0$.

Example 6.8 *The EDM of the simplex $D = \gamma(E - I)$ has strength 2. This follows since $D^2 = -\gamma D + \gamma^2(n-1)E$. Thus, D satisfies (6.22) with $\alpha = \gamma$ and $\lambda = \gamma(n-1)$. Note that $\lambda/\alpha = n - 1$ is an integer. In fact, $D = \gamma(E - I)$ has strength t for all positive integers t. This follows since $[D]^k = \gamma^{k-1}D$ for all positive integers k. Thus, $[D]^k[D]^0 = \gamma^{k-1}DE = \gamma^k(n-1)E$ and $[D]^k[D]^l = \gamma^{k+l-2}D^2 = -\gamma^{k+l-1}D + \gamma^{k+l}(n-1)E$ for all positive integers k and l.*

We saw in the beginning of this section how to construct a new EDM D' from an old EDM D using the smallest eigenvalue of D. The following theorem shows that this construction preserves the strength t for $t \leq 2$.

Theorem 6.10 ([152]) *Let D be an $n \times n$ EDM and let its smallest eigenvalue be $\lambda_n = -\alpha < 0$. Let $D' = \alpha(E - I) - D$. Then*

1. *D' has strength 1 iff D has strength 1.*
2. *D' has strength 2 iff D has strength 2.*

Proof. $D'E = \alpha(n-1)E - DE$. Thus, $D'E = \lambda'E$ iff $DE = (\alpha(n-1) - \lambda')E = \lambda E$, where $\lambda' + \lambda = \alpha(n-1)$. This proves Statement 1.

To prove Statement 2, note that $(D')^2 = -\alpha D' + \alpha^2(n-1)E - 2\alpha\lambda E + D^2 + \alpha D$, where we substituted $\alpha I = \alpha E - D - D'$. But

$$\lambda'(\lambda' + \alpha)/n = \alpha^2(n-1) - 2\alpha\lambda + \lambda(\lambda + \alpha)/n.$$

Thus,

$$(D')^2 = -\alpha D' + \lambda'(\lambda' + \alpha)E/n - \lambda(\lambda + \alpha)E/n + D^2 + \alpha D.$$

Hence, $(D')^2 = -\alpha D' + \lambda'(\lambda' + \alpha)E/n$ if and only if $D^2 + \alpha D - \lambda(\lambda + \alpha)E/n = 0$. Therefore, Statement 2 holds by setting $\alpha' = \alpha$.

□

A similar result for D with strength ≥ 3 is not true in general [152]. The fact that $\alpha' = \alpha$ should come as no surprise since $V^T D'V = -\alpha I_{n-1} - V^T DV$. But the nonpositive eigenvalues of D are exactly the eigenvalues of $V^T DV$. Therefore, if D has eigenvalues $(-\alpha)$ with multiplicity k and zero with multiplicity $n - 1 - k$, then D' has eigenvalues 0 with multiplicity k and $(-\alpha)$ with multiplicity $n - 1 - k$.

Example 6.9 *Let* $D = \begin{bmatrix} 0 & 2 & 4 & 2 \\ 2 & 0 & 2 & 4 \\ 4 & 2 & 0 & 2 \\ 2 & 4 & 2 & 0 \end{bmatrix}$ *be the EDM considered in Example 4.7. D is*

regular with eigenvalues $8, 0, -4, -4$ *and with Hoffman polynomial* $f(x) = x(x + 4)/24$. *Thus,* $D^2 + 4D = 24E$ *and therefore D satisfies (6.22) with* $\alpha = 4$ *and* $\lambda = 8$. *Hence, D has strength 2.*

Now $D' = \alpha(E - I) - D = \begin{bmatrix} 0 & 2 & 0 & 2 \\ 2 & 0 & 2 & 0 \\ 0 & 2 & 0 & 2 \\ 2 & 0 & 2 & 0 \end{bmatrix}$. *Hence, D' is also a regular EDM with*

eigenvalues $4, 0, 0, -4$. *Note that* $\alpha' = \alpha$ *and* $\lambda' = 4$. *The Hoffman polynomial of D'* *is* $f'(x) = x(x + 4)/8$. *Therefore,* $(D')^2 + 4D' = 8E$. *As a result, D' also has strength 2.*

Next, we turn to combinatorial designs. Let $\mathscr{P}$ be a collection of subsets of $\{1, \ldots, n\}$ called *blocks*, and let l be a positive integer and k be an integer such that $2 \le k \le n - 1$. Then $\mathscr{P}$ is called an (n, k, l) *2-design* if the following two conditions hold.

(i) Each block has cardinality k.
(ii) Each pair i and j is contained in exactly l blocks.

For example,

$$\mathscr{P} = \{\{1,2,4\}, \{2,3,5\}, \{3,4,6\}, \{4,5,7\}, \{1,5,6\}, \{2,6,7\}, \{1,3,7\}\}$$

is a $(7, 3, 1)$ 2-design. Also, the edges of the complete graph K_n is an $(n, 2, 1)$ 2-design. In this case, there are $n(n-1)/2$ blocks, $k = 2$ since each edge contains exactly two vertices and $l = 1$ since each pair of vertices is contained in exactly one edge.

Let b be the number of blocks. Then the $n \times b$, $(0-1)$ matrix A, where $a_{ij} = 1$ iff i is in block j, is called the *incidence matrix* of $\mathscr{P}$. Now $A^T e_n = k e_b$ since each block has cardinality k. Also, it can be proven that $A e_b = r e_n$ where r is a positive integer to be determined next. Thus, $e_b^T A^T e_n = kb$ and hence $b = nr/k$. Furthermore,

$$AA^T = lE + (r - l)I \qquad (6.24)$$

since

$$(AA^T)_{ij} = \sum_k a_{ik} a_{jk} = \begin{cases} l & \text{if } i \neq j, \\ r & \text{if } i = j. \end{cases}$$

To find r, we calculate $AA^T e_n$ in two ways. On the one hand, $A^T e_n = k e_b$ and thus $AA^T e_n = k A e_b = k r e_n$. On the other hand, it follows from (6.24) that $AA^T e_n = ((n-1)l + r)e_n$. Thus $r = l(n-1)/(k-1)$. Note that $r \neq l$ since $k \neq n$. Consequently, AA^T is nonsingular. Hence, null(A^T) is trivial and thus A^T has full column rank, i.e., rank$(A^T) = n$. Therefore, $n \le b$. For instance, in case of K_n, $r = n - 1$ since each vertex is in $n - 1$ edges.

Therefore,

$$D = rE - AA^T = (r-l)(E-I) \tag{6.25}$$

is a standard simplex EDM with $\gamma = r - l$. Note that $(A^T A)_{ii} = \sum_k a_{ki} = k$. Thus $\mathrm{diag}(A^T A) = k e_b$. Hence, $D' = kE - A^T A$ is a spherical EDM since $kE - D' = A^T A$ is PSD.

6.5 EDMs with Two or Three Distinct Eigenvalues

The eigenvalues of a nonzero EDM D cannot all be equal since $\mathrm{trace}(D) = 0$. As a result, EDMs have two or more distinct eigenvalues. EDMs with two distinct eigenvalues are easily characterized as those corresponding to the standard simplex.

Theorem 6.11 *Let D be an $n \times n$ EDM. Then D has exactly two distinct eigenvalues if and only if $D = \gamma(E - I)$ for some positive scalar γ.*

Proof.   The sufficiency part is immediate since the eigenvalues of $E - I$ are $n - 1$ with multiplicity 1 and (-1) with multiplicity $n - 1$. To prove the necessity part, let D have two distinct eigenvalues, say $\lambda > 0$ and $-\gamma < 0$. Then λ is the Perron eigenvalue and $(-\gamma)$ has multiplicity $n - 1$. Moreover, $\lambda = (n-1)\gamma$ since $\mathrm{trace}(D) = 0$. Let x be the normalized Perron eigenvector of D. Then $D = \gamma((n-1)xx^T - W_1 W_1^T)$ is the spectral decomposition of D. But $W_1 W_1^T = I - xx^T$, thus $D = \gamma(nxx^T - I)$. Furthermore, since $\mathrm{diag}(D) = \mathbf{0}$, it follows that $nx_i^2 = 1$ for $i = 1 \dots, n$ and hence $x = e/\sqrt{n}$. Therefore, $D = \gamma(E - I)$.

$\square$

There is no complete characterization of EDMs with exactly three distinct eigenvalues. Nevertheless, few partial results are known.

Theorem 6.12 *Let D be an $n \times n$ EDM. Then D is singular with strength 2 if and only if the distinct eigenvalues of D are $\lambda = e^T De/n$, 0 with multiplicity $n - 1 - k$ and $(-\alpha)$ with multiplicity k, where $1 \leq k \leq n - 2$.*

Proof.   Assume that the distinct eigenvalues of D are $\lambda = e^T De/n$, 0 with multiplicity $n - 1 - k$ and $(-\alpha)$ with multiplicity k, where $1 \leq k \leq n - 2$. Then $\lambda = k\alpha$ and D is obviously regular and singular. Moreover, the spectral decomposition of D is $D = \lambda E/n - V\Lambda V^T$, where $\Lambda = \begin{bmatrix} \alpha I_k & \mathbf{0} \\ \mathbf{0} & \mathbf{0} \end{bmatrix}$. Therefore, $D^2 = -\alpha D + \lambda(\lambda + \alpha)E/n$ and hence D satisfies (6.22).

To prove the other direction, assume that D is singular with strength 2. Then the spectral decomposition of D is $D = \lambda E/n - V\Lambda V^T$, where $\Lambda \succeq \mathbf{0}$. Moreover, (6.22) implies that $V\Lambda^2 V^T = \alpha V\Lambda V^T$ and hence, $\Lambda^2 = \alpha\Lambda$, i.e., $\Lambda_i(\Lambda_i - \alpha) = 0$ for $i = 1, \dots, n$. But $\Lambda \neq \mathbf{0}$ and $\Lambda \neq \alpha I$ since $\mathrm{diag}(D) = \mathbf{0}$ and D is singular. Therefore, the nonpositive eigenvalues of D are $(-\alpha)$ with multiplicity k and 0 with multiplicity $n - 1 - k$ for some $k : 1 \leq k \leq n - 2$.

$\square$

Recall that the standard simplex EDM $D = E - I$ has strength t for all positive integer t. Hence, it has strength 2 and has two distinct eigenvalues namely, $n - 1$ and (-1). However, D is nonsingular. The following theorem follows from a more general result of Neumaier. Recall that the degree of an EDM is the number of its distinct off-diagonal entries.

Theorem 6.13 (Neumaier [152]) *Let D be an $n \times n$ EDM of strength 2 and of degree 2 or 3. Then any two rows (columns) of D are obtained from each other by a permutation.*

Proof. Assume that the degree of D is 3 and let its distinct off-diagonal entries be β_1, β_2, and β_3. The proof of the case where the degree is 2 is similar.
  For $k = 1, 2, 3$, let $A_k = (a_{ij}^k)$ be the $(0 - 1)$ matrix such that

$$a_{ij}^k = \begin{cases} 1 \text{ if } d_{ij} = \beta_k, \\ 0 \text{ otherwise.} \end{cases}$$

Thus, $D = \beta_1 A_1 + \beta_2 A_2 + \beta_3 A_3$ and $A_1 + A_2 + A_3 = E - I$. Moreover, since D has strength 2, we have $De = \lambda e$ and

$$[D]^2 e = \operatorname{diag}(D^2) = \frac{1}{n} \lambda (\lambda + \alpha) e. \tag{6.26}$$

Thus,

$$\begin{aligned}
(E - I)e &= A_1 e + A_2 e + A_3 e &&= (n-1)e, \\
De &= \beta_1 A_1 e + \beta_2 A_2 e + \beta_3 A_3 e &&= \lambda e, \\
[D]^2 e &= \beta_1^2 A_1 e + \beta_2^2 A_2 e + \beta_3^2 A_3 e &&= (\lambda(\lambda + \alpha)/n)e.
\end{aligned}$$

Therefore, by grouping together the ith equation from each of the above three systems we obtain for $i = 1, \ldots, n$

$$\begin{bmatrix} 1 & 1 & 1 \\ \beta_1 & \beta_2 & \beta_3 \\ \beta_1^2 & \beta_2^2 & \beta_3^2 \end{bmatrix} \begin{bmatrix} (A_1 e)_i \\ (A_2 e)_i \\ (A_3 e)_i \end{bmatrix} = \begin{bmatrix} n - 1 \\ \lambda \\ \lambda(\lambda + \alpha)/n \end{bmatrix}.$$

This system is Vandermonde since the β's are distinct and hence its solution is unique. Therefore, for $k = 1, 2, 3$, $(A_k e)_i$ is independent of i and thus $A_k e = \gamma_k e$. That is, for $k = 1, 2, 3$, the entry β_k appears exactly γ_k times in each row. Note that $\gamma_1 + \gamma_2 + \gamma_3 = n - 1$.

$\square$

Example 6.10 *Let D be the regular EDM considered in Example 6.9. The eigenvalues of D are $8, 0, -4, -4$. Thus, $\alpha = 4$ and $\lambda = 8$. Moreover, $\beta_1 = 2$, $\beta_2 = 4$ and*

$$A_1 = \begin{bmatrix} 0 & 1 & 0 & 1 \\ 1 & 0 & 1 & 0 \\ 0 & 1 & 0 & 1 \\ 1 & 0 & 1 & 0 \end{bmatrix} \text{ and } A_2 = \begin{bmatrix} 0 & 0 & 1 & 0 \\ 0 & 0 & 0 & 1 \\ 1 & 0 & 0 & 0 \\ 0 & 1 & 0 & 0 \end{bmatrix}.$$

Clearly, $A_1 e = 2e$ and $A_2 e = e$.

To illustrate the proof of Theorem 6.13 for this case, note that

$$\begin{bmatrix} 1 & 1 \\ 2 & 4 \end{bmatrix} \begin{bmatrix} (A_1 e)_i \\ (A_2 e)_i \end{bmatrix} = \begin{bmatrix} 3 \\ 8 \end{bmatrix} \ \text{and thus} \ \begin{bmatrix} (A_1 e)_i \\ (A_2 e)_i \end{bmatrix} = \frac{1}{2} \begin{bmatrix} 4 & -1 \\ -2 & 1 \end{bmatrix} \begin{bmatrix} 3 \\ 8 \end{bmatrix} = \begin{bmatrix} 2 \\ 1 \end{bmatrix}.$$

So far, we considered regular EDMs with three distinct eigenvalues. Next, we use Theorem 3.17 to construct nonregular EDMs with three distinct eigenvalues from the simplex EDM $D = E - I$.

Theorem 6.14 *Let $t \geq (n-1)/2n$, $t \neq 1$. Then*

$$D' = \begin{bmatrix} 0 & te^T \\ te & E_n - I_n \end{bmatrix}$$

is a nonregular EDM with exactly three distinct eigenvalues, namely μ_1, μ_2, each with multiplicity 1, and (-1) with multiplicity $n-1$, where

$$\mu_1 = \frac{1}{2}\left(n - 1 + \sqrt{(n-1)^2 + 4nt^2}\right)$$

and

$$\mu_2 = \frac{1}{2}\left(n - 1 - \sqrt{(n-1)^2 + 4nt^2}\right).$$

Proof. By Theorem 3.17, D' is an EDM for all $t\sqrt{n} \geq (n-1)/2\sqrt{n}$, i.e., for all $t \geq (n-1)/2n$. Let

$$Q = \begin{bmatrix} 0 & 1 & 0 \\ e/\sqrt{n} & 0 & V \end{bmatrix}.$$

Then

$$Q^T D' Q = \begin{bmatrix} n-1 & tn^{1/2} & 0 \\ tn^{1/2} & 0 & 0 \\ 0 & 0 & -I_{n-1} \end{bmatrix}$$

and the result follows.

$\square$

Observe that if $t = 1$ in the previous theorem, then D' is the standard simplex EDM of order $n + 1$. Also, observe that in this case, D' has two distinct eigenvalues since $\mu_1 = n$ and $\mu_2 = -1$.

6.6 The EDM Inverse Eigenvalue Problem

Let (C) be a given class of matrices and let $\lambda_1, \ldots, \lambda_n$ be given scalars. The (C) inverse eigenvalue problem [55] is the problem of constructing a matrix in (C) with eigenvalues $\lambda_1, \ldots, \lambda_n$, or proving that no such matrix exists. In particular, if C is the EDM cone $\mathscr{D}^n$, then the corresponding problem is known as the EDM inverse eigenvalue problem (EDMIEP). There is no satisfactory solution for the EDMIEP for all n. More specifically, if n is an integer for which a Hadamard matrix of order n exists, then there is an elegant simple solution for such order [105]. However, for other orders, only partial results, based on Theorems 3.17, 3.18, 3.19 and 3.20, are known [105, 117, 151]. In what follows, we present the Hadamard matrix based solution.

A *Hadamard matrix* H_n is a $(1, -1)$-matrix of order n that satisfies

$$H_n^T H_n = nI.$$

For example, $H_2 = \begin{bmatrix} 1 & 1 \\ 1 & -1 \end{bmatrix}$ and $H_4 = H_2 \otimes H_2 = \begin{bmatrix} H_2 & H_2 \\ H_2 & -H_2 \end{bmatrix}$ are two Hadamard matrices. In fact, it is an immediate consequence of the definition that the Kronecker product of two Hadamard matrices is a Hadamard matrix. Consequently, Hadamard matrices of orders 2^k exist for every nonnegative integer k.

Another consequence of the definition is that Hadamard matrices are closed under the multiplication of any column or row by (-1). As a result, we can assume wlog that e is the first column of H_n. This implies that $n = 1, 2$ or $4k$, for some positive integer k, is a necessary condition for the existence of H_n. Whether this condition is also sufficient is an open problem. However, it has been a long-standing open conjecture that there exists a Hadamard matrix H_n for every $n = 4k$. The smallest n for which the existence of H_n is in doubt is $n = 668$ [120].

The EDM inverse eigenvalue problem is solved for all n for which a Hadamard matrix of order n exists.

Theorem 6.15 (Hayden et al. [105]) *Assume that n is a positive integer for which a Hadamard matrix of order n exists. Let $\lambda_1 > 0 \geq \lambda_2 \geq \cdots \geq \lambda_n$ be scalars such that $\sum_{i=1}^n \lambda_i = 0$. Then there exists a regular EDM with eigenvalues $\lambda_1, \ldots, \lambda_n$.*

Proof. Assume that $H_n = [e \ \bar{H}]$ is a Hadamard matrix and let $Q = H_n / \sqrt{n}$. Let $\Lambda = \text{Diag}(\lambda_1, \ldots, \lambda_n)$ and let $D = Q \Lambda Q^T = (\lambda_1 e e^T + \bar{H} \bar{\Lambda} \bar{H}^T)/n$ where $\bar{\Lambda} = \text{Diag}(\lambda_2, \ldots, \lambda_n)$. Then obviously D and Λ are cospectral since Q is orthogonal. Observe that $D_{ii} = \sum_{k=1}^n \lambda_k (q_{ik})^2 = (\sum_{k=1}^n \lambda_k)/n = 0$ and thus $\text{diag}(D) = \mathbf{0}$. Moreover, $\bar{H} = VA$ for some nonsingular A since the columns of $\bar{H}$ is a basis of $e^\perp$. Consequently, $\mathcal{T}_V(D) = -V^T(\lambda_1 e e^T + VA\bar{\Lambda}A^T V^T)V/(2n) = -A\bar{\Lambda}A^T/(2n)$ is PSD. Moreover, $e^T De/n = \lambda_1$. As a result, D is a regular EDM with eigenvalues $\lambda_1, \ldots, \lambda_n$.

$\square$

Chapter 7
The Entries of EDMs

This chapter focuses on two problems concerning the individual entries of an EDM. The first problem is how to determine a missing or an unknown entry of an EDM. We present two methods for solving this problem, the second of which yields a complete closed-form solution. The second problem is how far an entry of an EDM can deviate from its current value, assuming all other entries are kept unchanged, before the matrix stops being an EDM. We present explicit formulas for the intervals, within which, entries can vary, one at a time, if the matrix is to remain an EDM. Moreover, we present a characterization of those entries whose intervals have zero length; i.e., those entries where any deviation from their current values renders the matrix non-EDM.

7.1 Determining One Missing Entry of an EDM

Suppose that one entry of an EDM of order $n + 2$ is missing or unknown. By re-labelling the points if necessary, we can assume wlog that the missing entry is in positions $(1, n+2)$ and $(n+2, 1)$. More precisely, suppose that

$$\tilde{D} = \begin{bmatrix} 0 & d^T & \alpha \\ d & D & c \\ \alpha & c^T & 0 \end{bmatrix}$$

is an $(n+2) \times (n+2)$ EDM, where D is of order n, and where α is an unknown scalar. We first assume that $n \geq 2$ and later we consider the case $n = 1$. Let

$$D_1 = \begin{bmatrix} 0 & d^T \\ d & D \end{bmatrix} \quad \text{and} \quad D_2 = \begin{bmatrix} D & c \\ c^T & 0 \end{bmatrix}.$$

Then obviously, D, D_1, and D_2 are EDMs of embedding dimensions, say, r, r_1, and r_2, respectively. Clearly, $r \leq r_1, r_2 \leq r + 1$.

© Springer Nature Switzerland AG 2018
A.Y. Alfakih, *Euclidean Distance Matrices and Their Applications in Rigidity Theory*, https://doi.org/10.1007/978-3-319-97846-8_7

We present two methods for determining α. The first one is algorithmic and rather intuitive. However, in case there are multiple values of α, this method has the disadvantage of yielding only one such value. The second method is rather cumbersome, but it obtains a complete closed-form solution of the problem of finding α. It should be pointed out that a proof of the existence of a solution for this problem is given in [31] together with a condition for its uniqueness. Also, it should be pointed out that this problem has interesting consequences for the EDM completion problem to be discussed in the next chapter.

7.1.1 The First Method for Determining α

Assume that $D \neq 0$ and let

$$B_1 = -\frac{1}{2}(I - e(e^2)^T)D_1(I - e^2 e^T)$$

and

$$B_2 = -\frac{1}{2}(I - e(e^1)^T)D_2(I - e^1 e^T)$$

be the Gram matrices of D_1 and D_2, where e^1 and e^2 are the first two standard unit vectors in $\mathbb{R}^{n+1}$. Further, let

$$P_1 = \begin{bmatrix} (p^1)^T \\ (p^2)^T \\ \vdots \\ (p^{n+1})^T \end{bmatrix} = \begin{bmatrix} (p^1)^T \\ P \end{bmatrix} \text{ and } P_2 = \begin{bmatrix} (p'^2)^T \\ \vdots \\ (p'^{n+1})^T \\ (p'^{n+2})^T \end{bmatrix} = \begin{bmatrix} P' \\ (p'^{n+2})^T \end{bmatrix}$$

be $(n+1) \times r_1$ and $(n+1) \times r_2$ configuration matrices of D_1 and D_2 obtained by factorizing $B_1 = P_1 P_1^T$ and $B_2 = P_2 P_2^T$. Observe that, as a result of the above choice of the projection matrix on $e^\perp$, the origin, in both systems of coordinates, is fixed at the second point, i.e., $p^2 = 0$ and $p'^2 = 0$.

Note that P and P' are two "configuration matrices"[1] of D in $\mathbb{R}^{r_1}$ and $\mathbb{R}^{r_2}$, where $D = \mathcal{K}(PP^T) = \mathcal{K}(P'P'^T)$. Also, note that r_1, r_2, and r may be different. As a result, we consider three cases:

Case 1: $r_1 = r_2 = r$. In this case, p^1 lies in the affine span of $\{p^2, \ldots, p^{n+1}\}$ and p'^{n+2} lies in the affine span of $\{p'^2, \ldots, p'^{n+1}\}$. Since the origin is fixed at the second point, it follows that $P' = PQ$ for some orthogonal matrix Q. That is, P' is obtained from P by an orthogonal transformation. Consequently,

---

[1] Recall that a configuration of D is obtained by factorizing $\mathcal{T}(D)$, the Gram matrix of D. Hence, all configuration matrices of D have a full column rank which is equal to the rank of $\mathcal{T}(D)$. As a result, P and P' are not configuration matrices of D in the technical sense. Nonetheless, $D = \mathcal{K}(PP^T)$ and $D = \mathcal{K}(P'P'^T)$.

$$P_1 Q = \begin{bmatrix} (p^1)^T Q \\ PQ \end{bmatrix} = \begin{bmatrix} (p^1)^T Q \\ P' \end{bmatrix}$$

is a configuration matrix of D_1. Therefore, aligning the common points in both configuration matrices implies that

$$\tilde{P} = \begin{bmatrix} (p^1)^T Q \\ P' \\ (p'^{n+2})^T \end{bmatrix}$$

is a configuration matrix of $\tilde{D}$. Consequently,

$$\alpha = ||p^1||^2 + ||p'^{n+2}||^2 - 2(p^1)^T Q p'^{n+2}. \tag{7.1}$$

Observe that $||p^1||^2 = \tilde{D}_{2,1} = d_1$ and $||p'^{2}||^2 = \tilde{D}_{2,(n+1)} = c_1$ since the origin is fixed at the second point.

Case 2: $r_1 \neq r_2$. Wlog assume that $r_1 < r_2$. Then obviously, $r_1 = r$ and $r_2 = r + 1$. Hence, in this case, p^1 lies in the affine span of $\{p^2, \ldots, p^{n+1}\}$, while p'^{n+2} is not in the affine span of $\{p'^2, \ldots, p'^{n+1}\}$.

Let U' be the matrix whose columns form an orthonormal basis of $\text{null}(P')$ and let W' be the matrix such that $[W' \ U']$ is orthogonal. Observe that U' is $n \times 1$. Since configuration matrices are closed under multiplication from the right with an orthogonal matrix, it follows that the $n \times r$ matrix $P'W'$ is a configuration matrix of D. Hence, there exists an orthogonal matrix Q such that $P'W' = PQ$. As a result,

$$P_1 Q = \begin{bmatrix} (p^1)^T Q \\ PQ \end{bmatrix} = \begin{bmatrix} (p^1)^T Q \\ P'W' \end{bmatrix} \text{ and } P_2[W' \ U'] = \begin{bmatrix} P'W' & 0 \\ (p'^{n+2})^T W' & (p'^{n+2})^T U' \end{bmatrix}$$

are configuration matrices of D_1 and D_2, respectively. Again, aligning the common points in both configuration matrices implies that

$$\tilde{P} = \begin{bmatrix} (p^1)^T Q & 0 \\ P'W' & 0 \\ (p'^{n+2})^T W' & (p'^{n+2})^T U' \end{bmatrix}.$$

is a configuration matrix of $\tilde{D}$. Consequently, α is given by

$$\alpha = ||p^1||^2 + ||p'^{n+2}||^2 - 2(p^1)^T Q W'^T p'^{n+2}. \tag{7.2}$$

Case 3: $r_1 = r_2 = r + 1$. Let U and U' be the matrices whose columns form orthonormal bases of $\text{null}(P)$ and $\text{null}(P')$, respectively. Observe that U and U' are $n \times 1$ since the embedding dimension of D is r. Let W and W' be the matrices such that the matrices $[W \ U]$ and $[W' \ U']$ are orthogonal. Hence, the $n \times r$ matrices PW and $P'W'$ are configuration matrices of D. Therefore, there exists an orthogonal matrix Q such that $P'W' = PWQ$. Consequently,

$$P_1 [W \ U] \begin{bmatrix} Q & 0 \\ 0 & 1 \end{bmatrix} = \begin{bmatrix} (p^1)^T WQ & (p^1)^T U \\ PWQ & 0 \end{bmatrix} = \begin{bmatrix} (p^1)^T WQ & (p^1)^T U \\ P'W' & 0 \end{bmatrix}$$

and

$$P_2 [W' \ U'] = \begin{bmatrix} P'W' & 0 \\ (p'^{n+2})^T W' & (p'^{n+2})^T U' \end{bmatrix}$$

are configuration matrices of D_1 and D_2, respectively. Again, by aligning the common points in both configuration matrices, it follows that

$$\tilde{P} = \begin{bmatrix} (p^1)^T WQ & (p^1)^T U \\ P'W' & 0 \\ (p'^{n+2})^T W' & (p'^{n+2})^T U' \end{bmatrix}.$$

is a configuration matrix of $\tilde{D}$. Thus, α is given by

$$\alpha = ||p^1||^2 + ||p'^{n+2}||^2 - 2(p^1)^T (WQW'^T + UU'^T) p'^{n+2}. \tag{7.3}$$

Example 7.1 *Consider the EDM*

$$\tilde{D} = \begin{bmatrix} 0 & 1 & 2 & 5 & \alpha \\ 1 & 0 & 1 & 4 & 1 \\ 2 & 1 & 0 & 1 & 2 \\ 5 & 4 & 1 & 0 & 5 \\ \alpha & 1 & 2 & 5 & 0 \end{bmatrix},$$

where α is unknown. Then the embedding dimensions of D_1, D_2, and D are $r_1 = 2$, $r_2 = 2$, and $r = 1$, respectively. Hence, Case 3 applies. Then

$$P_1 = \begin{bmatrix} (p^1)^T \\ P \end{bmatrix} = \begin{bmatrix} 0 & 1 \\ 0 & 0 \\ 1 & 0 \\ 2 & 0 \end{bmatrix} \text{ and } P_2 = \begin{bmatrix} P' \\ (p^5)^T \end{bmatrix} \begin{bmatrix} 0 & 0 \\ -1/\sqrt{2} & -1/\sqrt{2} \\ -\sqrt{2} & -\sqrt{2} \\ -1/\sqrt{2} & 1/\sqrt{2} \end{bmatrix}.$$

Consequently,

$$[W \ U] = I_2 \text{ and } [W' \ U'] = \frac{1}{\sqrt{2}} \begin{bmatrix} 1 & 1 \\ 1 & -1 \end{bmatrix}.$$

Therefore,

$$PW = \begin{bmatrix} 0 \\ 1 \\ 2 \end{bmatrix}, P'W' = \begin{bmatrix} 0 \\ -1 \\ -2 \end{bmatrix} \text{ and thus } Q = -1.$$

As a result,

$$\tilde{P} = \begin{bmatrix} 0 & 1 \\ 0 & 0 \\ -1 & 0 \\ -2 & 0 \\ 0 & -1 \end{bmatrix}.$$

Therefore, $\alpha = 4$. Note that α is not unique. In fact, α can be any scalar between 0 and 4.

7.1.2 The Second Method for Determining α

This second method, unlike the first, obtains a complete closed-form solution of the problem of finding α. To this end, Theorem 3.3 implies that D_1 is an EDM iff $ed^T + de^T - D$ is PSD, where, as always, e denotes the vector of all 1's in $\mathbb{R}^n$. Recall that, in the first method, the origin was fixed at the second point. Here we fix the origin at the centroid of the generating points of D. Thus, $B = -JDJ/2$ and $X = V^T B V$ are the Gram and the projected Gram matrices of D. Let $X = W \Lambda W^T$ be the spectral decomposition of X where Λ is the diagonal matrix consisting of the r positive eigenvalues of X. Also, let U be the matrix whose columns form an orthonormal basis of null(X). Note that $\Lambda = W^T V^T P P^T V W$, where $P = V W \Lambda^{1/2}$ is a configuration matrix of D. Define the two orthogonal matrices

$$Q_1 = [V \ \frac{e}{\sqrt{n}}] \text{ and } Q_2 = \begin{bmatrix} W & U & 0 \\ 0 & 0 & 1 \end{bmatrix}.$$

Then $Q_2^T Q_1^T (ed^T + de^T - D) Q_1 Q_2$ is equal to

$$\begin{bmatrix} 2\Lambda & 0 & \sqrt{n} \, W^T V^T (d - De/n) \\ 0 & 0 & \sqrt{n} \, U^T V^T (d - De/n) \\ \sqrt{n} \, (d - De/n)^T V W & \sqrt{n} \, (d - De/n)^T V U & 2 e^T d - e^T De/n \end{bmatrix}.$$

Therefore, since D_1 is an EDM and in light of Lemma 3.8, it follows that

$$Z^T (d - \frac{De}{n}) = 0, \tag{7.4}$$

where Z is a Gale matrix of D, and

$$2e^T d - \frac{e^T De}{n} - \frac{n}{2} (d - \frac{De}{n})^T B^\dagger (d - \frac{De}{n}) \geq 0, \tag{7.5}$$

where we have used Schur complement and the fact that $B^\dagger = V W \Lambda^{-1} W^T V^T$. It is important to note that equality holds in (7.5) if and only if $r_1 = r$.

Similarly, since D_2 is an EDM, it follows that

$$Z^T (c - \frac{De}{n}) = 0, \tag{7.6}$$

and

$$2e^T c - \frac{e^T De}{n} - \frac{n}{2} (c - \frac{De}{n})^T B^\dagger (c - \frac{De}{n}) \geq 0, \tag{7.7}$$

where equality holds in (7.7) if and only if $r_2 = r$. An immediate consequence of (7.4) and (7.6) is that

$$Z^T (d - c) = 0.$$

Now Theorem 3.3 implies that $\tilde{D}$ is an EDM iff

$$\begin{bmatrix} de^T + ed^T - D & d - c + \alpha e \\ d^T - c^T + \alpha e^T & 2\alpha \end{bmatrix} \succeq 0.$$

Define the two orthogonal matrices

$$Q_1' = \begin{bmatrix} V & e/\sqrt{n} & 0 \\ 0 & 0 & 1 \end{bmatrix} \text{ and } Q_2' = \begin{bmatrix} W & U & 0 \\ 0 & 0 & I_2 \end{bmatrix}.$$

Then, $\tilde{D}$ is an EDM if and only if

$$Q_2'^T Q_1'^T \begin{bmatrix} de^T + ed^T - D & d - c + \alpha e \\ d^T - c^T + \alpha e^T & 2\alpha \end{bmatrix} Q_1' Q_2' \succeq 0. \tag{7.8}$$

But (7.8), using Schur complement, is equivalent to

$$\begin{bmatrix} nf(d,d) & \sqrt{n}\,(\alpha - g) \\ \sqrt{n}\,(\alpha - g) & 2\alpha - h \end{bmatrix} \succeq 0, \tag{7.9}$$

where

$$f(x,y) = \frac{e^T (x+y)}{n} - \frac{e^T De}{n^2} - \frac{1}{2}(x - \frac{De}{n})^T B^\dagger (y - \frac{De}{n}), \tag{7.10}$$

$$g = \frac{1}{2}(d - \frac{De}{n})^T B^\dagger (d - c) - \frac{e^T (d - c)}{n},$$

and

$$h = \frac{1}{2}(d - c)^T B^\dagger (d - c). \tag{7.11}$$

The function $f(x,y)$ plays an important role in determining α. Moreover, it is straightforward to verify that g and h can be expressed in terms of $f(x,y)$ as follows.

$$g = f(d,c) - f(d,d). \tag{7.12}$$

$$h = 2f(d,c) - f(d,d) - f(c,c). \tag{7.13}$$

An immediate consequence of (7.5) and (7.7) is that $f(d,d) \geq 0$ and $f(c,c) \geq 0$.
    Similar to the first method, we have to consider three cases.

Case 1: $r_1 = r_2 = r$. In this case, it follows from (7.5) and (7.7) that $f(d,d) = f(c,c) = 0$. Thus $g = h/2$. As a result, it follows from (7.9) that $\tilde{D}$ is an EDM if and only if

$$\alpha = g = \frac{h}{2} = f(d,c).$$

Therefore, in this case, α is unique and the matrix in (7.9) is zero. Hence, the embedding dimension of $\tilde{D} = r$.

Case 2: $r_1 = r$ and $r_2 = r+1$. Hence, in this case, $f(d,d) = 0$ and $f(c,c) > 0$. Thus $g > h/2$. Consequently, $\tilde{D}$ is an EDM if and only if

$$\alpha = g = f(d,c).$$

Therefore, in this case, α is also unique and the matrix in (7.9) has rank 1. Hence, the embedding dimension of $\tilde{D} = r+1$.

Case 3: $r_1 = r_2 = r+1$. Hence, in this case $f(d,d) > 0$ and $f(c,c) > 0$. Thus, by Schur complement, (7.9) holds iff

$$2\alpha - h - \frac{1}{f(d,d)}(\alpha - g)^2 \geq 0$$

or

$$\alpha^2 - 2\alpha(g + f(d,d)) + g^2 + f(d,d)h \leq 0. \qquad (7.14)$$

In light of (7.12) and (7.13), the quadratic inequality (7.14) reduces to

$$\alpha^2 - 2\alpha f(d,c) + f(d,c)^2 - f(d,d)f(c,c) \leq 0.$$

Therefore, the roots are:

$$\alpha_l = f(d,c) - \sqrt{f(d,d)f(c,c)} \text{ and } \alpha_u = f(d,c) + \sqrt{f(d,d)f(c,c)}.$$

Observe that (7.11) implies that $h \geq 0$. Thus $2f(c,d) \geq f(d,d) + f(c,c)$. Moreover, by the arithmetic mean-geometric mean inequality, we have

$$f(c,d) \geq \sqrt{f(d,d)f(c,c)}$$

Hence, $\alpha_l \geq 0$. Therefore, $\tilde{D}$ is an EDM if and only if $\alpha_l \leq \alpha \leq \alpha_u$.

So far, we have considered the case where D is of order $n \geq 2$. We now consider the case $n = 1$. Hence, $D = 0$ and thus

$$\tilde{D} = \begin{bmatrix} 0 & d & \alpha \\ d & 0 & c \\ \alpha & c & 0 \end{bmatrix}$$

is 3×3 and d and c are scalars. Using triangular inequality, it is immediate that $\tilde{D}$ is an EDM iff

$$(\sqrt{d} - \sqrt{c})^2 \leq \alpha \leq (\sqrt{d} + \sqrt{c})^2.$$

However, in this case, $B^\dagger = 0$ and hence $f(d,c) = c+d$, $f(d,d) = 2d$ and $f(c,c) = 2c$. Consequently,

$$f(d,c) - \sqrt{f(d,d)f(c,c)} = (\sqrt{d} - \sqrt{c})^2 \text{ and } f(d,c) + \sqrt{f(d,d)f(c,c)} = (\sqrt{d} + \sqrt{c})^2.$$

As a result, we have proved the following theorem.

Theorem 7.1 *Let*

$$D_1 = \begin{bmatrix} 0 & d^T \\ d & D \end{bmatrix} \text{ and } D_2 = \begin{bmatrix} D & c \\ c^T & 0 \end{bmatrix}$$

be two given $(n+1) \times (n+1)$ *EDMs of embedding dimensions* r_1 *and* r_2. *Assume that* D *is of order* n *and of embedding dimension* r. *Let* $B = -JDJ/2$ *be the Gram matrix of* D *and let*

$$\tilde{D} = \begin{bmatrix} 0 & d^T & \alpha \\ d & D & c \\ \alpha & c^T & 0 \end{bmatrix}.$$

Further let

$$f(x,y) = \frac{e^T(x+y)}{n} - \frac{e^T De}{n^2} - \frac{1}{2}(x - \frac{De}{n})^T B^\dagger (y - \frac{De}{n}).$$

Then $\tilde{D}$ *is an EDM if and only if* $\alpha_l \le \alpha \le \alpha_u$, *where*

$$\alpha_l = f(d,c) - \sqrt{f(d,d)f(c,c)} \text{ and } \alpha_u = f(d,c) + \sqrt{f(d,d)f(c,c)}.$$

Moreover, (i) the embedding dimension of $\tilde{D} = r$ *if and only if* $f(d,d) = f(c,c) = 0$, *(ii)* $f(d,d) = 0$ *if and only if* $r_1 = r$, *and (iii)* $f(c,c) = 0$ *if and only if* $r_2 = r$.

Example 7.2 *Consider the matrix* $\tilde{D}$ *of Example 7.1. Then* $c = d$ *and* $r_1 = r_2 = r+1$.

$$B^\dagger = \frac{1}{4} \begin{bmatrix} 1 & 0 & -1 \\ 0 & 0 & 0 \\ -1 & 0 & 1 \end{bmatrix}.$$

Therefore, $f(c,d) = f(d,d) = f(c,c) = 2$ *and hence* $\alpha_l = 0$ *and* $\alpha_u = 4$.

7.2 Yielding and Nonyielding Entries of an EDM

We saw in the previous section how to recover a missing entry of an EDM. Now suppose that a given entry of an EDM D is allowed to vary, while all other entries are kept fixed. In this section, we are interested in determining the interval within which this entry can vary if D is to remain an EDM. Clearly, depending on D and the given entry, this interval can have a zero or a nonzero length.

More precisely, let E^{ij} denote the $n \times n$ symmetric matrix with 1's in the (i,j) and (j,i) positions and zeros elsewhere. Further, let $l_{ij} \le 0$ and $u_{ij} \ge 0$ be the two scalars such that $D + tE^{ij}$ is an EDM if and only if $l_{ij} \le t \le u_{ij}$. That is, D remains an EDM iff its entry in the (i,j) and (j,i) positions varies between $d_{ij} + l_{ij}$ and $d_{ij} + u_{ij}$, while all other entries are kept unchanged. The closed interval $[l_{ij}, u_{ij}]$ is called the *yielding interval* of entry d_{ij}. Furthermore, entry d_{ij} is said to be *unyielding* if its yielding interval has zero length, i.e., if $l_{ij} = u_{ij} = 0$. Otherwise, if the yielding interval of an entry d_{ij} has a nonzero length, i.e., if $l_{ij} \ne u_{ij}$, then d_{ij} is said to be *yielding*. Gale transform, which we revisit next, plays a crucial role in establishing whether a given entry of an EDM is yielding or unyielding.

7.2.1 The Gale Matrix Z Revisited

In addition to the properties of Gale transform discussed in Chap. 3, more useful properties are given in this subsection. We begin, first, with the following simple proposition and its corollary.

Proposition 7.1 *Let D be an $n \times n$, $n \geq 3$, EDM of embedding dimension $r \leq n - 2$. Let Z and P be, respectively, a Gale matrix and a configuration matrix of D, where $P^T e = 0$. Then*

$$null(P^T) \cap null(Z^T) = col(e).$$

Proof.   This is an immediate consequence of the definition of Z.

□

Corollary 7.1 (Alfakih [15]) *Let D be an $n \times n$, $n \geq 3$, EDM of embedding dimension $r \leq n - 2$. Let Z and P be, respectively, a Gale matrix and a configuration matrix of D, where $P^T e = 0$. Let i and j be two distinct indices in $\{1, \ldots, n\}$.*

1. *If $z^i = 0$, then $p^i \neq 0$ and p^i is not in the affine hull of $\{p^1, \ldots, p^n\} \backslash \{p^i\}$.*
2. *If $z^i = z^j = 0$, then $p^i \neq 0$, $p^j \neq 0$ and $p^i - cp^j \neq 0$ for all scalars c.*
3. *If $z^i \neq 0$, $z^j \neq 0$ and $z^i = c'z^j$ for some nonzero scalar c', then $p^i - c'p^j \neq 0$.*

Proof.   To prove part 1, assume that $z^i = 0$. Then e^i, the ith standard unit vector in $\mathbb{R}^n$, lies in $null(Z^T)$. Now, by way of contradiction, assume that $p^i = 0$. Then e^i is also in $null(P^T)$ and hence $e^i \in null(P^T) \cap null(Z^T)$, a contradiction since $e^i \neq e$ ($n \geq 3$).

For ease of notation and wlog assume that $i = 1$, i.e., $z^1 = 0$. Now assume, to the contrary, that p^1 lies in the affine hull of $\{p^2, \ldots, p^n\}$. Then there exist scalars $\lambda_2, \ldots, \lambda_n$ such that

$$\begin{bmatrix} p^1 \\ 1 \end{bmatrix} = \sum_{i=2}^{n} \lambda_i \begin{bmatrix} p^i \\ 1 \end{bmatrix}.$$

Let $x = [-1 \ \lambda_2 \ \cdots \ \lambda_n]^T$. Thus, $P^T x = 0$ and $e^T x = 0$ and hence there exists $\xi \neq 0$ in $\mathbb{R}^{n-r-1}$ such that $x = Z\xi$. Consequently, $-1 = (z^1)^T \xi$, a contradiction. Therefore, p^1 is not in the affine hull of $\{p^2, \ldots, p^n\}$.

To prove part 2, assume wlog that $z^1 = z^2 = 0$. Then by part 1, $p^1 \neq 0$ and $p^2 \neq 0$. Assume, by way of contradiction, that $p^1 - cp^2 = 0$ for some scalar c and let x be the vector in $\mathbb{R}^n$ such that $x = [1 \ -c \ 0]^T$. Then x lies in $null(P^T) \cap null(Z^T)$ and $x \neq e$ since x has at least one zero entry ($n \geq 3$). Thus, we have a contradiction.

To prove part 3, assume wlog that $z^1 \neq 0$, $z^2 \neq 0$ and $z^1 = c'z^2$ for some scalar c'. Assume, by way of contradiction, that $p^1 - c'p^2 = 0$ and let x be the vector in $\mathbb{R}^n$ such that $x = [1 \ -c' \ 0]^T$. Then $x \in null(P^T) \cap null(Z^T)$ and hence we have a contradiction since $x \neq e$.

□

It should be pointed out that in part 3 of Corollary 7.1, p^i may be parallel to p^j, say $p^i = c p^j$, but c cannot be equal to c' as illustrated in the following example.

Example 7.3 *Consider the EDM* $D = \begin{bmatrix} 0 & 0 & 1 & 1 \\ 0 & 0 & 1 & 1 \\ 1 & 1 & 0 & 4 \\ 1 & 1 & 4 & 0 \end{bmatrix}$ *of embedding dimension* 1.

A configuration matrix of D is $P = \begin{bmatrix} 0 \\ 0 \\ -1 \\ 1 \end{bmatrix}$. *Thus, a Gale matrix of D is*

$Z = \begin{bmatrix} -2 & 0 \\ 0 & -2 \\ 1 & 1 \\ 1 & 1 \end{bmatrix}$. *Note that* $z^3 = z^4$ *and* $p^3 = -p^4$; *i.e, p^3 is parallel to p^4 but* $c \neq c'$.

Next, we characterize the yielding entries of an EDM in terms of Gale transform.

7.2.2 Characterizing the Yielding Entries

We consider two cases, depending on whether or not the generating points of D are affinely independent. Let r be the embedding dimension of the $n \times n$ EDM D.

We begin, first, with the case where $r = n - 1$; i.e., the case where the generating points of D are affinely independent.

Theorem 7.2 *Let D be an $n \times n$ EDM of embedding dimension $r = n - 1$. Then every entry of D is yielding.*

Proof.   Let $1 \leq k < l \leq n$. Then $D + tE^{kl}$ is an EDM iff $2X - tV^T E^{kl} V \succeq 0$, where X is the projected Gram matrix of D and V is as defined in (3.11). But, X is PD since X is of order $n - 1$ and $\text{rank}(X) = r = n - 1$. Thus, obviously, there exists $t \neq 0$ such that $2X - tV^T E^{kl} V \succeq 0$. Consequently, d_{kl} is yielding and the result follows.

□

Next, we consider the case where $r \leq n - 2$; i.e., the case where the generating points of D are affinely dependent. The following lemma is crucial for our results.

Lemma 7.1 *Let D be an $n \times n$ nonzero EDM of embedding dimension $r \leq n - 2$, and let Z and P be a Gale matrix and a configuration matrix of D, respectively, where $P^T e = 0$. Further, let X be the projected Gram matrix D. Then $2X - tV^T E^{kl} V \succeq 0$ iff*

$$\begin{bmatrix} 2(P^T P)^2 - t\left(p^k(p^l)^T + p^l(p^k)^T\right) & -t\left(p^k(z^l)^T + p^l(z^k)^T\right) \\ -t\left(z^k(p^l)^T + z^l(p^k)^T\right) & -t\left(z^k(z^l)^T + z^l(z^k)^T\right) \end{bmatrix} \succeq 0.$$

Proof. Let W and U be the two matrices whose columns form orthonormal bases of $\text{col}(X)$ and $\text{null}(X)$, respectively, and thus $Q = [W \ U]$ is orthogonal. Hence, $2X - tV^T E^{kl} V \succeq 0$ iff

$$Q^T (2X - tV^T E^{kl} V) Q = \begin{bmatrix} 2W^T XW - t\, W^T V^T E^{kl} VW & -t\, W^T V^T E^{kl} VU \\ -t\, U^T V^T E^{kl} VW & -t\, U^T V^T E^{kl} VU \end{bmatrix} \succeq 0.$$

But, it follows from Lemma 3.8 that $VU = ZA$ and $VW = PA'$, where A and A' are nonsingular. Hence, $2X - tV^T E^{kl} V \succeq 0$ iff

$$\begin{bmatrix} 2(P^T P)^2 - t\,(p^k(p^l)^T + p^l(p^k)^T) & -t\,(p^k(z^l)^T + p^l(z^k)^T) \\ -t\,(z^k(p^l)^T + z^l(p^k)^T) & -t\,(z^k(z^l)^T + z^l(z^k)^T) \end{bmatrix} \succeq 0.$$

□

Note that $(P^T P)^2$ is PD since P has full column rank. As the following theorem shows, the yielding entries of D are characterized in terms of Gale transform.

Theorem 7.3 (Alfakih [15]) *Let D be an $n \times n$ nonzero EDM of embedding dimension $r \leq n - 2$, and let $z^1, \ldots, z^n$ be Gale transforms of the generating points of D. Then entry d_{kl} is yielding if and only if z^k is parallel to z^l; i.e., iff there exists a nonzero scalar c such that $z^k = cz^l$.*

Proof. Let $1 \leq k < l \leq n$. Entry d_{kl} is yielding iff there exists $t \neq 0$ such that $D + tE^{kl}$ is an EDM or equivalently, iff $2X - tV^T E^{kl} V \succeq 0$.

Assume that z^k is parallel to z^l, i.e., $z^k = cz^l$ for some nonzero scalar c. Then $z^k(z^l)^T + z^l(z^k)^T = 2cz^l(z^l)^T \succeq 0$ and $p^k(z^l)^T + p^l(z^k)^T = (p^k + cp^l)(z^l)^T$. Hence, $\text{null}(z^l(z^l)^T) = \text{null}((z^l)^T) \subseteq \text{null}((p^k + cp^l)(z^l)^T)$. Consequently, it follows from Lemma 7.1 that there exists $t \neq 0$ such that $2X - tV^T E^{kl} V \succeq 0$ and therefore d_{kl} is yielding.

To prove the reverse direction, assume that z^k and z^l are not parallel and assume, to the contrary, that entry d_{kl} is yielding. Therefore, there exists $t \neq 0$ such that $2X - tV^T E^{kl} V \succeq 0$. Thus, it follows from Lemma 7.1 that there exists $t \neq 0$ such that $-tz^k(z^l)^T + z^l(z^k)^T \succeq 0$ and $\text{null}(z^k(z^l)^T + z^l(z^k)^T) \subseteq \text{null}(p^k(z^l)^T + p^l(z^k)^T)$. We consider two cases.

Case 1: $z^k = 0$ and $z^l \neq 0$. Thus $z^k(z^l)^T + z^l(z^k)^T = 0$. Moreover, by Corollary 7.1 (part 1), $p^k \neq 0$ and thus $p^k(z^l)^T + p^l(z^k)^T) = p^k(z^l)^T \neq 0$. Hence, we have a contradiction since $\text{null}(0) \not\subseteq \text{null}(p^k(z^l)^T)$.

Case 2: Both z^k and z^l are nonzero. Again, in this case we have a contradiction since Proposition 1.2 implies that $z^k(z^l)^T + z^l(z^k)^T$ is indefinite. As a result, d_{kl} is unyielding.

□

Example 7.4 *Let $\tilde{D}$ be the EDM considered in Example 7.1 with $\alpha = 4$. Then a Gale matrix of $\tilde{D}$ is*

$$Z = \begin{bmatrix} 0 & 1 \\ 1 & -2 \\ -2 & 0 \\ 1 & 0 \\ 0 & 1 \end{bmatrix}.$$

Therefore, entries d_{15} and d_{34} are yielding, while all other entries are unyielding.

When the generating points of D are in general position, Theorem 7.3 implies the following two corollaries.

Corollary 7.2 ([15]) *Let D be an $n \times n$ EDM of embedding dimension $r = n - 2$. Then D is in general position in $\mathbb{R}^r$ if and only if every entry of D is yielding.*

Proof.   In this case, $z^1, \ldots, z^n$ are scalars since $\bar{r} = n - r - 1 = 1$. Assume that D is in general position. Then it follows from Corollary 3.1 that $z^1, \ldots, z^n$ are nonzero. Thus, z^k is parallel to z^l for all $1 \leq k < l \leq n$, and hence every entry of D is yielding.

To prove the other direction, assume that one entry of D say, d_{kl}, is unyielding. Then z^k is not parallel to z^l. Thus, either $z^k = 0$ or $z^l = 0$, but not both. Therefore, Corollary 3.1 implies that D is not in general position.

$\square$

Corollary 7.3 ([15]) *Let D be an $n \times n$ EDM of embedding dimension $r \leq n - 3$. If D is in general position in $\mathbb{R}^r$, then every entry of D is unyielding.*

Proof.   Assume, to the contrary, that one entry of D, say d_{kl}, is yielding. Thus, it follows from Theorem 7.3 that $z^k = cz^l$ for some nonzero scalar c. Note that in this case, $\bar{r} = n - r - 1 \geq 2$. Hence, any $\bar{r} \times \bar{r}$ submatrix of Z containing $(z^k)^T$ and $(z^l)^T$ is singular. This contradicts Corollary 3.1 and the proof is complete.

$\square$

Observe that if a matrix is nonsingular, then, obviously, every two of its columns (rows) are linearly independent; i.e., nonparallel. However, the converse is not true; i.e., if every two columns (rows) of a matrix are linearly independent, then this matrix may be singular. Consequently, the converse of Corollary 7.3 is not true.

Example 7.5 *Consider the EDM* $D = \begin{bmatrix} 0 & 1 & 4 & 9 & 1 \\ 1 & 0 & 1 & 4 & 0 \\ 4 & 1 & 0 & 1 & 1 \\ 9 & 4 & 1 & 0 & 4 \\ 1 & 0 & 1 & 4 & 0 \end{bmatrix}$ *of embedding dimension 1. A configuration matrix and a Gale matrix of D are*

$$P = \frac{1}{5} \begin{bmatrix} -7 \\ -2 \\ 3 \\ 8 \\ -2 \end{bmatrix} \text{ and } Z = \begin{bmatrix} 1 & 0 & 0 \\ -2 & 1 & -1 \\ 1 & -2 & 0 \\ 0 & 1 & 0 \\ 0 & 0 & 1 \end{bmatrix}.$$

Obviously, D is not in general position in $\mathbb{R}^1$ since $p^2 = p^5$. However, every entry of D is unyielding.

Finally, the following important consequence of Theorem 7.2 and Corollaries 7.2 and 7.3 is worth pointing out. If an $n \times n$ EDM D of embedding dimension r is in general position, then the entries of D are either all yielding (if $n = r+1$ or $n = r+2$) or all unyielding (if $n \geq r+3$).

We determine, next, yielding intervals of the yielding entries of an EDM.

7.2.3 Determining Yielding Intervals

Let D be a given EDM and let $B = -JDJ/2$ be its Gram matrix. Further, let P be a configuration matrix of D and hence $P^T e = 0$. It is easy to verify that $B^\dagger = P(P^T P)^{-2} P^T$. Let $B^\dagger = SS^T$; i.e., let $S = P(P^T P)^{-1}$. As we show next, the yielding intervals of D can be expressed in terms of S.

Let d_{kl} be a yielding entry of D. Then either $r = n - 1$ or z^k is parallel to z^l. Consequently, to determine the yielding interval of d_{kl}, we have to consider the following three cases: (i) $r = n - 1$; (ii) $r \leq n - 2$ and $z^k = z^l = 0$; and (iii) $r \leq n - 2$, $z^k \neq 0$, $z^l \neq 0$ and $z^k = cz^l$ for some nonzero scalar c. As we will see below, in the first two cases, 0 lies in the interior of the yielding interval, while in the third case, 0 is an endpoint.

Proposition 7.2 *Let D be an $n \times n$ ($n \geq 3$) EDM matrix of embedding dimension $r = n - 1$ and let P be a configuration matrix of D such that $P^T e = 0$. Let $S = P(P^T P)^{-1}$ and let $(s^i)^T$ be the ith row of S; i.e., $s^i = (P^T P)^{-1} p^i$. Then s^k and s^l are nonzero and nonparallel for all $k \neq l$.*

Proof. Assume, to the contrary, that $s^k = 0$. Then $p^k = 0$ and thus $P^T e^k = 0$, where e^k is the kth standard unit vector in $\mathbb{R}^n$. Since $P^T e = 0$ and since e and e^k are linearly independent, this implies that rank$(PP^T) = r \leq n - 2$, a contradiction. To complete the proof, assume, to the contrary, that $s^k = cs^l$ for some nonzero scalar c, where $k \neq l$. Then $p^k = cp^l$ and thus $P^T(e^k - ce^l) = 0$, and again we have a contradiction.

$\square$

The following theorem establishes the yielding interval in case (i), where $r = n - 1$.

Theorem 7.4 (Alfakih [15]) *Let D be an $n \times n$ ($n \geq 3$) EDM of embedding dimension $r = n - 1$ and let P be a configuration matrix of D such that $P^T e = 0$. Further, let $S = P(P^T P)^{-1}$ and let $(s^i)^T$ be the ith row of S. Then, the yielding interval of entry d_{kl} is given by*

$$\left[\frac{2}{\lambda_r}, \frac{2}{\lambda_1} \right],$$

where $\lambda_1 = (s^k)^T s^l + ||s^k|| \, ||s^l||$ and $\lambda_r = (s^k)^T s^l - ||s^k|| \, ||s^l||$.

Proof. Let $1 \leq k < l \leq n$ and let X be the projected Gram matrix of D. Let $X = W\Lambda W^T$ be the spectral decomposition of X. Thus, Λ is PD and W is orthogonal since $r = n - 1$. Hence, $D + tE^{kl}$ is an EDM if and only if $2X - tV^T E^{kl} V \succeq 0$ iff $2W^T XW - tW^T V^T E^{kl} VW \succeq 0$. But $W^T XW = W^T V^T PP^T VW$. Thus, it follows from Lemma 3.8 that $2W^T XW - tW^T V^T E^{kl} VW \succeq 0$ iff

$$2(P^T P)^2 - tP^T E^{kl} P \succeq 0. \tag{7.15}$$

But Eq. (7.15) holds iff

$$2I_{n-1} - t(P^T P)^{-1} P^T E^{kl} P(P^T P)^{-1} = 2I_{n-1} - tS^T E^{kl} S \succeq 0.$$

In light of Propositions 1.2 and 7.2, let $\lambda_1 > 0 > \lambda_r$ be the nonzero eigenvalues of $S^T E^{kl} S = s^k (s^l)^T + s^l (s^k)^T$. Thus, $2I_{n-1} - tS^T E^{kl} S \succeq 0$ iff $2 - t\lambda_1 \geq 0$ and $2 - t\lambda_r \geq 0$. As a result, $D + tE^{kl}$ is an EDM iff $2/\lambda_r \leq t \leq 2/\lambda_1$.

$\square$

Note that $\lambda_1 = B_{kl}^\dagger + \sqrt{B_{kk}^\dagger B_{ll}^\dagger}$. Consequently, the yielding interval in Theorem 7.4 can also be expressed as

$$\frac{2}{B_{kk}^\dagger B_{ll}^\dagger - B_{kl}^{\dagger 2}} \left[-\sqrt{B_{kk}^\dagger B_{ll}^\dagger} - B_{kl}^\dagger , \sqrt{B_{kk}^\dagger B_{ll}^\dagger} - B_{kl}^\dagger \right], \tag{7.16}$$

where $B = -JDJ/2$. Observe that this yielding interval contains 0 in its interior.

Example 7.6 *Let $D = E - I$ be the EDM of the standard simplex of order n. Then $B = J/2$ and hence $B^\dagger = 2J$. Thus, for any $1 \leq k < l \leq n$, we have $B_{kk}^\dagger = B_{ll}^\dagger = 2(n-1)/n$ and $B^\dagger_{kl} = -2/n$. Consequently, the yielding interval of entry d_{kl} is equal to $[-1 , n/(n-2)]$. As a result,*

$$0 \leq d_{kl} \leq 2\frac{n-1}{n-2}. \tag{7.17}$$

Observe that Interval (7.17) could have been calculated using Theorem 7.1. In fact, in this case $f(c,d) = f(d,d) = f(c,c) = (n-1)/(n-2)$, where we replaced n in Theorem 7.1 with $(n-2)$ to agree with the notation of this example.

Example 7.7 *Let $D = \begin{bmatrix} 0 & 1 & 5/2 \\ 1 & 0 & 5/2 \\ 5/2 & 5/2 & 0 \end{bmatrix}$. Then D is an EDM of embedding dimension 2. A configuration matrix of D is $P = \begin{bmatrix} -1/2 & -1/2 \\ 1/2 & -1/2 \\ 0 & 1 \end{bmatrix}$. Thus, $S = P(P^T P)^{-1} = \begin{bmatrix} -1 & -1/3 \\ 1 & -1/3 \\ 0 & 2/3 \end{bmatrix}$. Hence, $||s^1||^2 = ||s^2||^2 = 10/9$ and $||s^3||^2 = 4/9$. Moreover, $(s^1)^T s^2 = -8/9$ and $(s^1)^T s^3 = (s^2)^T s^3 = -2/9$. Note that*

$$B^\dagger = SS^T = \frac{1}{9} \begin{bmatrix} 10 & -8 & -2 \\ -8 & 10 & -2 \\ -2 & -2 & 4 \end{bmatrix}.$$

Consequently, the yielding interval of d_{12} is

$$[-1\,,\,9],$$

while the yielding interval of d_{13} is equal to the yielding interval of d_{23} is equal to

$$[-\sqrt{10}+1\,,\,\sqrt{10}+1].$$

Note that, in this case, these intervals could have been calculated using triangular inequalities.

Next, we turn to case (ii), where $r \leq n-2$ and $z^k = z^l = \mathbf{0}$.

Theorem 7.5 (Alfakih [15]) *Let D be an $n \times n$ $(n \geq 4)$ EDM of embedding dimension $r \leq n-2$ and let Z and P be a Gale matrix and a configuration matrix of D respectively, where $P^T e = \mathbf{0}$. Further, let $S = P(P^T P)^{-1}$ and let $(s^i)^T$ be the ith row of S. If $z^k = z^l = \mathbf{0}$, then the yielding interval of entry d_{kl} is given by*

$$\left[\frac{2}{\lambda_r}\,,\,\frac{2}{\lambda_1} \right],$$

where $\lambda_1 = (s^k)^T s^l + ||s^k||\,||s^l||$ and $\lambda_r = (s^k)^T s^l - ||s^k||\,||s^l||$.

Proof. It follows from Corollary 7.1 (part 2) that s^k and s^l are nonzero and nonparallel. Moreover, in this case

$$\begin{bmatrix} 2(P^T P)^2 - t\,(p^k(p^l)^T + p^l(p^k)^T) & -t\,(p^k(z^l)^T + p^l(z^k)^T) \\ -t\,(z^k(p^l)^T \mid z^l(p^k)^T) & -t\,(z^k(z^l)^T + z^l(z^k)^T) \end{bmatrix}$$

reduces to

$$\begin{bmatrix} 2(P^T P)^2 - t\,P^T E^{kl} P & \mathbf{0} \\ \mathbf{0} & \mathbf{0} \end{bmatrix}.$$

The proof proceeds along the same line as in the proof of Theorem 7.4.

□

Finally, we turn to case (iii), where $r \leq n-2$, $z^k \neq \mathbf{0}, z^l \neq \mathbf{0}$ and $z^k = cz^l$ for some nonzero scalar c.

Theorem 7.6 (Alfakih [15]) *Let D be an $n \times n$ $(n \geq 3)$ EDM of embedding dimension $r \leq n-2$ and let Z and P be a Gale matrix and a configuration matrix of D respectively, where $P^T e = \mathbf{0}$. Further, let $S = P(P^T P)^{-1}$ and let s^i be the ith row of S, i.e., $s^i = (P^T P)^{-1} p^i$. If both z^k and z^l are nonzero and $z^k = cz^l$ for some nonzero scalar c, then the yielding interval of entry d_{kl} is given by*

$$\left[\frac{-4c}{||s^k - cs^l||^2}\,,\,0 \right] \text{ if } c > 0,$$

and

$$\left[0\,,\ \frac{4\,|c|}{||s^k - cs^l||^2}\right] \ \text{if } c < 0.$$

Proof. Assume that z^k and z^l are nonzero and $z^k = cz^l$, where $c \neq 0$. Let $1 \leq k < l \leq n$ and let X be the projected Gram matrix of D. Then $D + tE^{kl}$ is an EDM iff $2X - tV^T E^{kl} V \succeq 0$. In light of Lemma 7.1, $D + tE^{kl}$ is an EDM iff

$$\begin{bmatrix} 2(P^T P)^2 - t\,(p^k(p^l)^T + p^l(p^k)^T) & -t\,(p^k + cp^l)(z^l)^T) \\ -t\,(z^l(p^k + cp^l)^T) & -t\,2cz^l(z^l)^T \end{bmatrix} \succeq 0. \qquad (7.18)$$

Assume that $r = n - 2$, i.e., $\bar{r} = n - r - 1 = 1$. Then z^l is a nonzero scalar. Now Schur complement implies that Eq. (7.18) holds iff

$$tc \leq 0 \text{ and } 2(P^T P)^2 + \frac{t}{2c}(p^k - cp^l)(p^k - cp^l)^T \succeq 0, \qquad (7.19)$$

which is equivalent to

$$tc \leq 0 \text{ and } 2I_r + \frac{t}{2c}(s^k - cs^l)(s^k - cs^l)^T \succeq 0.$$

This in turn is equivalent to

$$tc \leq 0 \text{ and } 2 + \frac{t}{2c}||s^k - cs^l||^2 \geq 0.$$

The result follows from Corollary 7.1 (part 3) since $s^k - cs^l \neq 0$.

Now assume that $r \leq n - 3$, i.e., $\bar{r} \geq 2$. Let $Q' = [\frac{z^l}{||z^l||} \ M]$ be an $\bar{r} \times \bar{r}$ orthogonal matrix and define the $(n-1) \times (n-1)$ matrix $Q = \begin{bmatrix} I_r & 0 \\ 0 & Q' \end{bmatrix}$. Thus, obviously Q is orthogonal. By multiplying the LHS of Eq. (7.18) from the left with Q^T and from the right with Q we get that $D + tE^{kl}$ is an EDM iff

$$\begin{bmatrix} 2(P^T P)^2 - t\,(p^k(p^l)^T + p^l(p^k)^T) & -t\,(p^k + cp^l)\,||z^l||) \\ -t\,||z^l||\,(p^k + cp^l)^T) & -t\,2c\,||z^l||^2 \end{bmatrix} \succeq 0. \qquad (7.20)$$

Again, using Schur complement we arrive at Eq. (7.19) and thus the proof is complete.
                                                                                                  □

The following example illustrates cases (ii) and (iii).

Example 7.8 *Consider the EDM* $D = \begin{bmatrix} 0 & 0 & 1 & 1 \\ 0 & 0 & 1 & 1 \\ 1 & 1 & 0 & 2 \\ 1 & 1 & 2 & 0 \end{bmatrix}$ *of embedding dimension* 2. *A configuration matrix, a Gale matrix, and the Moore–Penrose inverse of the Gram matrix of D are given by*

$$P = \frac{1}{4} \begin{bmatrix} -1 & -1 \\ -1 & -1 \\ 3 & -1 \\ -1 & 3 \end{bmatrix}, Z = \begin{bmatrix} -1 \\ 1 \\ 0 \\ 0 \end{bmatrix} \text{ and } B^\dagger = \frac{1}{2} \begin{bmatrix} 1 & 1 & -1 & -1 \\ 1 & 1 & -1 & -1 \\ -1 & -1 & 2 & 0 \\ -1 & -1 & 0 & 2 \end{bmatrix}.$$

Thus, entries d_{12} and d_{34} are yielding, while all other entries are unyielding. To calculate the yielding intervals, note that

$$s^1 = s^2 = \frac{1}{2} \begin{bmatrix} 3 & 1 \\ 1 & 3 \end{bmatrix} \frac{1}{4} \begin{bmatrix} -1 \\ -1 \end{bmatrix} = \frac{1}{2} \begin{bmatrix} -1 \\ -1 \end{bmatrix}.$$

Moreover, $z^2 = -z^1$, i.e., $c = -1$. Therefore, the yielding interval of d_{12} is $[0, 2]$. On the other hand, $(s^3)^T s^4 - ||s^3|| \, ||s^4|| = B^\dagger_{34} - \sqrt{B^\dagger_{33} B^\dagger_{44}} = -1$ and $(s^3)^T s^4 + ||s^3|| \, ||s^4|| = 1$. Therefore, the yielding interval of d_{34} is $[-2, 2]$.

We conclude this chapter by remarking that, for two unyielding entries in a column (row), one can define the notion of jointly yielding entries. Such notion is defined and several results are presented in [15].

Chapter 8
EDM Completions and Bar Frameworks

This chapter has three parts. Part one addresses the problem of EDM completions. Part two is an introduction to the theory of bar-and-joint frameworks. Such frameworks, which are interesting in their own right, are particularly useful in the study of various uniqueness notions of EDM completions. In the third part, we discuss stress matrices, which play a pivotal role in the theory of bar-and-joint frameworks. The chapter concludes with the classic Maxwell–Cremona theorem. We begin first with EDM completions.

8.1 EDM Completions

Let $G = (V, E)$ be a simple incomplete connected graph and let $|V(G)| = n$ and $|E(G)| = m$. An $n \times n$ matrix $A = (a_{ij})$ is said to be *symmetric G-partial* if the entry a_{ij} is defined (or specified) if and only if $\{i, j\} \in E(G)$, and $a_{ij} = a_{ji}$ for all $\{i, j\} \in E(G)$. A *G-partial EDM* A is a symmetric G-partial matrix such that for each maximal clique K of G, the principal submatrix of A indexed by the nodes of K is an EDM. Evidently, the diagonal entries of a G-partial EDM are all zeros.

Let A be a given G-partial EDM. Then matrix D is said to be an *EDM completion* of A if: (i) D is an EDM and (ii) $d_{ij} = a_{ij}$ for all $\{i, j\} \in E(G)$; i.e, $\pi(D) = \pi(A)$, where $\pi : \mathscr{S}^n \to \mathbb{R}^m$ is the linear transformation defined in (5.3). The problem of finding an EDM completion of A, or showing that no such completion exists, is called the *EDM completion problem (EDMCP)*. Let r be a given positive integer. The *rEDM completion problem (rEDMCP)* is the EDMCP with the additional requirement that the embedding dimension of $D = r$. The EDMCP is closely related to the positive semidefinite matrix completion problem [98, 128, 129, 130].

Let $G = (V, E, a)$ be a given edge-weighted simple graph, where edge $\{i, j\}$ has a positive weight a_{ij}. A realization of G in $\mathbb{R}^r$ is a mapping of the nodes of G to points in $\mathbb{R}^r$, where node i is mapped to point p^i, such that

$$||p^i - p^j||^2 = a_{ij} \text{ for all } \{i, j\} \in E(G).$$

© Springer Nature Switzerland AG 2018
A.Y. Alfakih, *Euclidean Distance Matrices and Their Applications in Rigidity Theory*, https://doi.org/10.1007/978-3-319-97846-8_8

The problem of finding a realization of G in $\mathbb{R}^r$ is known as the *r-graph realization problem (rGRP)*. Likewise, the problem of finding a realization of G in some Euclidean space is known as the *graph realization problem (GRP)*. Clearly, the *r*EDMCP is equivalent to the *r*GRP and the EDMCP is equivalent to the GRP.

Saxe [164] proved that the *r*GRP, for graphs with positive integer weights, is NP-hard for all $r \geq 1$. Observe that if $G = C_n$, the cycle on n nodes, then the *partition problem* reduces to the 1GRP. This establishes the NP-hardness of the 1GRP since the partition problem is well known to be NP-hard [85]. Saxe also proved that the 1GRP remains NP-hard even if the weights a_{ij}'s are restricted to 1 and 2. It should be pointed out that the NP-hardness of the *r*GRP was independently proven by Yemini [199].

The complexity of the EDMCP is unknown and its membership in NP is open. However, for chordal graphs, the EDMCP can be solved exactly. On the other hand, for general graphs, the EDMCP can be formulated as a semidefinite programming problem (SDP) and thus can be solved approximately, up to any given accuracy, in polynomial time. For more details on the complexity of the EDMCP, see, e.g., [130]. Next, we discuss the EDMCP for chordal graphs.

8.1.1 Exact Completions

Let G be a connected chordal graph. Then there exists a sequence of connected chordal graphs $G = G_0, G_1, \ldots, G_{\bar{m}} = K_n$ such that, for $i = 1, \ldots, \bar{m}$, G_i is obtained from G_{i-1} by adding one new edge [98]. To prove this, assume that $1, \ldots, n$ is a perfect elimination ordering of the nodes of G and assume that $G \neq K_n$. Let $j = \max\{i \in V(G) : \{i,k\} \notin E(G) \text{ for some } k > i\}$. Hence, the set $\{i \in V(G) : i > j\}$, which obviously contains k, induces a clique in G. Let

$$N_G^+(j) = \{i \in V(G) : \{j,i\} \in E(G), i > j\}.$$

Then $N_G^+(j)$ is not empty since G is connected. Moreover, k is adjacent to all nodes in $N_G^+(j)$. Consequently, $G_1 = G \cup \{j,k\}$ is a chordal graph since $1, \ldots, n$ is also a perfect elimination ordering for G_1. Clearly, this process can be repeated until the complete graph K_n is obtained. An example of such sequence is given in Fig. 8.1.

Theorem 8.1 (Bakonyi and Johnson [31]) *Let A be a G-partial EDM, where G is a connected chordal graph. Then A admits an EDM completion.*

Proof.   Assume, wlog, that $1, \ldots, n$ is a perfect elimination ordering of G and let $G = G_0, G_1, \ldots, G_{\bar{m}} = K_n$ be a sequence of chordal graph such that G_i is obtained from G_{i-1} by adding the new edge $\{j_i, k_i\}$ as discussed above. The unspecified entries of A will be determined, one at a time, in the following order: $A_{j_1 k_1}, \ldots, A_{j_{\bar{m}} k_{\bar{m}}}$.

To determine $A_{j_1 k_1}$, wlog, assume that the partial submatrix of A indexed by the nodes $\{j_1\} \cup N_G^+(j_1) \cup \{k_1\}$ is

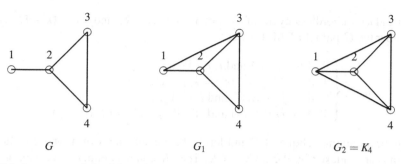

Fig. 8.1 A sequence of chordal graphs

$$\tilde{D} = \begin{bmatrix} 0 & d^T & \alpha \\ d & D & c \\ \alpha & c^T & 0 \end{bmatrix},$$

where α is the unspecified entry $A_{j_1 k_1}$. Then α is determined by using Theorem 7.1. The other unspecified entries $A_{j_2 k_2}, \ldots, A_{j_{\bar{m}} k_{\bar{m}}}$ are determined similarly.

□

Example 8.1 *Let* $A = \begin{bmatrix} 0 & 1 & & \\ 1 & 0 & 1 & 1 \\ & 1 & 0 & 2 \\ & 1 & 2 & 0 \end{bmatrix}$ *be a G-partial EDM, where G is the chordal graph depicted in Fig. 8.1. Then the unspecified entries of A are determined in the following order:* A_{13}, A_{14}. *Applying Theorem 7.1 to the submatrix of A whose rows and columns are induced by* $\{1,2,3\}$, *namely* $\begin{bmatrix} 0 & 1 & \alpha \\ 1 & 0 & 1 \\ \alpha & 1 & 0 \end{bmatrix}$, *yields that* $0 \leq \alpha \leq 4$.

Choose $\alpha = 2$. *Next, we apply Theorem 7.1 to the* G_1-*partial EDM* $\begin{bmatrix} 0 & 1 & 2 & \beta \\ 1 & 0 & 1 & 1 \\ 2 & 1 & 0 & 2 \\ \beta & 1 & 2 & 0 \end{bmatrix}$,

where β *is the unspecified entry* $A_{j_2 k_2}$. *In this case,* $B = \frac{1}{4} \begin{bmatrix} 1 & -1 \\ -1 & 1 \end{bmatrix}$ *and thus* $B^\dagger = \begin{bmatrix} 1 & -1 \\ -1 & 1 \end{bmatrix}$. *Hence,* $f(c,d) = f(d,d) = f(c,c) = 2$. *Consequently,* $0 \leq \beta \leq 4$ *and*

hence $D = \begin{bmatrix} 0 & 1 & 2 & 4 \\ 1 & 0 & 1 & 1 \\ 2 & 1 & 0 & 2 \\ 4 & 1 & 2 & 0 \end{bmatrix}$ *is an EDM completion of A. Obviously, A has no unique EDM completion.*

Bakonyi and Johnson [31] also presented the following example, which shows that for any nonchordal graph G, there exists a G-partial EDM A such that A has no EDM completion. Let G be a nonchordal graph and assume that $1, 2, \ldots, k$ ($k \geq 4$), is an induced chordless cycle of G. Let $V_1 = \{1,\ldots,k\}$ and $V_2 = \{k+1,\ldots,n\}$. Consider the G-partial EDM A, where

$$
a_{ij} = \begin{cases}
0 & \text{if } |i-j| = 1 \text{ and } i,j \in V_1, \\
1 & \text{if } (i=1, j=k \text{ or } i=k, j=1), \\
0 & \text{if } \{i,j\} \in E(G) \text{ and } i,j \in V_2, \\
1 & \text{if } \{i,j\} \in E(G) \text{ and } (i \in V_1, j \in V_2 \text{ or } i \in V_2, j \in V_1).
\end{cases}
$$

Let K be a maximal clique of G and let A_K be the submatrix of A whose rows and columns are indexed by the nodes of K. Then K contains either zero, one, or two nodes of V_1. Now if K does not contain any node of V_1, then $A_K = \mathbf{0}$. On the other hand, if K contains exactly one node in V_1, then $A_K = \begin{bmatrix} 0 & e^T \\ e & \mathbf{0} \end{bmatrix}$. Furthermore, if K contains exactly two nodes in V_1, then either $\{1,k\}$ is a subset of clique K, in which case $A_K = \begin{bmatrix} 0 & 1 & e^T \\ 1 & 0 & e^T \\ e & e & \mathbf{0} \end{bmatrix}$; or $\{1,k\} \not\subset K$, in which case $A_K = \begin{bmatrix} 0 & 0 & e^T \\ 0 & 0 & e^T \\ e & e & \mathbf{0} \end{bmatrix}$. It is easy to see that in all these cases A_K is an EDM. Consequently, A is a G-partial EDM.

Now the only nonzero specified entry in the kth leading principal submatrix of A is 1 in the $(1,k)$ and $(k,1)$ positions. Consequently, for any EDM completion of A, we must have $d_{12} = d_{23} = \cdots = d_{k-1k} = 0$ and $d_{1k} = 1$, an impossibility. Thus, A has no EDM completion.

Next, we consider the EDMCP and the rEDMCP for general graphs.

8.1.2 Approximate Completions

Let A be a G-partial EDM . A fairly intuitive approach for solving the rEDMCP is to formulate it as a global optimization problem. In particular, the rEDMCP can be posed as a global minimization problem where the objective function is

$$
f(P) = \sum_{ij:\{i,j\} \in E(G)} (\|p^i - p^j\|^2 - a_{ij})^2, \tag{8.1}
$$

where $p^1,\ldots,p^n$ are the unknown points in $\mathbb{R}^r$. Clearly, $\bar{P}$ solves the rEDMCP if and only if $f(\bar{P}) = 0$. The disadvantage of this approach is that finding a global minimum of f is an intractable problem since f is not convex and has a large number of local minima. For more details on this approach, see, e.g., [66, 106, 147, 148, 190, 77, 135, 100, 66]. In what follows, we focus on SDP approaches for the r-EDMCP and the EDMCP.

SDP Formulations of the EDMCP

Let A be a G-partial EDM and let H be the adjacency matrix of graph G. Observe that if G is disconnected, then the problem breaks down into at least two independent problems of lower dimensions. As a result, assume that G is connected. Wlog, assign 0 to all unspecified entries of A. Thus $A = \pi^*(\pi(A))$, where π and its adjoint π^* are the linear transformations defined in (5.3) and (5.4).

The following is an intuitive formulation of the EDMCP as an SDP.

$$\begin{aligned} \min \quad & \mathbf{0} \\ \text{subject to} \quad & H \circ \mathscr{K}_V(X) = H \circ A, \\ & X \succeq \mathbf{0} \end{aligned} \qquad (8.2)$$

where $\mathscr{K}_V$ is defined in (3.16) and ($\circ$) denotes the Hadamard product. Note that the feasible region of this problem is convex. SDP problems can be solved, up to any given precision, in polynomial time by interior-point methods [198]. Since the objective function is $\mathbf{0}$, any feasible solution of Problem (8.2) is optimal. Furthermore, if X^* is an optimal solution of Problem (8.2), then $D = \mathscr{K}_V(X^*)$ is an approximate solution of the EDMCP. It should be pointed out that Slater's condition may fail in Problem (8.2) which warrants the use of facial reduction. An SDP formulation for the rEDMCP is obtained by adding the constraint $\text{rank}(X) = r$ to Problem (8.2). Unfortunately, the presence of this rank constraint, in general, makes the feasible region nonconvex and renders the problem intractable.

The following quadratic formulation of the EDMCP, where Slater's condition is guaranteed to hold, was given in [21].

Let $B = \mathscr{T}_V(A)$ where $\mathscr{T}_V$ is defined in (3.17); and for $X \in \mathscr{S}^{n-1}$, let

$$f(X) = \|H \circ (A - \mathscr{K}_V(X))\|_F^2 = \|H \circ \mathscr{K}_V(B - X)\|_F^2, \qquad (8.3)$$

where $\|.\|_F$ denotes the Frobenius norm. Then, by Theorem 3.2, A has an EDM completion if and only if there exists $X \succeq \mathbf{0}$ such that $f(X) = 0$. As a result, the EDMCP can be formulated as the following SDP problem

$$\begin{aligned} \text{(P)} \quad \mu^* = \min \ & f(X) \\ \text{subject to} \quad & X \succeq \mathbf{0}. \end{aligned}$$

Evidently, $\mu^* = 0$ iff A has an EDM completion. Observe that the feasible region of this problem is rather simple since it is precisely $\mathscr{S}_+^{n-1}$, the cone of PSD matrices of order $n-1$. Put differently, all the complications of this problem lie in the objective function. The optimality conditions of Problem (P) are derived next. Let

$$L(X, \Lambda) = f(X) - \text{trace}(X\Lambda) \qquad (8.4)$$

denote the *Lagrangian* of (P). It is easy to see that (P) is equivalent to

$$\mu^* = \min_{X \succeq \mathbf{0}} \max_{\Lambda \succeq \mathbf{0}} L(X, \Lambda) = \min_{X} \max_{\Lambda \succeq \mathbf{0}} L(X, \Lambda).$$

Note that the semidefinite constraint on X can be treated as redundant since the inner max problem is unbounded unless $X \succeq 0$. Also, note that Slater's condition trivially holds for (P). Thus, strong duality holds and

$$\mu^* = \max_{\Lambda \succeq 0} \min_{X \succeq 0} L(X, \Lambda) = \max_{\Lambda \succeq 0} \min_{X} L(X, \Lambda),$$

and μ^* is attained for some $\Lambda \succeq 0$. To obtain the dual problem, since the inner minimization is unconstrained, we differentiate with respect to X to get the equivalent problem

$$\mu^* = \max_{\Lambda \succeq 0, \nabla f(X) - \Lambda = 0.} f(X) - \text{trace}(X\Lambda).$$

Therefore, the dual problem is

$$\text{(D) } \mu^* = \max\ f(X) - \text{trace}(X\Lambda)$$
$$\text{subject to } \nabla f(X) - \Lambda = 0,$$
$$\Lambda \succeq 0 (X \succeq 0).$$

For any two matrices X and Y, $\text{trace}(YH \circ \mathcal{K}_V(X)) = \sum_{\{i,j\} \in E(G)} Y_{ij}(\mathcal{K}_V(X))_{ij} = \text{trace}(H \circ Y \mathcal{K}_V(X)) = \text{trace}(\mathcal{K}_V^*(H \circ Y)X))$. Therefore,

$$\nabla f(X) = 2\mathcal{K}_V^*(H \circ \mathcal{K}_V(X - B)),$$

where $\mathcal{K}_V^*$ is given in Lemma 3.5. Next, we show that Slater's condition holds for the dual problem.

Lemma 8.1 ([21]) *Let H be the adjacency matrix of a connected graph G. Then*

$$\mathcal{K}_V^*(H \circ \mathcal{K}_V(I)) \succ 0$$

Proof.　Clearly, $\mathcal{K}_V(I) = 2(E - I)$ and thus $H \circ \mathcal{K}_V(I) = 2H$. Hence, $\mathcal{K}_V^*(H \circ \mathcal{K}_V(I)) = 4V^T LV$, where $L = \text{Diag}(He) - H$ is the Laplacian of G. But L is PSD and $Le = 0$. Moreover, $\text{rank}(L) = n - 1$ iff H is connected. Therefore, it follows from the spectral decomposition of L that $L = V\Phi V^T$, where Φ is PD. Hence, $\mathcal{K}_V^*(H \circ \mathcal{K}_V(I)) = 4\Phi$.

$\square$

An immediate consequence of this lemma is that Slater's condition holds for the dual since there exists a positive scalar α such that $\bar{X} = B + \alpha I \succ 0$. Therefore, $\mathcal{K}_V^*(H \circ \mathcal{K}_V(\bar{X} - B)) = \bar{\Lambda}$ is PD. Consequently, the optimality conditions are:

$$X \succeq 0 \qquad\qquad \text{primal feasibility,}$$
$$2\mathcal{K}_V^*(H \circ \mathcal{K}_V(X - B)) - \Lambda = 0, \Lambda \succeq 0 \text{ dual feasibility,}$$
$$\text{trace}(X\Lambda) = 0 \qquad\qquad \text{complementarity slackness.}$$

Now consider the following related problem known as the *closest EDM problem (CEDMP)*: Given any matrix A', find the closest EDM to A' in Frobenius norm [21, 24, 25, 23]. It is easy to see that the CEDMP can be formulated as a special case of Problem (P), where $H = E - I$. In other words, the CEDMP can be formulated as

$$\mu^* = \min \ ||A' - \mathscr{K}_V(X)||_F^2$$
$$\text{subject to} \qquad X \qquad \succeq 0.$$

One special case of the EDMCP, which received a great deal of attention recently, is the sensor network localization problem. This problem is discussed next.

The Sensor Network Localization Problem

Consider an ad hoc wireless sensor network in $\mathbb{R}^r$ ($r = 2$ or $r = 3$) consisting of m *anchors* and n *sensors*. The sensors are allowed to move freely, while the anchors have fixed known locations. Hence, the distance between any two anchors is known. On the other hand, the distance between any two sensors or between a sensor and an anchor is known only if it is within a given range. The problem of determining the positions of all the sensors is known as the *sensor network localization problem (SNLP)*. Clearly, the SNLP is a special case of the rEDMCP.

Next, we present two approaches, based on SDP relaxation, for finding an approximate solution of the SNLP. The first approach [71, 122] makes no distinction between anchors and sensors. More precisely, it treats the SNLP as an rEDMCP, ($r = 2$ or 3) where G is a graph on $m + n$ nodes and contains a clique of size m induced by the nodes corresponding to anchors. In other words, the only role anchors play in this approach is to induce a clique in G. Clearly, the presence of a clique in G results in the failure of Slater's condition. This failure is turned into an advantage, via facial reduction, by reducing the size of the problem.

In the second approach [43, 179], the nodes corresponding to the anchors are pinned down, and thus the only coordinates to be considered as those of the sensors. Let $c^1, \ldots, c^m$ be the known coordinates of the anchors and let $p^{m+1}, \ldots, p^{m+n}$ be the unknown coordinates of the sensors. Let $C^T = [c^1 \ \cdots \ c^m]$ and $P^T = [p^{m+1} \ \cdots \ p^{m+n}]$. Assume that the origin is fixed at the centroid of the anchors; i.e., $C^T e_m = 0$. Then the Gram matrix of the anchors and sensors is

$$\begin{bmatrix} C \\ P \end{bmatrix} \begin{bmatrix} C^T & P^T \end{bmatrix} = \begin{bmatrix} C & 0 \\ 0 & I_n \end{bmatrix} \begin{bmatrix} I_r & P^T \\ P & PP^T \end{bmatrix} \begin{bmatrix} C^T & 0 \\ 0 & I_n \end{bmatrix}. \tag{8.5}$$

Let Y be an $(r+n) \times (r+n)$ symmetric matrix partitioned as $Y = \begin{bmatrix} Y_{11} & Y_{12} \\ Y_{12}^T & Y_{22} \end{bmatrix}$, where Y_{11} is of order r. We will identify Y_{12} as P^T and Y_{22} as a relaxation of PP^T; i.e., $Y_{22} - PP^T \succeq 0$. Further, let $E_{12}(G) = \{\{i,j\} \in E(G) : i \leq m, j \geq m+1\}$ and let $E_{22}(G) = \{\{i,j\} \in E(G) : i, j \geq m+1\}$. Then, in this approach [43, 179], the SNLP is formulated as

$$\begin{aligned}
\text{min} \qquad & \text{trace}(\mathbf{0}\,Y) \\
\text{subject to} \qquad & Y_{11} = I_r, \\
& [(c^i)^T \ (-e^j)^T]Y \begin{bmatrix} c^i \\ -e^j \end{bmatrix} = a_{ij} \ \ \text{for all } \{i,j\} \in E_{12}(G), \\
& [\mathbf{0} \ (e^i - e^j)^T]Y \begin{bmatrix} \mathbf{0} \\ e^i - e^j \end{bmatrix} = a_{ij} \ \ \text{for all } \{i,j\} \in E_{22}(G), \\
& Y \succeq \mathbf{0},
\end{aligned} \qquad (8.6)$$

where e^i is the ith standard unit vector in $\mathbb{R}^n$. It should be pointed out that even though Y is of order $r+n$, Slater's condition may still fail in Problem (8.6), which necessitates the use of facial reduction. Also, note that the rank of the optimal solution of Problem (8.6) is $\geq r$. Moreover, the SNLP has an exact solution iff the rank of this optimal solution is r, in which case, $Y_{22} = Y_{12}^T Y_{12} = PP^T$.

The uniqueness of EDM completions [3, 134] is best studied in the context of the rigidity of bar-and-joint frameworks. The remainder of this chapter serves as an introduction to such frameworks, and the remaining chapters of this monograph are dedicated to various notions of rigidity of bar-and-joint frameworks.

8.2 Bar Frameworks

A bar-and-joint framework (a bar framework or a framework for short)[1] (G,p) in $\mathbb{R}^r$ is a simple incomplete connected graph G whose vertices are points $p^1,\ldots,p^n$ in $\mathbb{R}^r$, and whose edges are line segments between pairs of these points. We will refer to $p = (p^1,\ldots,p^n)$ as the *configuration* of (G,p). Framework (G,p) is r-dimensional if its configuration p affinely spans $\mathbb{R}^r$. A bar framework can be viewed as a mechanical *linkage* consisting of rigid bars (edges) and universal joints (vertices). An example of two frameworks is given in Fig. 8.2.

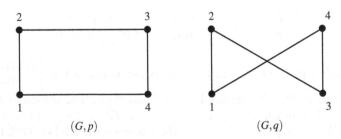

(G,p) $\qquad\qquad\qquad\qquad\qquad$ (G,q)

Fig. 8.2 Two equivalent two-dimensional bar frameworks (G,p) and (G,q) in the plane, where $G = C_4$, the cycle on four nodes

---

[1] In this monograph we are only interested in bar-and joint frameworks.

Observe that if two adjacent nodes of (G, p) coincide, then these two nodes can be merged into a single super node, with all possible multiple edges changed to single edges, to create a new framework with one less node. Thus, wlog, we make the following assumption:

Assumption 8.1 *In any bar framework (G, p), there are no edges of zero length; i.e., no two adjacent nodes coincide.*

Clearly, each framework (G, p) defines an EDM $D_p = (d_{ij} = ||p^i - p^j||^2)$. Furthermore, framework (G, p) also defines a G-partial EDM A in the natural way; i.e., a_{ij} is specified iff $\{i, j\} \in E(G)$, in which case, $a_{ij} = d_{ij}$. In this monograph, we always make the following assumption:

Assumption 8.2 *The configuration of every given r-dimensional bar framework (G, p) is in $\mathbb{R}^r$. That is, the configuration p of (G, p) lies in $\mathbb{R}^r$, where r is the embedding dimension of D_p.*

Two r-dimensional frameworks (G, p) and (G, q) are *congruent* if $D_p = D_q$; i.e., if the two configurations p and q are obtained from each other by a rigid motion (translation, rotation, or reflection). On the other hand, an r-dimensional framework (G, p) is said to be *equivalent* to an s-dimensional framework (G, q), s need not be equal to r, if $H \circ D_p = H \circ D_q$; i.e., if each edge of (G, p) has the same (Euclidean) length as the corresponding edge of (G, q). Observe that (G, p) and (G, q) are equivalent iff $\pi(D_p) = \pi(D_q)$, where π is the linear transformation defined in (5.3). An example of two equivalent two-dimensional frameworks is given in Fig. 8.2. It is a natural problem to characterize all frameworks that are equivalent to a given framework (G, p). This problem is discussed in the following subsection.

8.2.1 The Cayley Configuration Spectrahedron

Viewing framework (G, p) as a mechanical linkage, let the *Cayley configuration space* [176] of (G, p) denote the set of all possible distances between each pair of nonadjacent vertices of (G, p). Let A be the G-partial EDM defined by (G, p). Evidently, characterizing the Cayley configuration space of (G, p) is equivalent to characterizing all EDM completions of A.

Theorem 3.2 implies that (G, p') is equivalent to (G, p) iff

$$H \circ (D_{p'} - D_p) = \mathbf{0} \text{ iff } H \circ \mathcal{K}_V(X' - X) = \mathbf{0}, \tag{8.7}$$

where X' and X are the projected Gram matrices of (G, p') and (G, p), respectively. We begin, first, by finding a basis of the kernel of $H \circ \mathcal{K}_V$. For $i \neq j$, let E^{ij} be the $n \times n$ symmetric matrix with 1's in the (i, j) and (j, i) positions and zeros elsewhere, and let

$$M^{ij} = \mathcal{T}_V(E^{ij}) = -\frac{1}{2}V^T E^{ij} V. \tag{8.8}$$

We will find it convenient to work with $\bar{E}(\bar{G})$, the edge set of the complement graph $\bar{G}$, i.e., the set of missing edges of G.

Lemma 8.2 ([1]) *The set $\{M^{ij} : \{i,j\} \in \bar{E}(\bar{G})\}$ is a basis of the kernel of $H \circ \mathcal{K}_V$.*

Proof.   Lemma 3.6 implies that $\mathcal{K}_V(M^{ij}) = E^{ij}$ since $\mathrm{diag}(E^{ij}) = \mathbf{0}$. Therefore, $H \circ \mathcal{K}_V(M^{ij}) = \mathbf{0}$ for each $\{i,j\} \in \bar{E}(\bar{G})$.

Now let $\sum_{ij} \alpha_{ij} M^{ij} = \mathbf{0}$. Then $\mathcal{K}_V(\sum_{ij} \alpha_{ij} M^{ij}) = \sum_{ij} \alpha_{ij} E^{ij} = \mathbf{0}$. Hence, $\alpha_{ij} = 0$ for all $\{i,j\} \in \bar{E}(\bar{G})$. $\qquad\square$

Let X be the projected Gram matrix of a given framework (G,p). Let $\bar{m}$ denote the cardinality of $\bar{E}(\bar{G})$ and let $\mathcal{X} : \mathbb{R}^{\bar{m}} \to \mathcal{S}^{n-1}$ be the linear transformation such that

$$\mathcal{X}(y) = X + \sum_{\{i,j\}\in\bar{E}(\bar{G})} y_{ij} M^{ij}. \tag{8.9}$$

Let

$$\mathcal{F} = \{y \in \mathbb{R}^{\bar{m}} : \mathcal{X}(y) \succeq \mathbf{0}\}. \tag{8.10}$$

The set $\mathcal{F}$ is called the *Cayely configuration spectrahedron* of (G,p). As the following theorem shows, $\mathcal{F}$ is a translation of the Cayley configuration space of (G,p), and $\mathcal{X}(\mathcal{F})$ is the set of projected Gram matrices of all bar frameworks that are equivalent to (G,p). An important point to keep in mind is that all congruent bar frameworks have the same projected Gram matrix.

Theorem 8.2 (Alfakih [16]) *Let $\mathcal{F}$ be the Cayely configuration spectrahedron of a given r-dimensional bar framework (G,p) and let X' be the projected Gram matrix of bar framework (G,p'). Then (G,p') is equivalent to (G,p) if and only if*

$$X' \in \mathcal{X}(\mathcal{F}),$$

in which case, (G,p') is s-dimensional iff $\mathrm{rank}(X') = s$, and

$$\|p'^i - p'^j\|^2 = \|p^i - p^j\|^2 + y_{ij} \text{ for each } \{i,j\} \in \bar{E}(\bar{G}).$$

Proof.   The first part is an immediate consequence of Lemma 8.2 and Eq. (8.7). To prove the second part, note that

$$\|p'^i - p'^j\|^2 = (\mathcal{K}_V(X'))_{ij} = (\mathcal{K}_V(X))_{ij} + \sum_{\{k,l\}\in\bar{E}(\bar{G})} y_{kl}(\mathcal{K}_V(M^{kl}))_{ij}.$$

But $(\mathcal{K}_V(M^{kl}))_{ij} = (E^{kl})_{ij} = \delta_{ki}\delta_{lj}$. Consequently, $\|p'^i - p'^j\|^2 = \|p^i - p^j\|^2 + y_{ij}$ if $\{i,j\} \in \bar{E}(\bar{G})$. Obviously, $\|p'^i - p'^j\|^2 = \|p^i - p^j\|^2$ if $\{i,j\} \in E(G)$. $\qquad\square$

Evidently, $\mathcal{F}$ is a closed convex set that always contains $\mathbf{0}$ since $\mathcal{X}(\mathbf{0}) = X$ is PSD. Moreover, an immediate consequence of Theorem 8.2 is that $\mathcal{F}$ is bounded if and only G is connected. An example of set $\mathcal{F}$ is given in Fig. 8.3.

The definition of the Cayley configuration spectrahedron given in (8.10) is convenient for theoretical purposes. However, for pencil-and-paper calculations and as we remarked earlier, it is more convenient to use matrix U defined in (3.10). Recall that $V = US$, where S is a nonsingular matrix given in (3.13). Consequently, a simple calculation yields that the Cayely configuration spectrahedron of (G, p) is equivalently given by

$$\mathscr{F} = \{y \in \mathbb{R}^{\bar{m}} : -U^T(D_p + \sum_{\{i,j\} \in \bar{E}(\bar{G})} y_{ij}E^{ij})U \succeq \mathbf{0}, \tag{8.11}$$

where D_p is the EDM defined by framework (G, p).

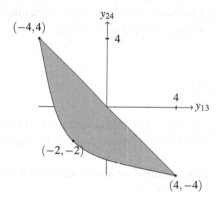

Fig. 8.3 The Cayley configuration spectrahedron of Example 8.2

Example 8.2 *Consider the framework (G, p) depicted in Fig. 8.2,*[2] *where*

$$D_p - \begin{bmatrix} 0 & 1 & 5 & 4 \\ 1 & 0 & 4 & 5 \\ 5 & 4 & 0 & 1 \\ 4 & 5 & 1 & 0 \end{bmatrix}.$$

Thus,

$$-U^T(D_p + y_{13}E^{13} + y_{24}E^{24})U = \begin{bmatrix} 2 & 2+y_{13} & -y_{24} \\ 2+y_{13} & 10+2y_{13} & 8+y_{13} \\ -y_{24} & 8+y_{13} & 8 \end{bmatrix}.$$

Since a symmetric matrix is PSD if and only if all of its principal minors are nonnegative, it is easy to see that the Cayley configuration spectrahedron of (G, p) is given by

---

[2] This example was discussed in Schoenberg [171] from a Cayley–Menger determinant point of view.

$$\mathscr{F} = \{y \in \mathbb{R}^2 : -4 \leq y_{13} \leq 4,$$
$$-4 \leq y_{24} \leq 4,$$
$$(y_{13} + y_{24})(y_{13}y_{24} + 5(y_{13} + y_{24}) + 16) \leq 0\}.$$

Thus $\mathscr{F}$, depicted in Fig. 8.3, is defined by

$$y_{24} \leq -y_{13}, \ and \ y_{24} \geq -\frac{5y_{13} + 16}{y_{13} + 5}.$$

It is worth pointing out that framework (G, q) in Fig. 8.2 corresponds to $y_{13} = y_{24} = -2$ shown in Fig. 8.3. Also, the points $(4, -4)$ and $(-4, 4)$ correspond to the two one-dimensional frameworks obtained by "flattening" (G, p). Finally, it is rather obvious that the origin corresponds to (G, p).

8.3 The Stress Matrix

Stresses and stress matrices play a key role in various rigidity problems of bar frameworks. Stress matrices resemble Laplacians and can be interpreted in various ways.

A *stress* (also called an *equilibrium stress*) of bar framework (G, p) is a real-valued function ω on $E(G)$ such that

$$\sum_{j:\{i,j\}\in E(G)} \omega_{ij}(p^i - p^j) = \mathbf{0} \text{ for each } i = 1, \ldots, n. \tag{8.12}$$

Clearly, the set of stresses is a subspace of $\mathbb{R}^m$. This fact will be elaborated on when we discuss the rigidity matrix in the next chapter. Note that if $\sum_{j:\{i,j\}\in E(G)} \omega_{ij} \neq 0$, then point p^i can be expressed as an affine combination of its neighbors.

Let ω be a stress of framework (G, p). Then the $n \times n$ symmetric matrix Ω, where

$$\Omega_{ij} = \begin{cases} -\omega_{ij} & \text{if } \{i, j\} \in E(G), \\ 0 & \text{if } \{i, j\} \in \bar{E}(\bar{G}), \\ \sum_{k:\{i,k\}\in E(G)} \omega_{ik} & \text{if } i = j, \end{cases} \tag{8.13}$$

is called a *stress matrix* of (G, p). A point to keep in mind is that if Ω is a stress matrix of (G, p), then so is $(-\Omega)$.

Example 8.3 *Consider the bar framework (G, p) depicted in Fig. 8.4. It is easy to see that $\omega_{12} = \omega_{23} = 2$, $\omega_{14} = \omega_{34} = 0$, and $\omega_{13} = -1$ is a stress of (G, p). Hence, the corresponding stress matrix is*

$$\Omega = \begin{bmatrix} 1 & -2 & 1 & 0 \\ -2 & 4 & -2 & 0 \\ 1 & -2 & 1 & 0 \\ 0 & 0 & 0 & 0 \end{bmatrix}.$$

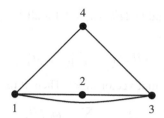

Fig. 8.4 The bar framework of Example 8.3. Edge $\{1,3\}$ is drawn as an arc to make edges $\{1,2\}$ and $\{2,3\}$ visible

It should be noted that the two frameworks depicted in Fig. 8.2 have no nonzero stress.

The following theorem is an immediate consequence of the definition of a stress matrix.

Theorem 8.3 *Let P be a configuration matrix of an r-dimensional bar framework (G, p) and let Ω be a symmetric matrix of order n. Then Ω is a stress matrix of (G, p) if and only if*

$$\Omega e = 0, \Omega P = 0 \text{ and } \Omega_{ij} = 0 \text{ for all } \{i, j\} \in \bar{E}(\bar{G}). \qquad (8.14)$$

Proof. Assume that Ω satisfies (8.14). Then, for all $i = 1, \ldots, n$, we have $\Omega_{ii} = -\sum_{k:\{i,k\}\in E(G)} \Omega_{ik}$. Hence,

$$(\Omega P)_{ij} = \Omega_{ii} p_j^i + \sum_{k:\{i,k\}\in E(G)} \Omega_{ik} p_j^k = \sum_{k:\{i,k\}\in E(G)} \Omega_{ik}(-p^i + p^k)_j = 0 \qquad (8.15)$$

for all $j = 1, \ldots, r$. As a result, $\omega = (\omega_{ij} = -\Omega_{ij})$ is a stress of (G, p) and $\Omega_{ii} = \sum_{k:\{i,k\}\in E(G)} \omega_{ik}$.

On the other hand, assume that Ω is a stress matrix of (G, p). Then the fact that $\Omega e = 0$ and $\Omega_{ij} = 0$ for all $\{i, j\} \in \bar{E}(\bar{G})$ follows immediately from (8.13). Furthermore, (8.13), (8.12), and (8.15) imply that $\Omega P = 0$ and the proof is complete.

□

An immediate consequence of Theorem 8.3 is that the maximum possible rank of a stress matrix is $n - r - 1$ since its null space contains e and the columns of P. Stress matrices of maximal rank play a pivotal role in the following chapters, where rigidity theory of bar frameworks is discussed.

The definition of projected Gram matrices necessitates the definition of projected stress matrices. Thus, the matrix $\Omega' = V^T \Omega V$ is called a *projected stress matrix*. Evidently, $\Omega = V\Omega'V^T$ and thus Ω is PSD of rank k iff Ω' is PSD of rank k. Let X and $B = VXV^T$ be the projected Gram and the Gram matrix of (G, p). Then $V\Omega'XV^T = \Omega B$ and $V^T\Omega BV = \Omega'X$. Hence, $\Omega'X = 0$ iff $\Omega B = 0$. As a result, a symmetric matrix Ω' is a projected stress matrix of (G, p) iff

$$\Omega'X = \mathbf{0} \text{ and } (V\Omega'V^T)_{ij} = 0 \text{ for all } \{i,j\} \in \bar{E}(\bar{G}). \tag{8.16}$$

Let

$$F^{ij} = (e^i - e^j)(e^i - e^j)^T, \tag{8.17}$$

where e^i is the ith standard unit vector in $\mathbb{R}^n$. Then, it readily follows that

$$\Omega = \sum_{\{i,j\}\in E(G)} \omega_{ij}F^{ij},$$

and

$$(\mathscr{K}(B))_{ij} = \text{trace}(F^{ij}B). \tag{8.18}$$

The first interpretation of the stress matrix was already alluded to in (5.6). More precisely, $\Omega = \text{Diag}(\pi^*(\omega)e) - \pi^*(\omega)$, where π^* is defined in (5.4). Hence,

$$\Omega = \frac{1}{2}\mathscr{K}^*(\pi^*(\omega)) \text{ and } \Omega' = \frac{1}{2}\mathscr{K}_V^*(\pi^*(\omega)). \tag{8.19}$$

The second interpretation of the stress matrix will be given after we discuss the relationship between the stress matrix and the Gale matrix.

8.3.1 The Stress Matrix and the Gale matrix

The stress matrix Ω of a bar framework (G,p) has two aspects: a geometric one dictated by the configuration p and a combinatorial one dictated by graph G. Equation (8.14) suggests a close connection between the geometric aspect of Ω and Gale matrix Z. As is shown in the following theorem, these two aspects can be separated by factorizing Ω as $\Omega = Z\Psi Z^T$, where the geometric aspect of Ω is captured in Z and the combinatorial one is captured in Ψ.

Theorem 8.4 (Alfakih [9]) *Let Z be a Gale matrix of an r-dimensional bar framework (G,p) on n nodes, $r \leq n-2$. Then Ω is a stress matrix of (G,p) if and only if*

$$\Omega = Z\Psi Z^T$$

for some symmetric matrix Ψ of order $n - r - 1$ such that

$$(z^i)^T\Psi z^j = 0 \text{ for all } \{i,j\} \in \bar{E}(\bar{G}).$$

Proof.   Assume that Ω is a stress matrix. Then, by Theorem 8.3, $\Omega = ZA$ for some matrix A. But Ω is symmetric. Hence, $\Omega = Z\Psi Z^T$ for some symmetric matrix Ψ. The reverse direction simply follows from Theorem 8.3. $\square$

An immediate consequence of Theorem 8.4 is that $\text{col}(\Omega) \subseteq \text{col}(Z)$. Hence, if Ω has maximal rank, i.e., if $\text{rank}(\Omega) = n-r-1$, then $\text{col}(\Omega) = \text{col}(Z)$ and thus any matrix whose columns form a basis of $\text{col}(\Omega)$ is a Gale matrix of framework (G,p).

Moreover, if in addition (G, p) is in general position, then we have the following lemma, which we will use in Chap. 10.

Lemma 8.3 *Let (G, p) be an r-dimensional bar framework with n nodes, $r \leq n - 2$. Let Ω be a stress matrix of (G, p) of rank $n - r - 1$. If (G, p) is in general position in $\mathbb{R}^r$, then any $n \times (n - r - 1)$ submatrix of Ω is a Gale matrix of (G, p).*

Proof. By the preceding remark, it suffices to show that any $n - r - 1$ columns of Ω are linearly independent. To this end, assume to the contrary that this is not the case and thus, wlog, assume that the first $n - r - 1$ columns of Ω are linearly dependent. Then there exists a nonzero $\lambda \in \mathbb{R}^{n-r-1}$ such that $\Omega x = Z\Psi Z^T x = \mathbf{0}$, where $x^T = [\lambda^T \; \mathbf{0}]$. But Z has full column rank and Ψ is nonsingular. Therefore, $Z^T x = \mathbf{0}$ and thus the first $n - r - 1$ rows of Z are linearly dependent, a contradiction to Corollary 3.1.

$\square$

8.3.2 Properties of PSD Stress Matrices

Evidently, a set of n points can affinely span a space of at most $n - 1$ dimensions, where the maximum dimensional space is obtained when these points are affinely independent. Let (G, p) be an r-dimensional bar framework on n nodes, where $r \leq n - 2$. A natural question to ask is whether there exists an $(n - 1)$-dimensional bar framework (G, q) that is equivalent to (G, p). In other words, it is of interest to know whether (G, p), when viewed as a mechanical linkage, can be flexed to a configuration in which its nodes are affinely independent. The following theorem uses stress matrices to answer this question.

Theorem 8.5 (Alfakih [10]) *Let (G, p) be an r-dimensional bar framework on n nodes, $r \leq n - 2$. Then there exists an $(n - 1)$-dimensional framework (G, q) that is equivalent to (G, p) if and only if there does not exist a nonzero positive semidefinite stress matrix Ω of (G, p).*

Proof. Let X be the projected Gram matrix of (G, p). By Theorem 8.2, there exists an $(n - 1)$-dimensional bar framework (G, q) that is equivalent to (G, p) iff there exists y such that $X + \sum_{\{i,j\} \in \bar{E}(\bar{G})} y_{ij} M^{ij} \succ \mathbf{0}$. But by Corollary 2.3, such y exists iff there does not exist $Y \succeq \mathbf{0}$, $Y \neq \mathbf{0}$, such that $\text{trace}(XY) = 0$ and $\text{trace}(YM^{ij}) = 0$ for all $\{i, j\} \in \bar{E}(\bar{G})$. Now $\text{trace}(XY) = 0$ iff $XY = \mathbf{0}$ iff $Y = U\Psi U^T$ for some $\Psi \succeq \mathbf{0}$, where U is the matrix whose columns form an orthonormal basis of $\text{null}(X)$. Consequently, (G, q) exists iff there does not exist a nonzero $\Psi \succeq \mathbf{0}$ such that $\text{trace}(U\Psi U^T M^{ij}) = 0$ for all $\{i, j\} \in \bar{E}(\bar{G})$. But by Lemma 3.8, VU is a Gale matrix of (G, p), i.e., $VU = Z$. Thus, $\text{trace}(U\Psi U^T M^{ij}) = -(Z\Psi Z^T)_{ij}/2$ and hence the result follows from Theorem 8.4.

$\square$

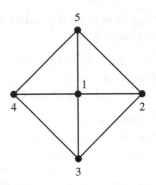

Fig. 8.5 The bar framework of Example 8.4. The set of missing edges is $\bar{E}(\bar{G}) = \{ \{2,4\}, \{3,5\} \}$

Example 8.4 *To illustrate Theorems 8.4 and 8.5, consider the framework* (G,p) *depicted in Fig. 8.5. A Gale matrix of* (G,p) *is*

$$Z = \begin{bmatrix} -2 & -2 \\ 1 & 0 \\ 0 & 1 \\ 1 & 0 \\ 0 & 1 \end{bmatrix}.$$

To find a stress matrix Ω, *we have to find* Ψ *such that* $(z^2)^T \Psi z^4 = (z^3)^T \Psi z^5 = 0$. *Hence,* $\Psi = \begin{bmatrix} 0 & 1 \\ 1 & 0 \end{bmatrix}$. *Consequently,*

$$\Omega = Z\Psi Z^T = \begin{bmatrix} 8 & -2 & -2 & -2 & -2 \\ -2 & 0 & 1 & 0 & 1 \\ -2 & 1 & 0 & 1 & 0 \\ -2 & 0 & 1 & 0 & 1 \\ -2 & 1 & 0 & 1 & 0 \end{bmatrix}.$$

Observe that Ω *is not PSD and* $\operatorname{rank}(\Omega) = \operatorname{rank}(\Psi) = 2$. *Therefore, there exists a four-dimensional bar framework* (G,q) *that is equivalent to* (G,p) *since* (G,p) *admits no nonzero PSD stress matrix. It is worth pointing out here that the Cayley configuration spectrahedron of* (G,p) *is a square; i.e., is polyhedral in this case.*

The following lemma establishes a connection between the stress matrix and the degrees of the node of graph G.

Lemma 8.4 *Let* (G,p) *be an r-dimensional bar framework with n nodes and assume that* (G,p) *is in general position in* $\mathbb{R}^r$. *Let* Ω *be a PSD stress matrix of* (G,p) *of rank* $n-r-1$. *Then* $\deg(i) \geq r+1$ *for every node i of G.*

Proof. By way of contradiction, assume that $\deg(v) \leq r$ for some node v and let $z^{j_1}, \ldots, z^{j_k}$ be Gale transforms of the nodes of (G,p) that are not adjacent to v. Then $k \geq n - 1 - r$ and thus, by Corollary 3.1, $z^{j_1},\ldots,z^{j_k}$ span $\mathbb{R}^{n-1-r}$. Hence, there exist scalars $\lambda_1,\ldots,\lambda_k$ such that $z^v = \sum_{i=1}^{k} \lambda_i z^{j_i}$. Now by Theorem 8.4, $\Omega = Z \Psi Z^T$ where Ψ is PD. Thus $(z^v)^T \Psi z^{j_i} = 0$ for all $i = 1,\ldots,k$. Consequently, $(z^v)^T \Psi z^v = \sum_{i=1}^{k} \lambda_i (z^v)^T \Psi z^{j_i} = 0$ which implies that $z^v \in \text{null}(\Psi)$, a contradiction. □

Lemma 8.4 will be strengthened in Chap. 10 (Lemma 10.5) by dropping the requirement that Ω is PSD.

Now we are ready to present the second interpretation of the stress matrix. Let X be the projected Gram matrix of (G, p) and consider the SDP problem

$$(P) \qquad \min \qquad \mathbf{0}$$
$$\text{subject to } X + \sum_{\{i,j\}\in \bar{E}(\bar{G})} y_{ij} M^{ij} \succeq 0.$$

Thus, the projected Gram matrix of every bar framework (G, q) that is equivalent to (G, p) is an optimal solution of (P) since the objective function is $\mathbf{0}$. The dual problem of (P) is

$$(D) \qquad \max \qquad -\text{trace}(XY)$$
$$\text{subject to } \text{trace}(M^{ij}Y) = 0 \text{ for all } \{i, j\} \in \bar{E}(\bar{G}),$$
$$Y \succeq \mathbf{0}.$$

Let Ω be a PSD stress matrix of (G, p). Then, by Theorem 8.4 and Lemma 3.8, $\Omega = V U \Psi U^T V^T$, where U is the matrix whose columns form an orthonormal basis of $\text{null}(X)$, and $\text{trace}(U \Psi U^T M^{ij}) = 0$ for all $\{i, j\} \in \bar{E}(\bar{G})$. Thus, the projected stress matrix $\Omega' = V^T \Omega V = U \Psi U^T$ is an optimal solution of (D) since $X \Omega' = 0$. A similar observation in the context of sensor networks was made by So and Ye in [178].

The following theorem presents another interesting property of PSD stress matrices.

Theorem 8.6 (Alfakih [10]) *Let Ω be a stress matrix of bar framework (G, p). If Ω is positive semidefinite, then Ω is a stress matrix of every bar framework (G, p') that is equivalent to (G, p).*

Proof. Let (G, p') be equivalent to (G, p) and let X and X' be the projected Gram matrices of (G, p) and (G, p'), respectively. Then $X' = X + \sum_{\{i,j\}\in \bar{E}(\bar{G})} y_{ij} M^{ij}$ for some $y = (y_{ij}) \in \mathbb{R}^{\bar{m}}$. Let $\Omega' = V^T \Omega V$ be the corresponding projected stress matrix. Then, by Eq. (8.16), it suffices to show that $\Omega'X' = \mathbf{0}$. To this end, we have $\Omega'X = \mathbf{0}$ and $\text{trace}(\Omega'M^{ij}) = -\text{trace}(\Omega E^{ij})/2 = -\Omega_{ij}/2 = 0$ for all $\{i, j\} \in \bar{E}(\bar{G})$. Therefore, $\text{trace}(\Omega'X') = 0$ and thus $\Omega'X' = \mathbf{0}$ since both matrices Ω' and X' are PSD. As a result, Ω is a stress matrix of (G, p'). □

The condition that Ω is positive semidefinite cannot be dropped in Theorem 8.6 as shown by the following example.

Example 8.5 Let (G, p) be the framework depicted in Fig. 8.5. Let (G, p') be the *two-dimensional framework obtained from* (G, p) *by folding* (G, p) *across the edges*

$\{1,2\}$ and $\{1,4\}$ so that points 3 and 5 coincide. Clearly, (G,p') is equivalent to (G,p) but Ω is not a stress matrix of (G,p').

Now let (G_2,p) and (G_2,p') be the frameworks obtained from (G,p) and (G,p') by adding edge $\{2,4\}$. Hence, $\{1,2,4\}$ is a clique of G_2. Clearly, (G_2,p) and (G,p) have the same configuration and hence the same Gale matrix Z. To find a stress matrix for (G_2,p), Ψ_2 has to satisfy only $(z^3)^T \Psi_2 z^5 = 0$. Therefore, (G_2,p) admits a PSD stress matrix $\Omega_2 = Z\Psi_2 Z^T = Z_{.1}Z_{.1}^T$ by choosing $\Psi_2 = \begin{bmatrix} 1 & 0 \\ 0 & 0 \end{bmatrix}$. Here, $Z_{.1}$ denotes the first column of Z. One can easily verify that Ω_2 is also a stress matrix of (G_2,p').

The following two corollaries are immediate consequences of Theorem 8.6.

Corollary 8.1 Let (G,p) be an r-dimensional bar framework on n vertices and let Ω be a positive semidefinite stress matrix of (G,p). Assume that $rank(\Omega) = k$ and let (G,p') be an s-dimensional bar framework that is equivalent to (G,p). Then $s \leq n-1-k$.

Proof. Let X' be the projected Gram matrix of (G,p') and let Ω' be the corresponding projected Gram matrix. Then $rank(\Omega') = k$ and $X'\Omega' = \mathbf{0}$. As a result, $rank(X') \leq n-1-k$.   □

Corollary 8.2 Let (G,p) be an r-dimensional bar framework on n vertices and let Ω' be a nonzero positive semidefinite projected stress matrix of (G,p). Further, let $\mathscr{F}$ be the Cayley configuration spectrahedron of (G,p). Then $\mathscr{X}(\mathscr{F})$ is contained in the hyperplane

$$H = \{A \in \mathscr{S}^{n-1} : trace(A\Omega') = 0\}.$$

Proof. Let $X \in \mathscr{X}(\mathscr{F})$, then X is the projected Gram matrix of a bar framework (G,p') that is equivalent to (G,p). Hence, $X\Omega' = \mathbf{0}$.   □

In the theorem that follows, we characterize frameworks with a positive semidefinite stress matrix of rank one.

Theorem 8.7 (Alfakih [2]) Let (G,p) be an r-dimensional bar framework. Then (G,p) admits a positive semidefinite stress matrix of rank one if and only if G has a clique whose nodes are affinely dependent.

Proof. Assume that G has a clique whose nodes are affinely dependent and wlog assume that this clique consists of the nodes $\{1,\ldots,k\}$. Thus, $k \geq r+2$. Let $\lambda = (\lambda_i)$ be a nonzero vector in $\mathbb{R}^k$ such that $\sum_{i=1}^k \lambda_i p^i = \mathbf{0}$ and $\sum_{i=1}^k \lambda_i = 0$. Further, let $\xi = \begin{bmatrix} \lambda \\ \mathbf{0} \end{bmatrix} \in \mathbb{R}^n$. Then, it is easy to see that $\Omega = \xi\xi^T$ is a PSD stress matrix of (G,p) of rank 1.

To prove the reverse direction, assume that $\Omega = \xi\xi^T$ is a nonzero stress matrix of (G,p) and let $\mathscr{I} = \{i : \xi_i \neq 0\}$. Then, $P^T\xi = \mathbf{0}$ and $e^T\xi = 0$ and thus $\sum_{i \in \mathscr{I}} \xi_i p^i = \mathbf{0}$ and $\sum_{i \in \mathscr{I}} \xi_i = 0$. Hence, the points $\{p^i : i \in \mathscr{I}\}$ are affinely dependent. Furthermore, $\{i, j\} \in E(G)$ for all $i, j \in \mathscr{I}$ since $\Omega_{ij} \neq 0$ for all $i, j \in \mathscr{I}$. Thus, the nodes of G induced by $\mathscr{I}$ form a clique and the proof is complete.

$\square$

Example 8.6 *Consider the framework* (G, p) *depicted in Fig. 8.4. Clearly, the nodes* $\{1, 2, 3\}$, *which induce a clique in* G, *are affinely dependent. It is also clear that* (G, p) *admits a PSD stress matrix of rank 1.*

It should be pointed out that a bar framework can admit a PSD stress matrix of rank ≥ 2 without admitting a PSD stress matrix of rank one. See, e.g., the framework (G, p) depicted in Fig. 10.4 of Chap. 10.

8.3.3 The Maxwell–Cremona Theorem

In this subsection, we assume that (G, p) is a three-connected two-dimensional planar bar framework, where no two of its nodes coincide, and where no two of its edges cross. Hence, wlog, we assume that all inner faces of (G, p), as well as the periphery, are convex polygons. Consequently, every edge separates exactly two distinct faces of (G, p). A *polyhedral terrain* is the image (graph) of a piece-wise linear continuous real function of two variables. That is, a polyhedral terrain is a surface in $\mathbb{R}^3$ consisting of connected polygonal faces, and thus it can be represented by a family of affine functions. The Maxwell–Cremona Theorem [141, 142, 192, 193, 65] establishes a correspondence between stressed two-dimensional planar bar frameworks and polyhedral terrains.

Theorem 8.8 (Maxwell–Cremona) *Every polyhedral terrain H that projects to a three-connected two-dimensional planar bar framework* (G, p) *defines a stress* ω *on* (G, p). *Conversely, every stressed three-connected two-dimensional planar bar framework* (G, p) *can be lifted to a polyhedral terrain H which is unique up to the addition of an affine function.*

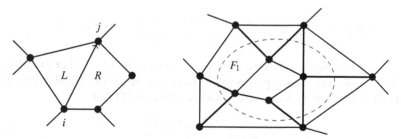

Fig. 8.6 A patch (i, j, L, R) and a Jordan curve in the interior of the faces of a face-cycle

The remainder of this subsection is dedicated to a proof of this theorem (see [193, 65, 111, 159]). First, we begin with a few definitions. A *face-path* of (G, p)

is a sequence of faces of (G, p), say $F_s, F_2, \ldots, F_k, F_t$, such that any two consecutive faces in this sequence have a common edge, and no face is repeated. A *face-cycle* of (G, p) is a face-path that begins and ends at the same face, i.e., $F_s = F_t$.

Evidently, every inner edge $\{i, j\}$ separates exactly two inner faces of (G, p). By orienting the edge $\{i, j\}$, we can denote these two faces by L (left) and R (right) as in Fig. 8.6. The ordered quadruple (i, j, L, R) is called a *patch*. Clearly, $(i, j, L, R) = (j, i, R, L)$. Therefore, a face-path consists of successive patches (i_k, j_k, L_k, R_k) where the common edges $\{i_k, j_k\}$ are properly oriented by the order of the faces in the face-path.

Assume that (G, p) is embedded in the plane $\{p \in \mathbb{R}^3 : p_3 = 1\}$ of $\mathbb{R}^3$. Then (G, p) can be lifted to a polyhedral terrain H by assigning a p_3-coordinate, i.e., a height $h(p^i)$, to each of its nodes such that the nodes of a face remain coplanar. Let H be a polyhedral terrain lifted from (G, p), and let $h(p) = (a^k)^T p + \alpha_k = (\bar{a}^k)^T \bar{p}$ be the restriction of H to face F_k, where

$$\bar{a}^k = \begin{bmatrix} a^k \\ \alpha_k \end{bmatrix} \in \mathbb{R}^3 \text{ and } \bar{p}^i = \begin{bmatrix} p^i \\ 1 \end{bmatrix} \in \mathbb{R}^3.$$

Then, since H is continuous on edge $\{i, j\}$ of patch (i, j, L, R) , it follows that

$$(a^R - a^L)^T (p^i - p^j) = 0, \tag{8.20}$$

that is, $a^R - a^L$ is proportional to $(p^i - p^j)^\perp$. As a result, for any patch (i, j, L, R), let

$$\bar{a}^R = \bar{a}^L + \omega_{ij}(\bar{p}^i \times \bar{p}^j), \tag{8.21}$$

where $(\times)$ denotes the usual cross product. Then $a^R - a^L = \omega_{ij} \begin{bmatrix} (p^i - p^j)_2 \\ -(p^i - p^j)_1 \end{bmatrix}$ and

thus satisfies (8.20). Moreover, let $\bar{p}^0 = \begin{bmatrix} p^0 \\ 1 \end{bmatrix}$ be an arbitrary point that is not collinear with any edge $\{i, j\}$, i.e., the points $\bar{p}^0, \bar{p}^i$ and $\bar{p}^j$ are affinely independent for all $i, j = 1, \ldots, n$. Recall that the signed area of the triangle defined by p^0, p^i, p^j is given by

$$\frac{1}{2} \det([\bar{p}^0 \ \bar{p}^i \ \bar{p}^j]) = -\frac{1}{2} \det\left(\begin{bmatrix} (p^0 - p^i)_1 & (p^0 - p^i)_2 \\ (p^j - p^i)_1 & (p^j - p^i)_2 \end{bmatrix} \right) = -\frac{1}{2}(p^0 - p^i) \times (p^j - p^i).$$

The sign of this area is determined by the usual *right-hand rule*. For example, in a patch (i, j, L, R), this area is positive if p^0 lies in the interior of L and negative if p^0 lies in the interior of R. As a result, it follows from (8.21) that

$$\omega_{ij} = \frac{(\bar{a}^R - \bar{a}^L)^T \bar{p}^0}{\det([\bar{p}^0 \ \bar{p}^i \ \bar{p}^j])}. \tag{8.22}$$

Note that ω_{ij} is the same for patches (i, j, L, R) and (j, i, R, L). Moreover, suppose that $\bar{q}^0$ is chosen in (8.22) instead of $\bar{p}^0$. Then (8.21) and (8.22) imply that

$$(\bar{a}^R - \bar{a}^L)^T \bar{q}^0 = \omega_{ij} \det([\bar{q}^0 \; \bar{p}^i \; \bar{p}^j]) = \frac{\det([\bar{q}^0 \; \bar{p}^i \; \bar{p}^j])}{\det([\bar{p}^0 \; \bar{p}^i \; \bar{p}^j])}(\bar{a}^R - \bar{a}^L)^T \bar{p}^0.$$

Consequently,

$$\frac{(\bar{a}^R - \bar{a}^L)^T \bar{q}^0}{\det([\bar{q}^0 \; \bar{p}^i \; \bar{p}^j])} = \frac{(\bar{a}^R - \bar{a}^L)^T \bar{p}^0}{\det([\bar{p}^0 \; \bar{p}^i \; \bar{p}^j])}.$$

That is, ω_{ij} is independent of the choice of $\bar{p}^0$.

Now let $\mathscr{C}$ be a face-cycle around a node i; i.e., i belongs to every face of $\mathscr{C}$ and let $F_k \in \mathscr{C}$. Then (8.21) implies that

$$\bar{a}^k = \bar{a}^k + \sum_{j:\{i,j\} \in E(G)} \omega_{ij}(\bar{p}^i \times \bar{p}^j).$$

Thus,

$$\sum_{j:\{i,j\} \in E(G)} \omega_{ij}(\bar{p}^i \times \bar{p}^j) = \left(\sum_{j:\{i,j\} \in E(G)} \omega_{ij}(\bar{p}^i - \bar{p}^j) \right) \times \bar{p}^i = \mathbf{0}$$

since $\bar{p}^i \times \bar{p}^j = -\bar{p}^j \times \bar{p}^i$ and $\bar{p}^i \times \bar{p}^i = \mathbf{0}$. But $(\bar{p}^i - \bar{p}^j)_3 = 0$ and $\bar{p}^i_3 = 1$. Therefore,

$$\sum_{j:\{i,j\} \in E(G)} \omega_{ij}(p^i - p^j) = \mathbf{0}.$$

As a result, every polyhedral terrain that projects to (G, p) induces a stress in (G, p) given by (8.22). This proves the first part of the Maxwell–Cremona Theorem.

To prove the second part, we need the following simple observation. Let V_1 be a subset of vertices of (G, p) of cardinality ≥ 2. Then since $\omega_{ij} = \omega_{ji}$, it follows that

$$\sum_{i \in V_1, \; j \in V_1:\{i,j\} \in E(G)} \omega_{ij}(\bar{p}^i \times \bar{p}^j) = \mathbf{0}. \tag{8.23}$$

Lemma 8.5 *Let ω be a stress of (G, p) and assume that $\bar{a}^1$, the vector associated with face F_1, is given. Then Eq. (8.21) consistently assigns vectors $\bar{a}^i$'s to all inner faces F_i's of (G, p).*

Proof. Let F_t be an inner face other than F_1. It suffices to show that the vector $\bar{a}^t$, calculated by successive application of (8.21), is independent of the face-path from F_1 to F_t we choose. Put differently, it suffices to show that $\bar{a}'^1 = \bar{a}^1$ where $\bar{a}'^1$ is the vector obtained by successively applying (8.21) to the face-cycle $\mathscr{C} = F_1, F_2, \ldots, F_1$.

To this end, let $\mathscr{J}$ be a Jordan curve through the interior of the faces of $\mathscr{C}$ and let V_1 and V_2 be the sets of nodes of (G, p) inside and outside $\mathscr{J}$, respectively (see Fig. 8.6). Then

$$\bar{a}'^1 = \bar{a}^1 + \sum_{i \in V_1, \; j \in V_2:\{i,j\} \in E(G)} \omega_{ij}(\bar{p}^i \times \bar{p}^j).$$

Now if $|V_1| = 1$, i.e., if $V_1 = \{i\}$, then

$$\bar{a}'^1 = \bar{a}^1 + \sum_{j:\{i,j\}\in E(G)} \omega_{ij}(\bar{p}^i \times \bar{p}^j)$$

$$= \bar{a}^1 + \left(\sum_{j:\{i,j\}\in E(G)} \omega_{ij}(\bar{p}^i - \bar{p}^j)\right) \times \bar{p}^i$$

$$= \bar{a}^1,$$

where the last equality follows from the definition of a stress ω. Therefore, assume that $|V_1| \geq 2$. But in this case, Eq. (8.23) implies that

$$\sum_{i\in V_1, j\in V_2:\{i,j\}\in E(G)} \omega_{ij}(\bar{p}^i \times \bar{p}^j) = \sum_{i\in V_1, j:\{i,j\}\in E(G)} \omega_{ij}(\bar{p}^i \times \bar{p}^j)$$

$$= \sum_{i\in V_1}\left(\sum_{j:\{i,j\}\in E(G)} \omega_{ij}(\bar{p}^i - \bar{p}^j)\right) \times \bar{p}^i$$

$$= \mathbf{0}.$$

Hence, $\bar{a}'^1 = \bar{a}^1$.

$\square$

Therefore, every two-dimensional planar bar framework with a nontrivial stress can be lifted, using Eq. (8.21), to a nontrivial polyhedral terrain. A polyhedral terrain H is trivial if all faces are coplanar; i.e., H is affine on $\mathbb{R}^2$. Moreover, H is unique up to the addition of an affine function. This proves the second part of the Maxwell–Cremona Theorem.

The following lemma establishes the connection between the signs of the stresses on the inner edges of (G, p) and the local convexity of H.

Lemma 8.6 *Let (i, j, L, R) be a patch and assume that $\omega_{ij} > 0$. Let p^l and p^r be two points in the interiors of L and R, respectively. Then*

$$h(p^l) = (\bar{a}^L)^T \bar{p}^l < (\bar{a}^R)^T \bar{p}^l$$

or equivalently

$$(\bar{a}^L)^T \bar{p}^r > (\bar{a}^R)^T \bar{p}^r = h(p^r),$$

where $\bar{p}^l = \begin{bmatrix} p^l \\ 1 \end{bmatrix}$ and $\bar{p}^r = \begin{bmatrix} p^r \\ 1 \end{bmatrix}$.

Proof. This follows from Eq. (8.21) since

$$(\bar{a}^R)^T \bar{p}^l - (\bar{a}^L)^T \bar{p}^l = \omega_{ij} \det([\bar{p}^l \ \bar{p}^i \ \bar{p}^j]) = -\omega_{ij}(p^l - p^i) \times (p^j - p^i) > 0$$

by the right-hand rule.

$\square$

As a result, the "mountain" ("valley") edges in H correspond to the inner edges of (G, p) with $\omega_{ij} > 0$ ($\omega_{ij} < 0$). It should be pointed out that in the literature, some authors define the stresses ω_{ij}'s with the opposite sign from our definition in (8.21). Consequently, their mountain (valley) stress correspondence is opposite to ours.

Chapter 9
Local and Infinitesimal Rigidities

This chapter focuses on the problems of local rigidity and infinitesimal rigidity of bar frameworks. These problems have a long and rich history going back at least as far as Cauchy [51]. The main tools in tackling these problems are the rigidity matrix R and the dual rigidity matrix $\bar{R}$. While R is defined in terms of the underlying graph G and configuration p, $\bar{R}$ is defined in terms of the complement graph $\bar{G}$ and Gale matrix Z. Nonetheless, both matrices R and $\bar{R}$ carry the same information. The chapter concludes with a discussion of generic local rigidity in dimension 2, where the local rigidity problem reduces to a purely combinatorial one depending only on graph G. The literature on the theory of local and infinitesimal rigidities is vast [59, 57, 66, 97, 194]. However, in this chapter, we confine ourselves to discussing only the basic results and the results pertaining to EDMs.

9.1 Local Rigidity

We start with the definition of local rigidity. Recall that D_p is the EDM defined by configuration p and H is the adjacency matrix of graph G. Also, recall that $(\circ)$ denotes the Hadamard product

Definition 9.1 *Let* (G, p) *be an r-dimensional bar framework. Then* (G, p) *is said to be* locally rigid *if there exists an* $\varepsilon > 0$ *such that there does not exist an r-dimensional bar framework* (G, q) *that satisfies: (i)* $\|q^i - p^i\| \le \varepsilon$ *for all* $i = 1, \ldots, n$, *(ii)* $H \circ D_q = H \circ D_p$ *and (iii)* $D_q \ne D_p$.

In other words, an r-dimensional bar framework (G, p) is locally rigid if there exists a neighborhood of p such that any r-dimensional bar framework (G, q) that is equivalent to (G, p) and within this neighborhood is actually congruent to (G, p). We say that (G, p) is *locally flexible* if it is not locally rigid. Evidently, local rigidity has two aspects: a combinatorial one dictated by graph G and a geometric one dictated by configuration p. Furthermore, it is equally evident that one can find a graph G

© Springer Nature Switzerland AG 2018
A.Y. Alfakih, *Euclidean Distance Matrices and Their Applications in Rigidity Theory*, https://doi.org/10.1007/978-3-319-97846-8_9

and two configurations p and q in $\mathbb{R}^r$ such that (G,p) is locally rigid while (G,q) is locally flexible.

Assume that $r = n - 1$, i.e., framework (G,p) is of dimension $n - 1$ and, as always, assume that $G \neq K_n$, where K_n denotes the complete graph on n nodes. Then X, the projected Gram matrix of (G,p), is PD. Hence, for a sufficiently small $\delta > 0$ and for some $y \in \mathbb{R}^{\bar{m}}$ such that $||y|| \leq \delta$, we have

$$\mathscr{X}(ty) = X + \sum_{\{i,j\} \in \bar{E}(\bar{G})} t y_{ij} M^{ij} \succ 0 \text{ for all } t : 0 \leq t \leq 1.$$

Therefore, (G,p) is locally flexible.

Let $\mathscr{S}_{nr}$ denote the set of real $n \times r$ matrices. Following Asimow and Roth [28, 29], for a graph G on n nodes and m edges, let $f_G = (f_G^{ij}) : \mathscr{S}^{nr} \to \mathbb{R}^m$ be the function defined by

$$f_G^{ij}(P) = ||p^i - p^j||^2 \text{ for each } \{i,j\} \in E(G),$$

where $(p^i)^T$ is the ith row of P. In other words, for framework (G,p),

$$f_G(P) = \pi(D_p),$$

where $\pi : \mathscr{S}^n \to \mathbb{R}^m$ is as defined in (5.3). f_G is called the *rigidity map* of (G,p) or the *edge function* of G. Hence, $f_G^{-1}(f_G(P)) = f_G^{-1}(\pi(D_p))$ is the set of all configurations q in $\mathbb{R}^r$ such that (G,q) is equivalent to (G,p). Moreover, it readily follows that $f_{K_n}^{-1}(f_{K_n}(P)) = f_{K_n}^{-1}(D_p)$ is the set of all configurations q in $\mathbb{R}^r$ such that (G,q) is congruent to (G,p). Clearly

$$f_{K_n}^{-1}(f_{K_n}(P)) \subseteq f_G^{-1}(f_G(P)).$$

Consequently, the structure of $f_G^{-1}(f_G(P))$ in a neighborhood of P is key to establishing the local rigidity or the local flexibility of (G,p). More precisely, (G,p) is locally rigid if and only if there exists a neighborhood W of P in $\mathscr{S}^{nr}$ such that

$$f_G^{-1}(f_G(P)) \cap W = f_{K_n}^{-1}(f_{K_n}(P)) \cap W.$$

We should point out that $f_{K_n}^{-1}(f_{K_n}(P))$ is a smooth manifold. Moreover, $f_G^{-1}(f_G(P))$ is a real algebraic variety and $f_{K_n}^{-1}(f_{K_n}(P))$ is a subvariety of $f_G^{-1}(f_G(P))$. Set S in $\mathscr{S}^{nr}$ is a *real algebraic variety* if it is the zero set of a finite number of polynomials with real coefficients.

Next, we present two other equivalent definitions of local rigidity and hence local flexibility. Let $\delta > 0$ be sufficiently small. A *continuous flex* of (G,p) is a continuous path $\gamma(t)$ for all $t : 0 \leq t \leq \delta$, such that: (i) $\gamma(t)$ is $n \times r$ and (ii) $\gamma(0) = P$. If, in addition, $\gamma(t)$ is analytic, then we say that $\gamma(t)$ is an *analytic flex* of (G,p). As a result, we have the following two definitions of local rigidity in terms of $\gamma(t)$.

Definition 9.2 *Bar framework (G,p) is locally rigid if every continuous flex of (G,p) in $f_G^{-1}(f_G(P))$ lies entirely in $f_{K_n}^{-1}(f_{K_n}(P))$.*

Definition 9.3 *Bar framework (G, p) is locally rigid if every analytic flex of (G, p) in $f_G^{-1}(f_G(P))$ lies entirely in $f_{K_n}^{-1}(f_{K_n}(P))$.*

Theorem 9.1 (Gluck [86]) *The above three definitions of local rigidity are equivalent.*

Proof. Clearly, Definition 9.2 implies Definition 9.3. Now assume that framework (G, p) is locally rigid by Definition 9.1 and assume that (G, p') is an r-dimensional framework such that $P' \in f_G^{-1}(f_G(P)) \backslash f_{K_n}^{-1}(f_{K_n}(P))$. Then $||p'^i - p^i|| > \varepsilon$ for some i. Hence, (G, p) is locally rigid by Definition 9.2.

On the other hand, assume that (G, p) is locally flexible by Definition 9.1. Then for every neighborhood W of P, there exists an r-dimensional framework (G, p') such that $P' \in W$ and $P' \in f_G^{-1}(f_G(P)) \backslash f_{K_n}^{-1}(f_{K_n}(P))$. But $f_G^{-1}(f_G(P))$ is a real algebraic variety and $f_{K_n}^{-1}(f_{K_n}(P))$ is a subvariety of $f_G^{-1}(f_G(P))$. Therefore, by the curve selection lemma (see, e.g., Wallace [191, Lemma 18.3] and Milnor [145, Lemma 3.1]), there exists an analytic flex of (G, p) in $f_G^{-1}(f_G(P)) \backslash f_{K_n}^{-1}(f_{K_n}(P))$ and thus framework (G, p) is locally flexible by Definition 9.3. Hence, Definition 9.3 implies Definition 9.1

$\square$

Figure 9.1 depicts two bar frameworks. Framework (a) is locally flexible, while Framework (b) is locally rigid. Note that Framework (a) is locally flexible since it can be continuously deformed into a family of rhombi.

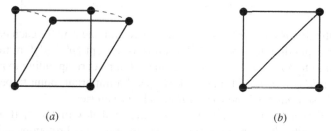

(a) $\qquad\qquad\qquad\qquad\qquad\qquad\qquad\qquad$ (b)

Fig. 9.1 An example of 2 two-dimensional bar frameworks. Framework (**a**) is locally flexible, while framework (**b**) is locally rigid

9.2 Infinitesimal Rigidity and the Rigidity Matrix R

The local rigidity problem turns out to be quite difficult. Therefore, instead of tackling this problem directly, it is sensible to consider the relatively simpler problem of infinitesimal rigidity. Infinitesimal rigidity is a linearized version of local rigidity which readily lends itself to linear algebraic tools.

Consider the process of smoothly deforming a bar framework (G, p) into a one-parameter family of equivalent bar frameworks $(G, q(t))$, with $q(0) = p$, where edges are allowed to pass through one another. It is helpful to think of the parameter t as time. Obviously, during such a process, $||q^i(t) - q^j(t)||^2$ must remain constant for each edge $\{i, j\}$. By differentiating with respect to t and setting $t = 0$, we get

$$(p^i - p^j)^T(\delta^i - \delta^j) = 0 \quad \text{for all } \{i, j\} \in E(G), \tag{9.1}$$

where we have substituted $q^i(0) = p^i$ and $q'^i(0) = \delta^i$. Any nonzero vector $\delta = [\delta^{1^T} \ \dots \ \delta^{n^T}]^T$ in $\mathbb{R}^{nr}$ that satisfies (9.1) is called an *infinitesimal flex* of (G, p). An infinitesimal flex is called *trivial* if it results from a rigid motion of (G, p) and is called *nontrivial* otherwise. Obviously, every bar framework (G, p) has trivial infinitesimal flexes. If (G, p) has only trivial infinitesimal flexes, then it is called *infinitesimally rigid*. Otherwise, i.e., if (G, p) has also nontrivial infinitesimal flexes, then it is called *infinitesimally flexible*.

System of Eq. (9.1) can be written in matrix form as $R\delta = \mathbf{0}$, where R is the $m \times nr$ matrix whose columns and rows are indexed, respectively, by the nodes and the edges of G such that the (i, j)th row is given by

$$\text{edge } \{i, j\} \begin{bmatrix} \vdots & \vdots & \vdots & \vdots & \vdots & \vdots \\ 0 \cdots 0 & (p^i - p^j)^T & 0 \cdots 0 & (p^j - p^i)^T & 0 \cdots 0 \\ \vdots & \vdots & \vdots & \vdots & \vdots \end{bmatrix}. \tag{9.2}$$

More specifically, R has r columns for each node and one row for each edge, where the row corresponding to edge $\{i, j\}$ has all zeros except $(p^i - p^j)^T$ in the columns corresponding to node i and $(p^j - p^i)^T$ in the columns corresponding to node j. R is called the *rigidity matrix* of framework (G, p). An important point to bear in mind is that, by Assumption 8.1, no row of R has all zero entries.

Clearly, δ is a (trivial or nontrivial) infinitesimal flex of (G, p) if and only if $\delta \in \text{null}(R)$. Evidently, there are r translations and $r(r-1)/2$ rotations in $\mathbb{R}^r$. Hence, trivial infinitesimal flexes form a subspace of $\mathbb{R}^{nr}$ of dimension $r(r+1)/2$. Consequently, $\dim(\text{null}(R)) \geq r(r+1)/2$ and thus

$$\text{rank}(R) \leq nr - \frac{1}{2}r(r+1).$$

As a result, (G, p) is infinitesimally rigid if and only if the null space of R consists only of trivial infinitesimal flexes; i.e., $\dim(\text{null}(R)) = r(r+1)/2$.

Theorem 9.2 *Let (G, p) be an r-dimensional bar framework on n nodes and let R be the rigidity matrix of (G, p). Then (G, p) is infinitesimally rigid if and only if*

$$rank(R) = nr - \frac{r(r+1)}{2}.$$

An immediate consequence of Theorem 9.2 is that if $m < nr - r(r+1)/2$, then framework (G, p) is infinitesimally flexible. This is intuitively clear since the fewer edges G has, the less likely that (G, p) is locally rigid.

Next, we establish the relationship between infinitesimal rigidity and local rigidity. The *Jacobian* of the rigidity map f_G at P, denoted by $df_G(P)$, is the $m \times nr$ matrix

$$df_G(P) = (\frac{\partial f^{ij}}{\partial p^k})_{ij=1,\ldots,m}^{k=1,\ldots,n} = 2R,$$

where R is the rigidity matrix. Let

$$k = \max\{\text{rank}(df_G(P)) : P \in \mathscr{S}^{nr}\}.$$

P in $\mathscr{S}^{nr}$ is called a *regular point* of f_G if $\text{rank}(df_G(P)) = \text{rank}(R) = k$ and is called a *singular point* otherwise. Let

$$g(P) = \sum\{(\det(R_{\mathscr{I}\mathscr{J}}))^2 : R_{\mathscr{I}\mathscr{J}} \text{ is a } k \times k \text{ submatrix of } R\}.$$

Thus, P is a regular of f_G if and only if $g(P) \neq 0$. Consequently, the set of regular points of f_G is open and dense in $\mathscr{S}^{nr}$. As a result, "almost all" points of $\mathscr{S}^{nr}$ are regular.

The following theorem is a special form of the well-known implicit function theorem.

Theorem 9.3 (Implicit Parameterization [30, p. 32]) *Let $f_G : \mathscr{S}^{nr} \to \mathbb{R}^m$ and assume that f_G^{ij}, for all $\{i, j\} \in E(G)$, are differentiable functions on a neighborhood W of the point P in $\mathscr{S}^{nr}$. Further, assume that $f_G(P) = \pi(D_p)$. If df_G has a constant rank k on W with $k < nr$. Then there exists a neighborhood U of $\mathbf{0} \in \mathbb{R}^{nr-k}$ and a differentiable mapping $\gamma : U \to W$ such that*

$$\gamma(\mathbf{0}) = P \text{ and } f_G(\gamma(y)) = \pi(D_p) \text{ for } y \in U.$$

Thus, the implicit parameterization theorem asserts that if P is a regular point of $\mathscr{S}^{nr}$, then by the lower semicontinuity of the rank function, $\text{rank}(df) = k$ on neighborhood W of P. Hence, $f_G^{-1}(f_G(P))$ is a smooth manifold of dimension $nr - k$ on W and $f_{K_n}^{-1}(f_{K_n}(P))$ is a submanifold of $f_G^{-1}(f_G(P))$.

As the following theorem shows, the notion of infinitesimal rigidity of a bar framework is stronger than that of local rigidity.

Theorem 9.4 (Gluck [86]) *If a bar framework (G, p) is infinitesimally rigid, then it is locally rigid.*

Proof. Assume that (G, p) is infinitesimally rigid and let P be a configuration matrix of (G, p). Then $\text{rank}(R) = \text{rank}(df_G(P)) = nr - r(r+1)/2$. Hence, P is a regular point and $\text{rank}(R)$ is constant on a neighborhood W of P. Moreover, by the Implicit Parameterization Theorem, $nr - k = r(r+1)/2$. Hence, $f_G^{-1}(f_G(P)) = f_{K_n}^{-1}(f_{K_n}(P))$ on W. That is, all bar frameworks in W that are equivalent to (G, p) are in fact congruent to it. Consequently, (G, p) is locally rigid.

$\square$

The converse of this theorem is not true as shown by the following example.

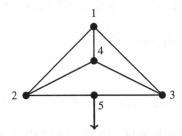

Fig. 9.2 An example of a two-dimensional bar framework which is both locally rigid and infinitesimally flexible

Example 9.1 *Consider the bar framework (G, p) depicted in Fig. 9.2. It is easy to see that $\delta_1 = \delta_2 = \delta_3 = \delta_4 = \mathbf{0}$ and $\delta_5 = [0, -1]^T$ is a nontrivial infinitesimal flex. Hence, (G, p) is infinitesimally flexible. On the other hand, it is also easy to see that (G, p) is locally rigid. Moreover, P is a singular point of f_G since $\mathrm{rank}(R) = 6$ and this rank increases to 7 if p^5 is slightly perturbed so that p^2, p^5, and p^3 are not collinear.*

As the following theorem shows, local rigidity and infinitesimal rigidity coincide at regular points.

Theorem 9.5 (Asimow and Roth [28]) *Let (G, p) be an r-dimensional bar framework on n nodes. Let P and R be a configuration matrix and the rigidity matrix of (G, p). Assume that P is a regular point of f_G in $\mathscr{S}^{nr}$. Then (G, p) is locally rigid if and only if*

$$\mathrm{rank}(R) = nr - \frac{r(r+1)}{2}.$$

As a result, (G, p) is either locally rigid on all regular points or locally flexible on all regular points [28].

The rigidity matrix R has more uses than just to establish the infinitesimal rigidity of a given bar framework (G, p). In fact, while the null space of R contains the space of infinitesimal flexes of (G, p), its left null space contains the space of stresses of (G, p). More precisely, the following theorem is a direct consequence of the definitions of a stress and R in (8.12) and (9.2).

Theorem 9.6 *Let (G, p) be an r-dimensional bar framework and let R be its rigidity matrix. Then $\omega \in \mathbb{R}^m$ is a stress of (G, p) if and only if ω lies in the left null space of R; i.e., iff*

$$\omega^T R = \mathbf{0}.$$

Note that the left null space of R is null(R^T). Consequently, by the definition of the rank of a matrix, we have that, for any bar framework (G,p), the dimension of the space of stresses of (G,p) is equal to

$$m - nr + \frac{1}{2}r(r+1) + \text{dim of the space of nontrivial infinitesimal flexes.} \qquad (9.3)$$

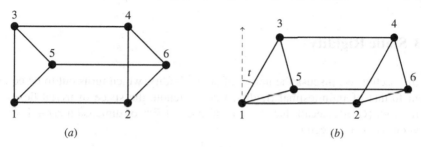

Fig. 9.3 The two-dimensional locally flexible bar framework of Example 9.2

Example 9.2 *Consider the framework (G,p) depicted in Fig. 9.3a, where*

$$p^1 = \begin{bmatrix} 0 \\ 0 \end{bmatrix}, p^2 = \begin{bmatrix} 3 \\ 0 \end{bmatrix}, p^3 = \begin{bmatrix} 0 \\ 2 \end{bmatrix}, p^4 = \begin{bmatrix} 3 \\ 2 \end{bmatrix}, p^5 = \begin{bmatrix} 1 \\ 1 \end{bmatrix} \text{ and } p^6 = \begin{bmatrix} 4 \\ 1 \end{bmatrix}.$$

(G,p) is locally flexible since $p^1 = \begin{bmatrix} 0 \\ 0 \end{bmatrix}$, $p^2 = \begin{bmatrix} 3 \\ 0 \end{bmatrix}$, $p^3 = \begin{bmatrix} 2\sin(t) \\ 2\cos(t) \end{bmatrix}$,

$$p^4 = \begin{bmatrix} 3 + 2\sin(t) \\ 2\cos(t) \end{bmatrix}, p^5 = \begin{bmatrix} \cos(t) + \sin(t) \\ \cos(t) - \sin(t) \end{bmatrix}, p^6 - \begin{bmatrix} 3 + \cos(t) + \sin(t) \\ \cos(t) - \sin(t) \end{bmatrix}$$

is a continuous deformation of (G,p) as shown in Fig. 9.3b.

The rigidity matrix R of (G,p) is 9×12 and of rank 8. Thus, the dimension of the space of stresses is 1 and the dimension of nontrivial infinitesimal flexes of (G,p) is also 1. Moreover, P is a singular point of f_G since any slight perturbation of the second coordinate of p^6 increases the rank of R to 9. Hence, (G,p'') is locally rigid on all regular points P'' of f_G.

Example 9.3 *Consider the framework (G,p') depicted in Fig. 9.4. Similar to framework (G,p) of Fig. 9.3a, rank$(R) = 8$ and thus P' is a singular point of f_G. But unlike (G,p), framework (G,p') is locally rigid since any potential continuous deformation of (G,p') would increase the distance between p'^5 and p'^6.*

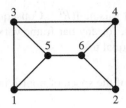

Fig. 9.4 The two-dimensional locally rigid bar framework of Example 9.3

9.3 Static Rigidity

In this section we discuss the notion of static rigidity, which turns out to be equivalent to that of infinitesimal rigidity. Let $\mathcal{L}$ denote the space of trivial flexes of framework (G, p). Recall that $\mathcal{L}$ is a subspace of $\mathbb{R}^{nr}$ of dimension $r(r+1)/2$. A *load* on (G, p) is a vector

$$F = \begin{bmatrix} F^1 \\ \vdots \\ F^n \end{bmatrix} \in \mathbb{R}^{nr}$$

where $F^1, \ldots, F^n \in \mathbb{R}^r$ are external forces acting on the nodes of (G, p). A load is said to be an *equilibrium load* if $F \in \mathcal{L}^\perp$, the orthogonal complement of $\mathcal{L}$ in $\mathbb{R}^{nr}$. To gain a better understanding of this condition, we first determine a basis of $\mathcal{L}$.

As always, let e^j denote the jth standard unit vector in $\mathbb{R}^r$ and let Q^{kl} be the $r \times r$ skew-symmetric matrix with 1 in the (k, l) position, (-1) in the (l, k) position and zeros elsewhere. Further, let τ^j and ρ^{kl} be the two vectors in $\mathbb{R}^{nr}$ such that

$$\tau^j = \begin{bmatrix} e^j \\ \vdots \\ e^j \end{bmatrix} \text{ and } \rho^{kl} = \begin{bmatrix} Q^{kl} p^1 \\ \vdots \\ Q^{kl} p^n \end{bmatrix}.$$

Then $\{\tau^1, \ldots, \tau^r\}$ and $\{\rho^{kl} : 1 \leq k < l \leq r\}$ are bases of the trivial infinitesimal flexes resulting from the r translations and the $r(r-1)/2$ rotations in $\mathbb{R}^r$, respectively. Consequently,

$$\{\tau^1, \ldots, \tau^r\} \cup \{\rho^{kl} : 1 \leq k < l \leq r\}$$

is a basis of $\mathcal{L}$. As a result, the condition $F \in \mathcal{L}^\perp$ amounts to two equalities. First,

$$(\tau^j)^T F = \sum_{i=1}^n (F^i)_j = 0 \text{ for all } j = 1, \ldots, r$$

and hence,

$$\sum_{i=1}^n F^i = \mathbf{0}.$$

That is, the net force exerted by F on (G, p) is zero. Second,

$$(\rho^{kl})^T F = \sum_{i=1}^{n} (p^i)^T Q^{kl} F^i = 0 \text{ for all } 1 \le k < l \le r. \tag{9.4}$$

Equation (9.4), for $r = 3$, is equivalent to the assertion

$$\sum_{i=1}^{n} p^i \times F^i = \mathbf{0},$$

where $(\times)$ denotes the usual cross product in $\mathbb{R}^3$. That is, the net torque exerted by F on (G, p) is zero. Therefore, a load is an equilibrium load if it results in a zero net force and a zero net torque.

An equilibrium load is *resolved* by framework (G, p) if there exist scalars w_{ij}, for all $\{i, j\} \in E(G)$, such that

$$F^i = \sum_{j : \{i,j\} \in E(G)} w_{ij}(p^i - p^j) \text{ for each } i = 1, \dots, n.$$

It should be pointed out here that a stress on framework (G, p) is a resolution of the trivial zero load. Framework (G, p) is said to be *statically rigid* if every equilibrium load is resolved by (G, p). That is, (G, p) is statically rigid if and only if the system

$$R^T w = F, \tag{9.5}$$

where R is the rigidity matrix of (G, p), has a solution $w \in \mathbb{R}^m$ for every equilibrium load F. Now System (9.5) has a solution for every $F \in \mathscr{L}^\perp$ if and only if $\mathscr{L}^\perp \subseteq \text{col}(R^T)$ if and only if $\text{null}(R) \subseteq \mathscr{L}$. But $\mathscr{L} \subseteq \text{null}(R)$. Therefore, framework (G, p) is statically rigid if and only if $\text{null}(R) = \mathscr{L}$. Hence, we have proven the following theorem.

Theorem 9.7 (Whiteley and Roth [195]) *A bar framework is statically rigid if and only if it is infinitesimally rigid.*

9.4 The Dual Rigidity Matrix $\bar{R}$

The rigidity matrix R is a function of $p^1, \dots, p^n$ and hence it is not invariant under rigid motions. Consequently, to establish the infinitesimal rigidity of a bar framework, one has to take into account the space of trivial infinitesimal flexes, and hence the factor $r(r+1)/2$ appears in Theorems 9.2 and 9.5. In this section, we discuss an alternative approach [7] to infinitesimal rigidity which circumvents the need to account for rigid motions. This approach is based on projected Gram matrices and leads to a dual rigidity matrix $\bar{R}$. Our presentation follows closely [7].

Let X be the projected Gram matrix of r-dimensional framework (G, p), $r \le n - 2$, and thus X is PSD of rank r. Recall from (8.10) and (8.9) that

$$\mathscr{F} = \{y \in \mathbb{R}^{\bar{m}} : \mathscr{X}(y) \succeq \mathbf{0}\}$$

is the Cayley configuration spectrahedron of (G, p), where

$$\mathscr{X}(y) = X + \sum_{\{i,j\} \in \bar{E}(\bar{G})} y_{ij} M^{ij}.$$

Also, recall that $\{\mathscr{X}(y) : y \in \mathscr{F}\}$ is the set of projected Gram matrices of all frameworks that are equivalent to (G, p). No attention will be paid to rigid motions since all frameworks that are congruent to (G, p) have the same projected Gram matrix X. Let

$$\mathscr{M}(y) = \sum_{\{i,j\} \in \bar{E}(\bar{G})} y_{ij} M^{ij}. \tag{9.6}$$

Let ζ be a sufficiently small neighborhood of zero in $\mathbb{R}^{\bar{m}}$. Since $X = \mathscr{X}(\mathbf{0})$, it follows that

$$\{\mathscr{X}(y) : y \in \zeta, \mathscr{X}(y) \succeq \mathbf{0} \text{ and } \mathrm{rank}(\mathscr{X}) = r\}$$

is the set of projected Gram matrices of all r-dimensional frameworks near (G, p) that are equivalent to (G, p). To characterize such frameworks, we need the following lemma which is an immediate consequence of Schur complement.

Lemma 9.1 *Let*

$$M = \begin{bmatrix} A & B \\ B^T & C \end{bmatrix}$$

be a symmetric matrix, where A is an $r \times r$ positive definite matrix. Then matrix M is positive semidefinite with rank r if and only if $C - B^T A^{-1} B = \mathbf{0}$.

Let U be the matrix whose columns form an orthonormal basis of $\mathrm{null}(X)$ and let $X = W \Lambda W^T$ be the spectral decomposition of X, where Λ is the diagonal matrix consisting of the r positive eigenvalues of X. Hence, $Q = [W \ U]$ is an orthogonal matrix of order $n - 1$. Moreover, $\mathscr{X}(y)$ is PSD with rank r if and only if

$$Q^T \mathscr{X}(y) Q = \begin{bmatrix} \Lambda + W^T \mathscr{M}(y) W & W^T \mathscr{M}(y) U \\ U^T \mathscr{M}(y) W & U^T \mathscr{M}(y) U \end{bmatrix}$$

is PSD of rank r. Now $\Lambda + W^T \mathscr{M}(y) W \succ 0$ for all $y \in \zeta$. Therefore, it follows from Lemma 9.1 that for $y \in \zeta$, $\mathscr{X}(y)$ is PSD with rank r if and only if

$$\Phi(y) = U^T \mathscr{M}(y) U - U^T \mathscr{M}(y) W (\Lambda + W^T \mathscr{M}(y) W)^{-1} W^T \mathscr{M}(y) U = \mathbf{0}. \tag{9.7}$$

Hence, the linearization of $\Phi(y)$ near $y = \mathbf{0}$ is given by

$$U^T \mathscr{M}(\bar{\delta}) U = \mathbf{0}. \tag{9.8}$$

Therefore, (G, p) is infinitesimally flexible if and only if there exists a nonzero $\bar{\delta}$ in $\mathbb{R}^{\bar{m}}$ satisfying Eq. (9.8). Let

$$\mathscr{E}(y) = \sum_{\{i,j\} \in \bar{E}(\bar{G})} y_{ij} E^{ij}. \tag{9.9}$$

Hence, $\mathscr{M}(y) = -V^T \mathscr{E}(y) V / 2$. The following theorem follows directly from Eq. (9.8), Lemmas 3.8 and (8.8). Recall that $(n-1)$-dimensional frameworks are locally flexible.

Theorem 9.8 (Alfakih [7]) *Let (G, p) be an r-dimensional bar framework with n nodes, $r \leq n - 2$. Let Z be a Gale matrix of (G, p). Then (G, p) is infinitesimally flexible if and only if there exists a nonzero $\bar{\delta}$ in $\mathbb{R}^{\bar{m}}$ such that*

$$Z^T \mathscr{E}(\bar{\delta}) Z = 0. \tag{9.10}$$

The dual rigidity matrix $\bar{R}$ is derived by writing Eq. (9.10) in matrix form. To this end, we need a few definitions.

Given an $n \times n$ symmetric matrix A, let svec(A) denote the $\frac{n(n+1)}{2}$ vector formed by stacking the columns of A from the main diagonal downwards after having multiplied the off-diagonal entries of A by $\sqrt{2}$. For example, if A is a 3×3 matrix, then

$$\text{svec}(A) = \begin{bmatrix} a_{11} \\ \sqrt{2}\, a_{21} \\ \sqrt{2}\, a_{31} \\ a_{22} \\ \sqrt{2}\, a_{32} \\ a_{33} \end{bmatrix}. \tag{9.11}$$

Let B be an $m \times n$ matrix and let A be an $n \times n$ symmetric matrix. The *symmetric Kronecker product* between B and itself, denoted by $B \otimes_s B$, is defined such that

$$(B \otimes_s B)\, \text{svec}(A) = \text{svec}(BAB^T). \tag{9.12}$$

For more details on the symmetric Kronecker product, see [27].

Definition 9.4 *Let Z be a Gale matrix of r-dimensional bar framework (G, p) and let $\bar{R}^T$ be the submatrix of $Z \otimes_s Z$ obtained by keeping only rows corresponding to $\bar{E}(\bar{G})$, the missing edges of G. Then the matrix $\bar{R}$ is called the dual rigidity matrix of (G, p).*

As always, let

$$\bar{r} = n - 1 - r.$$

Recall that z^i is a Gale transform of p^i given by the ith row of Z. Then the dual rigidity matrix $\bar{R}$ is the $\frac{\bar{r}(\bar{r}+1)}{2} \times \bar{m}$ matrix whose columns are indexed by $\bar{E}(\bar{G})$, where the (i, j)th column is equal to $\frac{1}{\sqrt{2}} \text{svec}(z^i z^{jT} + z^j z^{iT})$. For example, if $\bar{E}(\bar{G}) = \{(i_1, j_1), \ldots, (i_{\bar{m}}, j_{\bar{m}})\}$, then

$$\bar{R} = \frac{1}{\sqrt{2}} [\, \text{svec}(z^{i_1} z^{j_1 T} + z^{j_1} z^{i_1 T}) \quad \cdots \quad \text{svec}(z^{i_{\bar{m}}} z^{j_{\bar{m}} T} + z^{j_{\bar{m}}} z^{i_{\bar{m}} T}) \,]. \tag{9.13}$$

That is,

$$
\bar{R} = \begin{bmatrix}
\sqrt{2}\, z_1^{i_1} z_1^{j_1} & \sqrt{2}\, z_1^{i_2} z_1^{j_2} & \cdots & \sqrt{2}\, z_1^{i_{\bar{m}}} z_1^{j_{\bar{m}}} \\
z_2^{i_1} z_1^{j_1} + z_1^{i_1} z_2^{j_1} & z_2^{i_2} z_1^{j_2} + z_1^{i_2} z_2^{j_2} & \cdots & z_2^{i_{\bar{m}}} z_1^{j_{\bar{m}}} + z_1^{i_{\bar{m}}} z_2^{j_{\bar{m}}} \\
z_3^{i_1} z_1^{j_1} + z_1^{i_1} z_3^{j_1} & z_3^{i_2} z_1^{j_2} + z_1^{i_2} z_3^{j_2} & \cdots & z_3^{i_{\bar{m}}} z_1^{j_{\bar{m}}} + z_1^{i_{\bar{m}}} z_3^{j_{\bar{m}}} \\
\cdots & \cdots & \cdots & \cdots \\
\sqrt{2}\, z_{\bar{r}}^{i_1} z_{\bar{r}}^{j_1} & \sqrt{2}\, z_{\bar{r}}^{i_2} z_{\bar{r}}^{j_2} & \cdots & \sqrt{2}\, z_{\bar{r}}^{i_{\bar{m}}} z_{\bar{r}}^{j_{\bar{m}}}
\end{bmatrix}, \tag{9.14}
$$

where z_k^l denotes the kth coordinate of vector z^l. A justification of the definition of $\bar{R}$ is given in the following theorem.

Theorem 9.9 (Alfakih [7]) *Let $\bar{R}$ be the dual rigidity matrix of an r-dimensional bar framework (G,p). Then (G,p) is infinitesimally rigid if and only if $\bar{R}$ has a trivial null space, i.e., if and only if*

$$
rank\,(\bar{R}) = \bar{m}. \tag{9.15}
$$

Proof. This follows from Eq. (9.10) and the definition of $\bar{R}$ since $Z^T \mathscr{E}(\bar{\delta})Z = 0$ if and only if $\bar{R}\bar{\delta} = 0$.

□

Before presenting an example to illustrate the dual rigidity matrix $\bar{R}$, we make the following four observations. First, as the graph becomes denser, i.e., as the number of edges of G increases, the number of rows in the rigidity matrix R increases, while the number of columns of $\bar{R}$ decreases and vice versa. Second, $\bar{R}$ is a function of Gale matrix and thus is invariant under rigid motions. Hence, unlike Theorem 9.2, the term $r(r+1)/2$ is absent from Eq. (9.15). Third, $\bar{R}$ is in general sparse since the Gale matrix Z can be chosen sparse. Fourth, the rank of $\bar{R}$ does change if the factors of $\sqrt{2}$ are dropped from the definition of $\bar{R}$. These factors are kept in order to make the definition of $\bar{R}$ in terms of the symmetric Kronecker product simple.

Example 9.4 *Consider the framework (G,p) depicted in Fig. 9.2. A Gale matrix of (G,p) is given by*

$$
Z = \begin{bmatrix}
1 & 0 \\
0 & 1 \\
0 & 1 \\
-2 & 0 \\
1 & -2
\end{bmatrix}.
$$

Then

$$
\bar{R} = \begin{bmatrix}
\sqrt{2} & 0 & -2\sqrt{2} \\
-2 & 0 & 4 \\
0 & \sqrt{2} & 0
\end{bmatrix}
$$

where the columns of $\bar{R}$ are indexed in the order $\{1,5\},\{2,3\},\{4,5\}$. Note that the rigidity matrix R of (G,p) is 7×10. Also, note that $\bar{\delta} = [2\ 0\ 1]^T$ is a basis of the null space of $\bar{R}$ and $x = [2\ \sqrt{2}\ 0]^T$ is a basis of the left null space of $\bar{R}$. Therefore, (G,p) is infinitesimally flexible.

Example 9.5 *Consider the framework* (G,p) *depicted in Fig. 9.3a. A Gale matrix of* (G,p) *is given by*

$$Z = \begin{bmatrix} 3 & 0 & 0 \\ 0 & 3 & 0 \\ 0 & 0 & 3 \\ 3 & 3 & -3 \\ -5 & -2 & -3 \\ -1 & -4 & 3 \end{bmatrix}.$$

Then

$$\bar{R} = \begin{bmatrix} 9\sqrt{2} & -3\sqrt{2} & 0 & 0 & 0 & -15\sqrt{2} \\ 9 & -12 & 0 & -15 & 0 & -21 \\ -9 & 9 & 0 & 0 & -3 & 6 \\ 0 & 0 & 0 & -6\sqrt{2} & 0 & -6\sqrt{2} \\ 0 & 0 & 9 & -9 & -12 & -3 \\ 0 & 0 & 0 & 0 & 9\sqrt{2} & 9\sqrt{2} \end{bmatrix}.$$

The columns of $\bar{R}$ *are indexed in the order* $\{1,4\}, \{1,6\}, \{2,3\}, \{2,5\}, \{3,6\}, \{4,5\}$. *Note that the rigidity matrix* R *of* (G,p) *is* 9×12. *Also note that*

$$\bar{\delta} = [2 \ 1 \ -2 \ -1 \ -1 \ 1]^T$$

is a basis of the null space of $\bar{R}$ *and*

$$x = [1 \ 2\sqrt{2} \ 3\sqrt{2} \ -5 \ 0 \ 1]^T$$

is a basis of the left null space of $\bar{R}$. *Therefore,* (G,p) *is infinitesimally flexible. In fact, as we showed in Example 9.2,* (G,p) *is locally flexible.*

Example 9.6 *Consider the framework* (G,p) *depicted in Fig. 9.4. A Gale matrix of* (G,p) *is given by*

$$Z = \begin{bmatrix} 1 & 0 & 0 \\ 0 & 1 & 0 \\ 0 & 0 & 1 \\ 1 & 1 & -1 \\ -1 & 2 & -3 \\ -1 & -4 & 3 \end{bmatrix}.$$

Then

$$\bar{R} = \begin{bmatrix} \sqrt{2} & -\sqrt{2} & 0 & 0 & 0 & -\sqrt{2} \\ 1 & -4 & 0 & -1 & 0 & 1 \\ -1 & 3 & 0 & 0 & -1 & -2 \\ 0 & 0 & 0 & 2\sqrt{2} & 0 & 2\sqrt{2} \\ 0 & 0 & 1 & -3 & -4 & -5 \\ 0 & 0 & 0 & 0 & 3\sqrt{2} & 3\sqrt{2} \end{bmatrix}.$$

The columns of $\bar{R}$ are indexed in the order $\{1,4\},\{1,6\},\{2,3\},\{2,5\},\{3,6\},\{4,5\}.$
Note that

$$\delta = [2 \ \ 1 \ \ -2 \ \ -1 \ \ -1 \ \ 1]^T$$

is a basis of the null space of $\bar{R}$ and

$$x = [1 \ \ 2\sqrt{2} \ \ 3\sqrt{2} \ \ 1 \ \ 0 \ \ 1]^T$$

is a basis of the left null space of $\bar{R}$. Therefore, framework (G,p) is infinitesimally flexible. However, as we showed in Example 9.3, (G,p) is locally rigid.

9.4.1 Similarities and Dissimilarities Between R and $\bar{R}$

We begin with the following theorem characterizing the left null space of $\bar{R}$.

Theorem 9.10 (Alfakih [7]) *Let (G,p) be an r-dimensional bar framework and let $\bar{R}$ be its dual rigidity matrix. Further, let Z be a Gale matrix of (G,p). Then $\Omega = Z\Psi Z^T$ is a stress matrix of (G,p) if and only if $(svec(\Psi))^T \bar{R} = \mathbf{0}$.*

Proof.   By Theorem 8.4, $\Omega = Z\Psi Z^T$ is a stress matrix of (G,p) if and only if $(Z\Psi Z^T)_{ij} = 0$ for all $\{i,j\} \in \bar{E}(\bar{G})$. But, by the definition of the symmetric Kronecker product, $svec(Z\Psi Z^T) = (Z \otimes_s Z) svec(\Psi)$. Hence, Ω is a stress matrix of (G,p) if and only if $((Z \otimes_s Z) svec(\Psi))_{ij} = 0$ for all $\{i,j\} \in \bar{E}(\bar{G})$. However, by the definition of $\bar{R}$, this last statement is equivalent to $\bar{R}^T svec(\Psi) = \mathbf{0}$.   □

Example 9.7 *Let (G,p) be the framework depicted in Fig. 9.2. We found in Example 9.4 that $x = [2 \ \ \sqrt{2} \ \ 0]^T$ is a basis of the left null space of $\bar{R}$. Moreover, it can be shown that*

$$\omega = (\omega_{12} = -1, \omega_{13} = -1, \omega_{14} = 4, \omega_{24} = 2, \omega_{25} = -1, \omega_{34} = 2, \omega_{35} = -1)$$

is a stress of (G,p) with corresponding stress matrix

$$\Omega = \begin{bmatrix} 2 & 1 & 1 & -4 & 0 \\ 1 & 0 & 0 & -2 & 1 \\ 1 & 0 & 0 & -2 & 1 \\ -4 & -2 & -2 & 8 & 0 \\ 0 & 1 & 1 & 0 & -2 \end{bmatrix} = Z\Psi Z^T,$$

where $\Psi = \begin{bmatrix} 2 & 1 \\ 1 & 0 \end{bmatrix}$. Note that $svec(\Psi) = x = [2 \ \ \sqrt{2} \ \ 0]^T$.

The following theorem establishes the relationship between the null spaces and the left null spaces of R and $\bar{R}$.

Theorem 9.11 (Alfakih [7]) *Let R and $\bar{R}$ be, respectively, the rigidity and the dual rigidity matrices of r-dimensional bar framework (G,p). Then*

1. $null(\bar{R}^T)$ is isomorphic to $null(R^T)$.
2. $dim\ of\ null(\bar{R}) = dim\ of\ null(R) - \dfrac{r(r+1)}{2}$.

Proof. The left null space of R is isomorphic to the space of stress matrices of (G, p) which, by Theorem 9.10, is isomorphic to the left null space of $\bar{R}$. To prove Statement 2, note that

$$\dim \text{null}(R) = \dim \text{null}(R^T) + nr - m$$

and

$$\dim \text{null}(\bar{R}) = \dim \text{null}(\bar{R}^T) + \bar{m} - \frac{\bar{r}(\bar{r}+1)}{2}.$$

But, since $\dim \text{null}(R^T) = \dim \text{null}(\bar{R}^T)$, we have

$$\dim \text{null}(\bar{R}) - \dim \text{null}(R) = m + \bar{m} - nr - \frac{\bar{r}(\bar{r}+1)}{2} = -\frac{r(r+1)}{2}.$$

$\square$

Next, we show that for each infinitesimal flex of (G, p), i.e., for each vector in the null space of R, there corresponds a vector y in the null space of $\bar{R}$, which can be determined explicitly.

Theorem 9.12 (Alfakih [7]) *Let* $\delta \in \mathbb{R}^{nr}$ *be an infinitesimal flex of* (G, p) *and let* $\Delta^T = [\delta^1 \ \cdots \ \delta^n]$, *i.e.,* Δ^T *is an* $r \times n$ *matrix. Then, there exists a vector* $\bar{\delta}$ *in the null space of* $\bar{R}$ *such that*

$$\mathscr{E}(\bar{\delta}) = \mathscr{K}(P\Delta^T + \Delta P^T).$$

That is,

$$\mathscr{E}(\bar{\delta}) = diag(P\Delta^T + \Delta P^T)e^T + e(diag(P\Delta^T \mid \Delta P^T))^T - 2(P\Delta^T + \Delta P^T), \quad (9.16)$$

where P *is a configuration matrix of* (G, p) *and* $\mathscr{E}(y)$ *is as defined in (9.9).*

Proof. It is straightforward to verify that

$$2(p^i - p^j)^T(\delta^i - \delta^j) = (P\Delta^T + \Delta P^T)_{ii} + (P\Delta^T + \Delta P^T)_{jj} - 2(P\Delta^T + \Delta P^T)_{ij}.$$

Let $\mathscr{L}$ denote the space of $n \times n$ symmetric matrices $A = (a_{ij})$ such that $a_{ij} = 0$ if $i = j$ or if $\{i, j\} \in E(G)$. Then, since $(p^i - p^j)^T(\delta^i - \delta^j) = 0$ for all $\{i, j\} \in E(G)$, it follows that the right-hand side of Eq. (9.16) belongs to $\mathscr{L}$. Therefore, there exists $\bar{\delta} \in \mathbb{R}^{\bar{m}}$ that satisfies Eq. (9.16) since the set $\{E^{ij} : \{i, j\} \in \bar{E}(\bar{G})\}$ forms a basis of $\mathscr{L}$. Now multiplying Eq. (9.16) from left and right by Z^T and Z, respectively, yields $Z^T \mathscr{E}(\bar{\delta})Z = 0$. Thus, $\bar{\delta}$ belongs to $null(\bar{R})$.

$\square$

Now if δ is a trivial infinitesimal flex resulting from a translation, then $\Delta^T = e^j e_n^T$, where e^j is the jth standard unit vector in $\mathbb{R}^r$. On the other hand, if δ is a trivial infinitesimal flex resulting from a rotation, then $\Delta^T = Q^{kl} P^T$, where Q^{kl} is the skew-symmetric matrix defined above. It is easy to verify that in both of these cases, the right-hand side of Eq. (9.16) is identically zero. Consequently, if δ is a trivial infinitesimal flex of (G, p), then $\bar{\delta} = \mathbf{0}$. As a result, if $\bar{\delta}$ in Eq. (9.16) is nonzero, then δ is a nontrivial infinitesimal flex.

Example 9.8 *Let (G, p) be the framework depicted in Fig. 9.2, where $\delta_1 = \delta_2 = \delta_3 = \delta_4 = \mathbf{0}$ and $\delta_5 = [0, -1]^T$ is a nontrivial infinitesimal flex. Hence, Eq. (9.16) yields that $\bar{\delta} = [4\ 0\ 2]^T$, where the missing edges are listed in the order $\{1,5\}, \{2,3\}$ and $\{4,5\}$. This agrees with Example 9.4, where we found that $\bar{\delta} = [2\ 0\ 1]^T$ is a basis of $null(\bar{R})$.*

Example 9.9 *Consider the framework depicted in Fig. 9.4, where the missing edges are listed in the order $\{1,4\}, \{1,6\}, \{2,3\}, \{2,5\}, \{3,6\}, \{4,5\}$; and where*

$$p^1 = \begin{bmatrix} 0 \\ 0 \end{bmatrix}, p^2 = \begin{bmatrix} 3 \\ 0 \end{bmatrix}, p^3 = \begin{bmatrix} 0 \\ 2 \end{bmatrix}, p^4 = \begin{bmatrix} 3 \\ 2 \end{bmatrix}, p^5 = \begin{bmatrix} 1 \\ 1 \end{bmatrix} \text{ and } p^6 = \begin{bmatrix} 2 \\ 1 \end{bmatrix}.$$

We found, in Example 9.6, that

$$\bar{\delta} = [2\ 1\ -2\ -1\ -1\ 1]^T$$

is a basis of $null(\bar{R})$. On the other hand, one can verify that

$$\delta^1 = \begin{bmatrix} -7 \\ -3 \end{bmatrix}, \delta^2 = \begin{bmatrix} -7 \\ 3 \end{bmatrix}, \delta^3 = \begin{bmatrix} 7 \\ -3 \end{bmatrix}, \delta^4 = \begin{bmatrix} 7 \\ 3 \end{bmatrix} \delta^5 = \begin{bmatrix} 0 \\ -10 \end{bmatrix} \text{ and } \delta^6 = \begin{bmatrix} 0 \\ 10 \end{bmatrix}$$

is a nontrivial infinitesimal flex of (G, p). Consequently, Eq. (9.16) yields that

$$\bar{\delta} = 54[2\ 1\ -2\ -1\ -1\ 1]^T.$$

9.4.2 Geometric Interpretation of $\bar{R}$

Let (G, p) be an r-dimensional bar framework where $r \leq n - 2$. In this subsection, we assume that the Cayley configuration spectrahedron $\mathscr{F}$ of (G, p) is full dimensional; i.e., we assume that there exists $\hat{y} \in \mathscr{F}$ such that $\mathscr{X}(\hat{y})$ is PD. In other words, we assume that there exists an $(n-1)$-dimensional framework that is equivalent to (G, p). Then the rows of $\bar{R}$ have a geometric interpretation in terms of the normal cone of $\mathscr{F}$ at the origin [1, 2].

Lemma 9.2 ([1]) *Let (G, p) be an r-dimensional bar framework on n nodes, where $r \leq n - 2$, and let $\mathscr{F}$ be its Cayley configuration spectrahedron as defined in (8.10). Assume that there exists $\hat{y} \in \mathscr{F}$ such that $\mathscr{X}(\hat{y})$ is positive definite. Then the normal cone $N_{\mathscr{F}}(y^0)$ is given by*

$$N_{\mathscr{F}}(y^0) = \{c \in \mathbb{R}^{\bar{m}} : c_{ij} = -trace(M^{ij}Y), \text{ for some } Y \succeq 0 : trace(\mathscr{X}(y^0)Y) = 0\}.$$

Proof. Let $c = (c_{ij})$ in $\mathbb{R}^{\bar{m}}$, where $c_{ij} = -\text{trace}(M^{ij}Y)$ for some $Y \succeq 0$ such that $\text{trace}(\mathscr{X}(y^0)Y) = 0$, and let y be any point in $\mathscr{F}$. Then

$$c^T(y^0 - y) = \text{trace}((\mathscr{M}(y) - \mathscr{M}(y^0))Y) = \text{trace}((\mathscr{X}(y) - \mathscr{X}(y^0))Y) \geq 0$$

since $\text{trace}(\mathscr{X}(y^0)Y) = 0$, and since both matrices $\mathscr{X}(y)$ and Y are PSD. Therefore, $c \in N_{\mathscr{F}}(y^0)$.

To prove the reverse direction, let $c = (c_{ij}) \in N_{\mathscr{F}}(y^0)$ and consider the following pair of dual SDP problems:

(P) max $\quad c^T y$ $\qquad$ (D) min $\text{trace}(XY)$
$\quad$ s. t. $X + \mathscr{M}(y) \succeq 0$ $\qquad$ s. t. $-\text{trace}(M^{ij}Y) = c_{ij}$ for $\{i,j\} \in \bar{E}(\bar{G})$,
$\qquad\qquad\qquad\qquad\qquad\qquad Y \succeq 0$,

where X is the projected Gram matrix of framework (G, p). Hence, y^0 is an optimal solution of (P). Moreover, Slater's condition holds by our assumption. Consequently, by SDP strong duality, there exists $Y \succeq 0$ such that $c^T y^0 = \text{trace}(XY)$ and $c_{ij} = -\text{trace}(M^{ij}Y)$ for all $\{i,j\} \in \bar{E}(\bar{G})$. The result follows since $c^T y^0 - \text{trace}(XY) = -\text{trace}((\mathscr{M}(y^0) + X)Y) = -\text{trace}(\mathscr{X}(y^0)Y) = 0$. $\qquad\square$

The following corollary is an immediate consequence of Lemmas 9.2 and 3.8.

Corollary 9.1 ([2]) *Let (G, p) be an r-dimensional bar framework on n nodes, where $r \leq n - 2$. Assume that there exists $\hat{y} \in \mathscr{F}$ such that $\mathscr{X}(\hat{y})$ is positive definite. Let Z be a Gale matrix for (G, p). Then*

$$N_{\mathscr{F}}(0) = \{c \in \mathbb{R}^{\bar{m}} : c_{ij} = \text{trace}(Z^T E^{ij} Z \Phi) \text{ for some } \Phi \in \mathscr{S}_+^{\bar{r}}\}. \qquad (9.17)$$

Proof. Set $y^0 = 0$ in Lemma 9.2. Then $\text{trace}(XY) = 0$ implies that $XY = 0$ since both matrices X and Y are PSD. Let U be the matrix whose columns form an orthonormal basis of $\text{null}(X)$. Then $Y = U\Phi U^T$ for some $\bar{r} \times \bar{r}$ PSD matrix Φ. Therefore, $-\text{trace}(M^{ij}Y) = \text{trace}(V^T E^{ij} V U \Phi U^T)/2$. The result follows from Lemma 3.8. $\qquad\square$

As always, let $e^1, \ldots, e^{\bar{r}}$ denote the standard unit vectors in $\mathbb{R}^{\bar{r}}$. Then the following $\bar{r}(\bar{r}+1)/2$ matrices:

$$\psi^{kk} = e^k e^{k^T} \text{ for all } k = 1, \ldots, \bar{r},$$
$$\psi^{kl} = \frac{1}{\sqrt{2}}(e^k e^{l^T} + e^l e^{k^T}) + e^k e^{k^T} + e^l e^{l^T} \text{ for all } 1 \leq k < l \leq \bar{r}$$

are obviously symmetric PSD and linearly independent. Thus, their conic hull is a full-dimensional subset of $\mathscr{S}_+^{\bar{r}}$, the PSD cone of order $\bar{r}$; i.e., the dimension of their conic hull is $\bar{r}(\bar{r}+1)/2$. Moreover, the conic hull of the $\bar{r}(\bar{r}+1)/2$ vectors $c^{kl} = (c_{ij}^{kl}) \in \mathbb{R}^{\bar{m}}$, where

$$c_{ij}^{kl} = \frac{1}{\sqrt{2}} \, \text{trace}(Z^T E^{ij} Z \psi^{kl}),$$

is a subset of $N_{\mathscr{F}}(\mathbf{0})$. But

$$c_{ij}^{kl} = \begin{cases} \sqrt{2} \, z_k^i z_k^j & \text{if } k = l \\ z_k^i z_l^j + z_l^i z_k^j + \sqrt{2} z_k^i z_k^j + \sqrt{2} z_l^i z_l^j & \text{if } k \neq l, \end{cases}$$

where z_k^l denotes the kth coordinate of the lth Gale transform z^l. Hence,

$$c^{kl} = \begin{cases} \text{the } (k,k)\text{th row of } \bar{R} & \text{if } k = l, \\ \text{the sum of the } (k,l)\text{th, } (k,k)\text{th and } (l,l)\text{th rows of } \bar{R} & \text{if } k \neq l, \end{cases}$$

Hence, if $\text{rank}(\bar{R}) = \bar{m}$, then $\dim N_{\mathscr{F}}(\mathbf{0}) = \bar{m}$. As a result, if a bar framework (G, p), with full dimensional Cayley configuration spectrahedron $\mathscr{F}$, is infinitesimally rigid, then $N_{\mathscr{F}}(\mathbf{0})$ is full dimensional; i.e., $\dim N_{\mathscr{F}}(\mathbf{0}) = \bar{m}$.

Example 9.10 *Consider the two-dimensional framework (G, p) depicted in Fig. 8.2 and discussed in Example 8.2. In this case, $\bar{r} = 1$ and $Z = [1 \ -1 \ 1 \ -1]^T$. Then*

$$N_{\mathscr{F}}(\mathbf{0}) = \{c \in \mathbb{R}^2 : c_{13} = 2\psi, c_{24} = 2\psi, \text{ where } \psi \geq 0\}.$$

Obviously, $N_{\mathscr{F}}(\mathbf{0})$ is one-dimensional in $\mathbb{R}^2$ and (G, p) is infinitesimally flexible. Note that the dual rigidity matrix in this case is $\bar{R} = [\sqrt{2} \ \sqrt{2}]$.

Now consider the one-dimensional framework (G, p') corresponding to $y^0 = (y_{13} = 4$ and $y_{24} = -4)$ in Fig. 8.3. In this case, $\bar{r} = 2$ and

$$Z = \begin{bmatrix} 1 & 0 \\ -2 & 1 \\ 0 & 1 \\ 1 & -2 \end{bmatrix}.$$

Then

$$N_{\mathscr{F}}(y^0) = \{c \in \mathbb{R}^2 : c_{13} = \text{trace}\left(\begin{bmatrix} 0 & 1 \\ 1 & 0 \end{bmatrix} \psi\right), c_{24} = \text{trace}\left(\begin{bmatrix} -4 & 5 \\ 5 & -4 \end{bmatrix} \psi\right), \text{ where } \psi \succeq 0\}.$$

In this case,

$$c^{11} = \begin{bmatrix} 0 \\ -2\sqrt{2} \end{bmatrix}, c^{22} = \begin{bmatrix} 0 \\ -2\sqrt{2} \end{bmatrix} \text{ and } c^{12} = \begin{bmatrix} 1 \\ 5 - 4\sqrt{2} \end{bmatrix}.$$

Obviously, $N_{\mathscr{F}}(y^0)$ is two-dimensional and (G, p') is infinitesimally rigid. Note that the dual rigidity matrix in this case is

$$\bar{R} = \begin{bmatrix} 0 & -2\sqrt{2} \\ 1 & 5 \\ 0 & -2\sqrt{2} \end{bmatrix}.$$

9.5 Combinatorial Local Rigidity

Suppose we restrict ourselves to bar frameworks with "typical" or generic configurations, i.e., configurations possessing no special structure. Then in this case, as is shown in this section, the local rigidity of framework (G, p) depends only on graph G and not on configuration p. In other words, the local rigidity problem becomes a purely combinatorial one [139, 97]. More formally, bar framework (G, p) is said to be *generic* if the coordinates of $p^1, \ldots, p^n$ are algebraically independent over the rationals. That is, if $p^1, \ldots, p^n$ do not satisfy any nonzero polynomial with rational coefficients. Put differently, these coordinates can be treated as indeterminates. Evidently, generic bar frameworks are regular points of the rigidity map f_G. As a result, the notions of local rigidity and infinitesimal rigidity coincide for generic frameworks and such frameworks are either all locally flexible or all locally rigid. In other words, local rigidity is a generic property of bar frameworks.

A graph G is said to be *generically locally rigid in dimension r* if there exists a locally rigid r-dimensional generic bar framework (G, p). In this section, we characterize generically locally rigid graphs in dimensions one and two. We begin first with the one-dimensional case [128].

Theorem 9.13 *Let (G, p) be a one-dimensional bar framework on n nodes. Then (G, p) is locally rigid if and only if G is connected.*

Proof. If G is not connected, then each connected component of G can be moved relative to the other components and thus (G, p) is locally flexible.

To prove the reverse direction, assume that G is connected and note that the rigidity matrix R in this case has rank $\leq n - 1$. Let M be the node-edge incidence matrix of G. Hence, $\text{rank}(M) = n - 1$ since G is connected. Moreover, the rigidity matrix of (G, p) is given by $R = QM^T$, where Q is the $m \times m$ diagonal matrix, whose diagonal entries are indexed by the edges of G, where the diagonal entry corresponding to edge $\{i, j\}$ is equal to $p^i - p^j$, up to a minus sign. Now by Assumption 8.1, $p^i - p^j \neq 0$ for each $\{i, j\} \in E(G)$ and thus Q is nonsingular. Consequently, $\text{rank}(R) = n - 1$. As a result, (G, p) is infinitesimally rigid and hence is locally rigid since R has maximal rank.

$\square$

Next, we turn to the two-dimensional case. Let $G = (V, E)$ be a given graph and let $V' \subset V(G)$ and $E' \subset E(G)$. We say that V' *spans* E', or E' is spanned by V', if $E' = \{\{i, j\} \in E(G) : i, j \in V'\}$. Furthermore, we say that $G' = (V', E')$ is an *induced subgraph* of $G = (V, E)$ if V' spans E'. A graph G, with n nodes and m edges, is called a *Laman graph* if it satisfies the following two conditions:

  (i) $m = 2n - 3$.
  (ii) Every induced subgraph with $n' \geq 2$ nodes spans at most $2n' - 3$ edges.

We will refer to Conditions (i) and (ii) as Laman Conditions. An immediate consequence of this definition is that a Laman graph G on $n \geq 3$ nodes cannot have a leaf, i.e., a node of degree 1. This follows since if G has a leaf, say v, then the vertices of G other than v span $m - 1 > 2(n - 1) - 3$ edges, a contradiction. Moreover, the nodes of G cannot all have degrees ≥ 4. To see this, suppose that every node of G has degree ≥ 4. Then $m \geq 2n > 2n - 3$, also a contradiction. As a result, every Laman graph on $n \geq 3$ nodes has at least one node of degree either 2 or 3. For example, the complete graph K_n is a Laman graph if $n = 2$ or 3, but not if $n \geq 4$.

We say that graph G admits a *Henneberg construction* [107] if there exists a sequence of graphs $K_2 = G_0, G_1, \ldots, G_{k-1}, G_k = G$ such that G_{i+1} is obtained from G_i by either one of the following two steps:

H1:   Add one new node and two new edges connecting this new node to two existing nodes of G_i.

H2:   Delete an existing edge of G_i. Add one new node and three new edges, where two of these new edges connect the new node to the end nodes of the deleted edge, and the third new edge connects the new node to any other existing node of G_i.

The following lemma shows that Laman graphs are closed under H1 and H1 in reverse.

Lemma 9.3 *Let G_i be a Laman graph with $n \geq 3$ nodes. If G_{i+1} is a graph obtained from G_i by Henneberg step H1, then G_{i+1} is also a Laman graph. On the other hand, if G_i has a node, say v, of degree 2, then there exists a Laman graph G_{i-1} such that G_i is obtained from G_{i-1} by Henneberg step H1.*

Proof.   To prove the first part of the lemma, note that G_{i+1} clearly satisfies Laman Condition (i). Now let v be the node added by the H1 step. Then obviously $\deg(v) = 2$. Let G' be an induced subgraph of G_{i+1} where $|V(G')| \geq 2$. If $v \notin V(G')$, then G' is an induced subgraph of G_i and thus G' automatically satisfies Laman Condition (ii). Hence, assume that $v \in V(G')$. Here we have to consider three cases: (a) $V(G')$ contains v but none of its neighbors, (b) $V(G')$ contains v and one of its neighbors and (c) $V(G')$ contains v and both of its neighbors. Let us consider case (c) first. In this case, the nodes of G' other than v span at most $2(|V(G')| - 1) - 3$ edges. Thus, $V(G')$ spans at most $2(|V(G')| - 1) - 3 + 2 = 2|V(G')| - 3$ edges. By a similar argument, $V(G')$ spans at most $2|V(G')| - 5$ in case (a) and at most $2|V(G')| - 4$ in case (b). Hence, in all these three cases, G' satisfies Laman Condition (ii). Consequently, G_{i+1} is a Laman graph.

To prove the second part, let G_{i-1} be the graph obtained from G_i be deleting node v and the two edges incident with it. Then obviously, G_{i-1} satisfies Laman Condition (i). Moreover, any induced subgraph of G_{i-1} is also an induced subgraph of G_i and hence automatically satisfies Laman Condition (ii). Consequently, G_{i-1} is a Laman graph.

$\square$

Laman graphs are also closed under H2 and H2 in reverse.

Lemma 9.4 *Let G_i be a Laman graph with $n \geq 4$ nodes. If G_{i+1} is a graph obtained from G_i by Henneberg step H2, then G_{i+1} is also a Laman graph. On the other hand, if G_i has a node, say v, of degree 3, then there exists a Laman graph G_{i-1} such that G_i is obtained from G_{i-1} by Henneberg step H2.*

Proof.   To prove the first part of the lemma, note that G_{i+1} obviously satisfies Laman Condition (i). Now let v be the node added by the H2 step. Then obviously $\deg(v) = 3$. Let G' be an induced subgraph of G_{i+1} where $|V(G')| \geq 2$. If $v \notin V(G')$, then G' is an induced subgraph of G_i and hence G' automatically satisfies Laman Condition (ii). Hence, assume that $v \in V(G')$. Consider the case where all three neighbors of v are in $V(G')$. Then the nodes of G' other than v span at most $2(|V(G')| - 1) - 4$ edges since one edge between the neighbors of v is deleted. Thus, $V(G')$ spans at most $2(|V(G')| - 1) - 4 + 3 = 2|V(G')| - 3$. The cases where one or more of the neighbors of v are not in $V(G')$ also span at most $2|V(G')| - 3$. Thus, G_{i+1} satisfies Laman Condition (ii) and consequently, G_{i+1} is a Laman graph.

To prove the second part, we need the following two claims. For $i = 1, 2, 3$, let H_i be an induced subgraph of G_{i-1} with n_i nodes and m_i edges. Following [138], we say that $V(H_i)$ is *tight* if $m_i = 2n_i - 3$.

Claim 1: Let both $V(H_1)$ and $V(H_2)$ be tight and assume that $|V(H_1) \cap V(H_2)| \geq 2$. Then $V(H_1) \cup V(H_2)$ is tight.

A useful observation is that $V(H_1) \cap V(H_2)$ spans the edges of $E(H_1) \cap E(H_2)$, while $E(H_1) \cup E(H_2)$ is a subset of the edges of G_{i-1} spanned by $V(H_1) \cup V(H_2)$. This follows since an edge $\{i, j\}$ where $i \in V(H_1)$ and $j \in V(H_2)$ is in the latter set but not in the former one.

Proof of Claim 1:   Let $m' = |E(H_1) \cap E(H_2)|$, $n' = |V(H_1) \cap V(H_2)|$, $m = |E(H_1) \cup E(H_2)|$ and $n = |V(H_1) \cup V(H_2)|$. Then $m = m_1 + m_2 - m'$ and $n = n_1 + n_2 - n'$. Moreover, $m - 2n + 3 = -m' + 2n' - 3$. But, $m' \leq 2n' - 3$ since $n' \geq 2$ and thus $m \geq 2n - 3$. But $m \leq 2n - 3$ and hence $m = 2n - 3$.

Claim 2: Let i, j, k be the neighbors of v in G_i and assume that i, j are in $V(H_1)$, i, k are in $V(H_2)$ and j, k are in $V(H_3)$. Then, at least one of the sets $V(H_1), V(H_2)$ and $V(H_3)$ is not tight.

Proof of Claim 2: By way of contradiction, assume that $V(H_1), V(H_2)$, and $V(H_3)$ are all tight. We need to consider two cases:

Case 1: $|V(H_1) \cap V(H_2)| = |V(H_1) \cap V(H_3)| = |V(H_2) \cap V(H_3)| = 1$. Then $V(H_1) \cap V(H_2) \cap V(H_3) = \emptyset$. Let $m = |E(H_1) \cup E(H_2) \cup E(H_3)|$, then $m = m_1 + m_2 + m_3$. Now let $n = |V(H_1) \cup V(H_2) \cup V(H_3)|$, thus $n = n_1 + n_2 + n_3 - 3$. Since by assumption $m_i = 2n_i - 3$ for $i = 1, 2, 3$, it follows that $m = 2(n + 3) - 9 = 2n - 3$; i.e., $V(H_1) \cup V(H_2) \cup V(H_3)$ is tight. Consequently, the subgraph of G_i induced by $V(H_1) \cup V(H_2) \cup V(H_3) \cup \{v\}$ spans $\geq m + 3 = 2n > 2(n + 1) - 3$, a contradiction.

Case 2: At least one of the above three cardinalities, say $|V(H_1) \cap V(H_2)|$, is ≥ 2. Let $m = |E(H_1) \cup E(H_2)|$ and $n = |V(H_1) \cup V(H_2)|$. Then $\{i, j, k\} \subseteq V(H_1) \cup V(H_2)$. Now by assumption $V(H_1)$ and $V(H_2)$ are tight and thus, by claim 1, $V(H_1) \cup V(H_2)$ is tight. Therefore, the subgraph of G_i induced by $V(H_1) \cup V(H_2) \cup \{v\}$ spans $\geq m + 3 = 2n > 2(n + 1) - 3$, a contradiction. Hence, in both cases, we have a contradiction and the proof of Claim 2 is complete.

Now assume that H_1 is not tight and thus $|E(H_1)| < 2|V(H_1)| - 3$. By taking $V(H_1) = \{i, j\}$, we conclude that $\{i, j\} \notin E(G_i)$. Let G_{i-1} be the graph obtained from G_i be deleting node v and the three edges incident with it and adding edge $\{i,j\}$. Then obviously G_{i-1} satisfies Laman Condition (i). Now let G' be an induced subgraph of G_{i-1}. If either i or j is not in $V(G')$, then G' is an induced subgraph of G_i and hence automatically satisfies Laman Condition (ii). Therefore, assume that both i and j are in $V(G')$. Thus, $V(G') = V(H_1)$ and $E(G') = E(H_1) \cup \{i,j\}$. Consequently, $|E(G')| = |E(H_1)| + 1 \leq 2|V(G')| - 3$ and thus G' satisfies Laman Condition (ii). As a result, G_{i+1} is a Laman graph.

$\square$

As the following theorem shows, Laman graphs are precisely those graphs which admit a Henneberg construction.

Theorem 9.14 *A graph G is a Laman graph if and only if it admits a Henneberg construction.*

Proof. This follows from Lemmas 9.3 and 9.4 since every Laman graph on $n \geq 3$ nodes must have either a node of degree 2 or a node of degree 3.

$\square$

Next, we show that generically locally rigid Laman graphs are closed under H1.

Lemma 9.5 *Let (G,p) be a generic two-dimensional bar framework with n nodes and m edges, where G is a Laman graph. Assume that G is obtained from Laman graph G' by Henneberg step H1. Further, assume that (G',p') is locally rigid, where p' is the restriction of p to G'. Then (G,p) is locally rigid.*

Proof. Let R and R' be the rigidity matrices of (G,p) and (G',p'), respectively. Wlog assume that node 1 is the new node of G and that nodes 2 and 3 are adjacent to node 1. Also, wlog assume that the first two rows of R are indexed by the edges $\{1,2\}$ and $\{1,3\}$, respectively. Then

$$R = \begin{bmatrix} (p^1 - p^2)^T & (p^2 - p^1)^T & 0 & 0 & \cdots & 0 \\ (p^1 - p^3)^T & 0 & (p^3 - p^1)^T & 0 & \cdots & 0 \\ 0 & R'_{.2} & R'_{.3} & R'_{.4} & \cdots & R'_{.n} \end{bmatrix},$$

where $R'_{.j}$ denotes the two columns of R' associated with node j. Let $\omega^T = [\alpha \ \beta \ \lambda^T]$, where α and β are scalars and $\lambda \in \mathbb{R}^{m-2}$. Then $\omega^T R = 0$ implies, by examining the first two columns of R, that $\alpha = \beta = 0$ since p^1, p^2, p^3 are not collinear. Therefore, $\lambda = 0$ since the rows of R' are linearly independent. Consequently, the rows of R are linearly independent and thus $\text{rank}(R) = m$.

$\square$

Generically locally rigid Laman graphs are also closed under H2.

Lemma 9.6 *Let (G,p) be a generic two-dimensional bar framework with n nodes and m edges, where G is a Laman graph. Assume that G is obtained from Laman graph G' by Henneberg step H2. Further, assume that (G',p') is locally rigid, where p' is the restriction of p to G'. Then (G,p) is locally rigid.*

Proof. Let R and R' be the rigidity matrices of (G,p) and (G',p'), respectively. Wlog assume that nodes 2, 3 and 4 are adjacent to node 1, the new node of G; and that $\{2,3\}$ is the deleted edge. Further, assume, wlog, that the first three rows of R

are indexed by the edges $\{1,2\}$, $\{1,3\}$ and $\{1,4\}$. Following [138], let us consider configuration q where $q^1 = (q^2 + q^3)/2$ and $(q^1 - q^4)$ is not parallel to $q^2 - q^3$ (see Fig. 9.5). Obviously, q is nongeneric. To prove the lemma, it suffices to show that R for (G,q) has rank m since this would imply that $\mathrm{rank}(R) \geq m$ for any generic (G,p). To this end,

$$
R = \begin{bmatrix}
\frac{1}{2}(q^3 - q^2)^T & \frac{1}{2}(q^2 - q^3)^T & \mathbf{0} & \mathbf{0} & \mathbf{0} & \cdots & \mathbf{0} \\
\frac{1}{2}(q^2 - q^3)^T & \mathbf{0} & \frac{1}{2}(q^3 - q^2)^T & \mathbf{0} & \mathbf{0} & \cdots & \mathbf{0} \\
(q^1 - q^4)^T & \mathbf{0} & \mathbf{0} & (q^4 - q^1)^T & \mathbf{0} & \cdots & \mathbf{0} \\
\mathbf{0} & \bar{R}'_{.2} & \bar{R}'_{.3} & \bar{R}'_{.4} & \bar{R}'_{.5} & \cdots & \bar{R}'_{.n}
\end{bmatrix},
$$

where $\bar{R}'_{.j}$ denotes the two columns of R', after deleting the row of R' indexed by edge $\{2,3\}$, associated with node j. Let $\omega^T = [\alpha \ \beta \ \gamma \ \lambda^T]$, where α, β, and γ are scalars and $\lambda \in \mathbb{R}^{m-3}$. Then $\omega^T R = \mathbf{0}$ implies, by examining the first two columns of R, that $\alpha = \beta$ and $\gamma = 0$ since $(q^1 - q^4)$ is not parallel to $q^2 - q^3$. Therefore, $\omega^T R = \mathbf{0}$ reduces to

$$
[\alpha/2 \ \lambda^T] \begin{bmatrix}
(q^2 - q^3)^T & (q^3 - q^2)^T & \mathbf{0} & \mathbf{0} & \cdots & \mathbf{0} \\
\bar{R}'_{.2} & \bar{R}'_{.3} & \bar{R}'_{.4} & \bar{R}'_{.5} & \cdots & \bar{R}'_{.n}
\end{bmatrix} = [\alpha/2 \ \lambda^T] R' = \mathbf{0}.
$$

Hence, $\alpha = 0$ and $\lambda = \mathbf{0}$ since the rows of R' are linearly independent. Consequently, the rows of R are linearly independent and thus $\mathrm{rank}(R) = m$.

$\square$

It should be pointed out that the proofs of Lemmas 9.5 and 9.6 amount to showing that the considered frameworks admit only zero stresses. In fact, if (G,p) is a two-dimensional framework with n nodes and $m = 2n - 3$ edges, then (9.3) implies that the dimension of the space of stress of (G,p) is equal to the dimension of its space of nontrivial infinitesimal flexes. As a result, framework (G,p) is infinitesimally rigid if and only if it does not admit any nonzero stress.

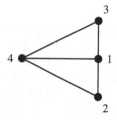

Fig. 9.5 The bar framework (G,q) of Example 9.11. Note that (G,q) has one missing edge namely $\{2,3\}$

Example 9.11 *To illustrate the proof of Lemma 9.6, let (G,q) be the two-dimensional nongeneric framework depicted in Fig. 9.5 and let R be its rigidity matrix. Then G is obtained from K_3 by an H2 step. Clearly, (G,q) admits no nonzero stress and hence $\mathrm{rank}(R) = 5$. Another way to see this is to note that the dual rigidity matrix of (G,q) is $\bar{R} = \sqrt{2}$, i.e., $\mathrm{rank}(\bar{R}) = \bar{m} = 1$.*

Theorem 9.15 (Laman [127]) *Let G be a graph with n nodes and $m = 2n - 3$ edges. Then G is generically locally rigid in dimension 2 if and only if G is a Laman graph.*

Proof.   To prove the "only if" part, assume that G is not a Laman graph and let (G, p) be a generic two-dimensional framework with rigidity matrix R. Then there exists an induced subgraph G' such that $|E(G')| > 2|V(G')| - 3$. Hence, the rows of R indexed by $E(G')$ are linearly dependent. Therefore, the rows of R are linearly dependent and thus rank$(R) < m$. Consequently, (G, p) is not locally rigid.

The "if" part is proved by induction on n with induction base $n = 2$ since K_2 is obviously locally rigid. The induction steps use Lemmas 9.3, 9.4, 9.5, and 9.6 and the fact that every Laman graph must have at least one node of degree either 2 or 3. $\square$

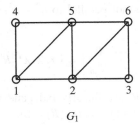 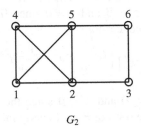

G_1 G_2

Fig. 9.6 The graphs of Example 9.12

Example 9.12 *Consider the two graphs depicted in Fig. 9.6. Graph G_1 is a Laman graph and thus is generically locally rigid in dimension 2, while graph G_2 is not a Laman graph and thus is generically locally flexible in dimension 2. The subgraph of G_2 induced by the nodes $V' = \{1, 2, 4, 5\}$ spans six edges and thus violates Laman Condition (ii).*

Let (G, p) be a two-dimensional locally rigid framework. We say that (G, p) is *minimally locally rigid* if it becomes locally flexible upon the deletion of any of its edges. Hence, if (G, p) is minimally locally rigid, then $m = 2n - 3$. Evidently, Laman graphs are the generically minimally locally rigid graphs in dimension 2. Using matroid theory, Lovász and Yemini [139] slightly generalized Laman theorem. Let (G, p) be a generic two-dimensional bar framework. The *generic degree of freedom* of a graph G, denoted by $\phi(G)$, is defined as

$$\phi(G) = 2n - 3 - \text{rank}(R),$$

where R is the rigidity matrix of (G, p). Hence, (G, p) is locally rigid if and only if $\phi(G) = 0$. A collection of nonempty subsets $E^1, \ldots, E^k$ of $E(G)$ is called a *partition* of $E(G)$ if $E^i \cap E^j = \emptyset$ whenever $i \neq j$ and $E^1 \cup \cdots \cup E^k = E(G)$.

Theorem 9.16 (Lovász and Yemini [139]) *The generic degree of freedom of graph G on n nodes, $n \geq 2$, is given by*

$$\phi(G) = 2n - 3 - \min \sum_{i=1}^{k} (2n_i - 3) \tag{9.18}$$

where the minimum is taken over all partitions $E^1, \ldots, E^k$ of $E(G)$ and n_i is the number of nodes incident to E^i.

Corollary 9.2 (Lovász and Yemini [139]) *Let (G, p) be a generic two-dimensional bar framework on n nodes. Then (G, p) is locally rigid if and only if*

$$2n - 3 \leq \sum_{i=1}^{k} (2n_i - 3) \tag{9.19}$$

for every partition $E^1, \ldots, E^k$ of $E(G)$, where n_i denotes the number of nodes incident to E^i.

Consider the partitioning of $E(G)$ into m subsets each consisting of a single edge. Then $\sum_{i=1}^{m} (2n_i - 3) = m$. As a result, any graph that satisfies Inequality (9.19) must have at least $2n - 3$ edges. Next, we show that Corollary 9.2 is equivalent to Laman Theorem. Let G have n nodes and $m = 2n - 3$ edges.

Now assume that (G, p) is locally rigid by Corollary 9.2 and let G' be an induced subgraph of G with n' nodes and m' edges. Consider the partition of $E(G)$ into $m - m' + 1$ subsets such that $E^1 = E'(G')$ and each of the remaining subsets $E^2, \ldots, E^{m-m'+1}$ consists of a single edge of $E(G) \backslash E'(G')$. Let n_1 denote the number of nodes incident to $E'(G')$, then $n_1 \leq n'$ since both end nodes of every edge in $E'(G')$ are in G'. Therefore, Inequality (9.19) implies that

$$m \leq 2n_1 - 3 + m - m'.$$

Therefore, $m' \leq 2n' - 3$ and thus (G, p) is locally rigid by Laman Theorem.

To prove the reverse direction, assume that (G, p) is locally rigid by Laman Theorem and let $E^1, \ldots, E^k$ be a partition of $E(G)$ where $|E^i| = m_i$. For $i = 1, \ldots, k$, let V'^i be the set of nodes incident to E^i and let $n_i = |V'^i|$. Let $G'^1, \ldots, G'^k$ be the subgraphs of G induced by $V'^1, \ldots, V'^k$, respectively, and let $m'_i = |E'(G'^i)|$. Then $E^i \subseteq E'(G'^i)$ and thus $m_i \leq m'_i$. Moreover, for each $i = 1, \ldots, k$, we have $m'_i \leq 2n_i - 3$. Therefore,

$$\sum_{i=1}^{k} (2n_i - 3) \geq \sum_{i=1}^{k} m'_i \geq m.$$

Hence, (G, p) is locally rigid by Corollary 9.2.

We conclude this chapter by noting that generically locally rigid graphs in dimension 2, i.e., Laman graphs, can be recognized by a fast algorithm due to Jacobs and Hendrickson [115] known as the *pebble game* (see also [133]). Laman graphs can also be efficiently recognized by matroidal methods [157, 83, 139]. On the other hand, a characterization of generically locally rigid graphs in dimension 3 is not known. In fact, the three-dimensional analogue of Laman Conditions namely:

  (i) $|E(G)| = 3|V(G)| - 6$,
  (ii) $|E(G')| \leq 3|V(G')| - 6$ for each induced subgraph G' with $|V(G')| \geq 3$,

are not sufficient for generic local rigidity in dimension 3 as shown by the so-called *double banana graph* [173, p. 14]. It should be pointed out that Condition (ii) trivially holds if $|V(G')| = 3$ or 4.

Chapter 10
Universal and Dimensional Rigidities

In this chapter, we study the universal rigidity problem of bar frameworks and the related problem of dimensional rigidity. The main tools in tackling these two problems are the Cayley configuration spectrahedron $\mathscr{F}$, defined in (8.10), and Ω, the stress matrix, defined in (8.13). The more general problem of universally linked pair of nonadjacent nodes is also studied and the results are interpreted in terms of the Strong Arnold Property and the notion of nondegeneracy in semidefinite programming.

10.1 Definitions and Basic Results

Recall that D_p is the EDM defined by configuration p and H is the adjacency matrix of graph G. Also, recall that $(\circ)$ denotes the Hadamard product.

Definition 10.1 *Let (G, p) be an r-dimensional bar framework with n nodes. Then (G, p) is said to be* universally rigid *if for any integer $s : 1 \leq s \leq n - 1$, there does not exist an s-dimensional bar framework (G, p') such that $H \circ D_{p'} = H \circ D_p$ and $D_{p'} \neq D_p$.*

In other words, (G, p) is universally rigid if every bar framework (G, p') that is equivalent to (G, p) is actually congruent to it. Note that the notion of universal rigidity is equivalent to the uniqueness of EDM completions, i.e., a given EDM completion is unique if and only if the corresponding bar framework is universally rigid. Moreover, from a geometric viewpoint, (G, p) is universally rigid if and only if its Cayley configuration spectrahedron is a singleton, i.e., $\mathscr{F} = \{0\}$. We should remark here that the notion of *unique localizability* of So and Ye [179] is very close, but not identical, to the notion of universal rigidity. Evidently, universal rigidity implies local rigidity but the converse is not true. As a result, universal rigidity is a stronger notion than local rigidity.

© Springer Nature Switzerland AG 2018

A.Y. Alfakih, *Euclidean Distance Matrices and Their Applications in Rigidity Theory*, https://doi.org/10.1007/978-3-319-97846-8_10

As will be shown below, the notions of affine motions and dimensional rigidity are indispensable for the study of universal rigidity. An *affine motion* in $\mathbb{R}^r$ is a map $f : \mathbb{R}^r \to \mathbb{R}^r$ of the form

$$f(p^i) = Ap^i + b \qquad (10.1)$$

for all $p^i \in \mathbb{R}^r$, where A is an $r \times r$ nonsingular matrix[1] and b is a vector in $\mathbb{R}^r$. Observe that if A is orthogonal, then the affine motion is a rigid motion. The following lemma is an immediate consequence of (10.1).

Lemma 10.1 *Let (G, p) and (G, p') be two r-dimensional bar frameworks on n nodes. Then configuration p' is obtained from configuration p by an affine motion if and only if the Gale spaces of (G, p) and (G, p') are equal.*

Proof. The "only if" part is obvious. To prove the "if" part, assume that (G, p) and (G, p') have the same Gale space and let P and P' be two configuration matrices of (G, p) and (G, p'), respectively. Then $\mathrm{col}([P'\ e]) = \mathrm{col}([P\ e])$. Hence, $P' = PA + eb^T$ for some $r \times r$ matrix A and some $b \in \mathbb{R}^r$. Wlog assume that $P^T e = P'^T e = \mathbf{0}$. Then $b = \mathbf{0}$ and thus A is nonsingular since both P and P' have rank r.   □

Let (G, p) be an r-dimensional bar framework. (G, p) is said to admit a *nontrivial affine flex* if there exists an r-dimensional bar framework (G, p') such that: (i) (G, p') is equivalent, but not congruent, to (G, p) and (ii) configuration p' is obtained from configuration p by an affine motion; in which case, we say that (G, p') is obtained from (G, p) be a nontrivial affine flex. For example, in Fig. 10.1, framework (a), unlike framework (b), admits a nontrivial affine flex.

Lemma 10.2 *Let (G, p) be an r-dimensional bar framework with n nodes, $r \leq n - 2$. Let X be the projected Gram matrix of (G, p) and let U be the matrix whose columns form an orthonormal basis of $\mathrm{null}(X)$. Then the following two statements are equivalent:*

(i) (G, p) admits a nontrivial affine flex.
(ii) There exists a nonzero $y \in \mathbb{R}^{\bar{m}}$ such that

$$\mathcal{M}(y)U = \mathbf{0}, \qquad (10.2)$$

where $\mathcal{M}(y)$ is defined in (9.6).

Proof. Assume that (G, p') is obtained from (G, p) by a nontrivial affine flex. Let X and X' be the projected Gram matrices of (G, p) and (G, p'), respectively. Then

$$X' = X + \mathcal{M}(y)$$

for some nonzero $y \in \mathbb{R}^{\bar{m}}$ and the Gale spaces of (G, p) and (G, p') are equal. Therefore, by Lemma 3.8, $\mathcal{M}(y)U = (X' - X)U = \mathbf{0}$ and thus Statement (i) implies Statement (ii).

---

[1] Affine motions are often defined without the nonsingularity assumption on A. However, this assumption is more convenient for our purposes.

To prove the other direction, assume that there exists a nonzero $y \in \mathbb{R}^{\bar{m}}$ such that $\mathcal{M}(y)U = 0$. Let $X = W\Lambda W^T$ be the spectral decomposition of X, where Λ is the diagonal matrix consisting of the positive eigenvalues of X. Thus $Q = [W \ U]$ is orthogonal. Let $X' = X + t\mathcal{M}(y)$, where t is a scalar. Now X' is the projected Gram matrix of an r-dimensional framework (G, p') that is equivalent, but not congruent, to (G, p) if and only if X' is PSD of rank r. But

$$Q^T X' Q = \begin{bmatrix} \Lambda + tW^T \mathcal{M}(y) W & 0 \\ 0 & 0 \end{bmatrix}.$$

Therefore, X' is PSD of rank r for a sufficiently small $t > 0$. Moreover, (G, p') is obtained from (G, p) by a nontrivial affine flex since $X'U = 0$.

□

In order to provide a geometric interpretation of Lemma 10.2, let $\mathcal{F}$ denote the Cayley configuration spectrahedron of (G, p) and recall that face$(x, \mathcal{F})$ denotes the smallest face of $\mathcal{F}$ containing x. Also, recall from Theorem 2.21 that the affine hull of face$(0, \mathcal{F})$ is given by

$$\text{aff}(\text{face}(0, \mathcal{F})) = \{y \in \mathbb{R}^{\bar{m}} : \mathcal{M}(y)U = 0\}.$$

Consequently, (G, p) admits a nontrivial affine flex if and only if the dimension of the affine hull of face$(0, \mathcal{F})$ is ≥ 1; i.e., aff$(\text{face}(0, \mathcal{F})) \neq \{0\}$.

Next, we turn to the notion of dimensional rigidity.

Definition 10.2 *Let (G, p) be an r-dimensional bar framework with n nodes, $r \leq n - 2$. Then (G, p) is said to be* dimensionally rigid *if for any integer $s : r + 1 \leq s \leq n - 1$, there does not exist an s-dimensional bar framework (G, p') such that $H \circ D_{p'} = H \circ D_p$.*

In other words, if an r-dimensional bar framework (G, p) is dimensionally rigid and if (G, p') is equivalent to (G, p), then (G, p') is of dimension r or less. Furthermore, from a geometric viewpoint, (G, p) is dimensionally rigid if and only if $r \geq \text{rank}(\mathcal{X}(y))$ for all $y \in \mathcal{F}$ if and only if 0 lies in the relative interior of $\mathcal{F}$ (Theorem 2.20). Clearly, universal rigidity implies dimensional rigidity but the converse is not true. For example, in Fig. 10.1, framework (a) is dimensionally rigid since it does not have an equivalent three-dimensional framework. Put differently, every framework that is equivalent to (a) is of dimension 2 or 1. On the other hand, framework (b) is not dimensionally rigid since it has an infinite number of equivalent three-dimensional frameworks.

As the following theorem shows, the universal rigidity problem, i.e., the problem of determining whether or not a given bar framework (G, p) is universally rigid, can be split into two independent problems: the affine flexing problem, i.e., whether or not (G, p) admits a nontrivial affine flex, and the dimensional rigidity problem, i.e., whether or not (G, p) is dimensionally rigid.

Theorem 10.1 (Alfakih [6]) *Let (G, p) be an r-dimensional bar framework with n nodes, $r \leq n - 2$. Then (G, p) is universally rigid if and only if the two following conditions hold.*

  (i) *(G,p) is dimensionally rigid.*
  (ii) *(G,p) does not admit a nontrivial affine flex.*

We present two proofs of this theorem, the second of which is geometric.

First proof. The "only if" part is obvious. To prove the "if" part assume, by way of contradiction, that Conditions (i) and (ii) hold and (G,p) is not universally rigid. Then, there exists a framework (G,p') such that (G,p') is equivalent, but not congruent, to (G,p). Let X and X' be the projected Gram matrices of (G,p) and (G,p'), respectively. Then $X' = \mathscr{X}(y) = X + \mathscr{M}(y)$ for some nonzero y in $\mathbb{R}^{\bar{m}}$. Now, for a sufficiently small $\delta > 0$, $\mathscr{X}(ty) = X + t\mathscr{M}(y)$ is PSD for all $t : 0 \le t \le \delta$. Moreover, by the lower semicontinuity of the rank function, $\text{rank}(\mathscr{X}(ty)) \ge r$ for all $t : 0 \le t \le \delta$. Hence, by Condition (i), $\text{rank}(\mathscr{X}(ty)) = r$ for all $t : 0 \le t \le \delta$.

Let U be the matrix whose columns form an orthonormal basis of $\text{null}(X)$ and let $X = W\Lambda W^T$ be the spectral decomposition of X, where Λ is the diagonal matrix consisting of the positive eigenvalues of X. Thus $Q = [W\ U]$ is orthogonal. Hence,

$$Q^T \mathscr{X}(ty) Q = \begin{bmatrix} \Lambda + t\, W^T \mathscr{M}(y) W & t\, W^T \mathscr{M}(y) U \\ t\, U^T \mathscr{M}(y) W & t\, U^T \mathscr{M}(y) U \end{bmatrix}$$

is PSD and of rank r for all $t : 0 \le t \le \delta$. Therefore, $U^T \mathscr{M}(y) U = \mathbf{0}$ and thus $W^T \mathscr{M}(y) U = \mathbf{0}$. Consequently, $\mathscr{M}(y) U = \mathbf{0}$, which, in light of Lemma 10.2, contradicts Condition (ii).

$\square$

Second proof. Again the "only if" part is obvious. Now Condition (i) implies that $\mathbf{0}$ lies in the relative interior of $\mathscr{F}$, where $\mathscr{F}$ denotes the Cayley configuration spectrahedron of (G,p). Hence, by Theorem 2.20, $\text{face}(\mathbf{0}, \mathscr{F}) = \text{face}(\mathscr{F}, \mathscr{F}) = \mathscr{F}$. On the other hand, Condition (ii) implies that $\text{aff}(\text{face}(\mathbf{0}, \mathscr{F})) = \{\mathbf{0}\}$. Therefore, $\mathscr{F} = \{\mathbf{0}\}$ and hence, (G,p) is universally rigid.

$\square$

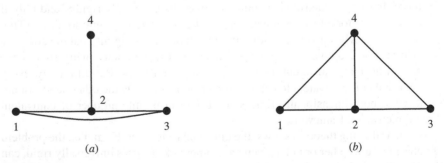

Fig. 10.1 Two two-dimensional bar frameworks. Edge $\{1,3\}$ in framework (**a**) is drawn as an arc to make edges $\{1,2\}$ and $\{2,3\}$ visible. Framework (**a**) is dimensionally rigid and admits a nontrivial affine flex. Framework (**b**) is not dimensionally rigid and does not admit a nontrivial affine flex

Consider the frameworks of Fig. 10.1. Both frameworks are not universally rigid since framework (*a*) admits a nontrivial affine flex, while framework (*b*) is not dimensionally rigid.

Theorem 10.1 allows us to tackle the universal rigidity problem by tackling the affine flexing problem and the dimensional rigidity problem separately.

10.2 Affine Flexes

We present two characterizations of bar frameworks that admit nontrivial affine flexes. These characterizations are then used to identify four cases where a given bar framework does not admit a nontrivial affine flex. As a result, in these cases, universal rigidity coincides with dimensional rigidity. The first of these characterizations, given in the following lemma, is in terms of $E(G)$, the edges of the underlying graph, and the points $p^1, \ldots, p^n$ of configuration p.

Lemma 10.3 (Connelly [61]) *Let (G, p) be an r-dimensional bar framework. Then the following two statements are equivalent:*

(i) (G, p) admits a nontrivial affine flex.
(ii) There exists a nonzero symmetric $r \times r$ matrix Φ such that

$$(p^i - p^j)^T \Phi (p^i - p^j) = 0 \text{ for all } \{i, j\} \in E(G). \tag{10.3}$$

Proof. Assume that Statement (i) holds and let (G, p') be a bar framework obtained from (G, p) be a nontrivial affine flex. Then

$$||p^i - p^j||^2 = ||p'^i - p'^j||^2 = (p^i - p^j)^T A^T A (p^i - p^j) \text{ for all } \{i, j\} \in E(G).$$

Therefore, $(p^i - p^j)^T \Phi (p^i - p^j) = 0$ for all $\{i, j\} \in E(G)$, where $\Phi = I - A^T A$. Note that Φ is symmetric and nonzero since A is not orthogonal.

On the other hand, assume that Statement (ii) holds and observe that $I - \varepsilon\Phi$ is PD for a sufficiently small $\varepsilon > 0$. Then $I - \varepsilon\Phi$ can be factorized as $I - \varepsilon\Phi = A^T A$, where A is nonsingular. Note that A is not orthogonal since $\varepsilon > 0$. Let $p'^i = Ap^i$ for $i = 1, \ldots, n$. Then $||p'^i - p'^j||^2 = (p^i - p^j)^T (I - \varepsilon\Phi)(p^i - p^j) = ||p^i - p^j||^2$ for all $\{i, j\} \in E(G)$. Hence, (G, p') is obtained from (G, p) by a nontrivial affine flex and thus Statement (i) holds.

$\square$

Let C be the $m \times (r(r+1)/2)$ matrix whose rows are indexed by the edges of (G, p), where the row indexed by edge $\{i, j\}$ is given by

$$(p^i - p^j)^T \otimes_s (p^i - p^j)^T. \tag{10.4}$$

Evidently, there exists an $r \times r$ nonzero symmetric Φ satisfying (10.3) if and only if C has a nontrivial null space. Few remarks are in order here. First, if a bar framework satisfies Condition (ii) of Lemma 10.3, then we say that the edge directions of (G, p)

lie on a *conic at infinity* [61]. Second, we saw earlier that $(n-1)$-dimensional bar frameworks are locally flexible. Now by Lemma 10.3, such frameworks actually admit nontrivial affine flexes. This follows since for $r = n-1$ and for incomplete graphs, the null space of A is nontrivial since $m < n(n-1)/2$. Third, as a result of Assumption 8.1, any one-dimensional bar framework with at least one edge does not admit a nontrivial affine flex. As a result, universal rigidity and dimensional rigidity coincide on the line.

Using Lemma 10.3, we present, next, the first case where framework (G, p) does not admit a nontrivial affine flex.

Theorem 10.2 (Connelly [61]) *Let (G, p) be an r-dimensional bar framework. If (G, p) is generic and if $\deg(i) \geq r$ for every node i of G. Then (G, p) does not admit a nontrivial affine flex.*

Proof.   First, note that, by the premise of the theorem, the number of edges of G is $m \geq nr/2 \geq r(r+1)/2$. Thus, matrix C as defined in (10.4) has at least as many rows as columns. Also, note that it suffices to prove that the result is true for a particular framework, not necessarily generic. The proof is by induction on r. For $r = 2$, (G, p) has at least 3 edges. If no two of these edges are parallel, then $\text{null}(C) = \{0\}$ and the assertion of the theorem is true.

Now assume that the assertion of the theorem is true for $r = k$ and consider the $(k+1)$-dimensional bar framework (G, p), where $p^n = \mathbf{0}$ and $p^1, \ldots, p^{n-1}$ lie in a hyperplane, say [138] $H = \{p \in \mathbb{R}^{k+1} : p^T e^1 = 1\}$, where e^1 is the first standard unit vector in $\mathbb{R}^{k+1}$. Thus, $p^i = \begin{bmatrix} 1 \\ p'^i \end{bmatrix}$ for $i = 1, \ldots, n-1$. Let G' be the graph obtained from G by deleting node n and the edges incident to it. Wlog assume that the edges incident to node n are $\{1, n\}, \ldots, \{s, n\}$ where $s \geq k+1$. Let (G', p') be the k-dimensional bar framework where $p' = (p'^1, \ldots, p'^{n-1})$ and assume that (G', p') is generic. Obviously, $\deg(i) \geq k$ for each node i of G'.

Let $(p^i - p^j)^T \Phi (p^i - p^j) = 0$ for all $\{i, j\} \in E(G)$ and assume that Φ is partitioned as $\Phi = \begin{bmatrix} \sigma & \rho^T \\ \rho & \Phi' \end{bmatrix}$, where Φ' is $k \times k$, $\rho \in \mathbb{R}^k$ and σ is a scalar. Clearly, $E(G) = E(G') \cup \{\{1, n\}, \ldots, \{s, n\}\}$.

Then, for all $\{i, j\} \in E(G')$, the equation $(p^i - p^j)^T \Phi (p^i - p^j) = 0$ reduces to

$$[0 \ \ (p'^i - p'^j)^T] \begin{bmatrix} \sigma & \rho^T \\ \rho & \Phi' \end{bmatrix} \begin{bmatrix} 0 \\ p'^i - p'^j \end{bmatrix} = (p'^i - p'^j)^T \Phi' (p'^i - p'^j) = 0.$$

Hence, by the induction hypothesis, $\Phi' = \mathbf{0}$. Consequently, for edge $\{i, n\}$, the equation $(p^i - p^j)^T \Phi (p^i - p^j) = 0$ reduces to

$$[1 \ \ (p'^i)^T] \begin{bmatrix} \sigma & \rho^T \\ \rho & \mathbf{0} \end{bmatrix} \begin{bmatrix} 1 \\ p'^i \end{bmatrix} = [1 \ \ (p'^i)^T] \begin{bmatrix} \sigma \\ 2\rho \end{bmatrix} = 0.$$

As a result, for edges $\{1, n\}, \ldots, \{k+1, n\}$, the equation $(p^i - p^j)^T \Phi (p^i - p^j) = 0$ reduces to

$$\begin{bmatrix} 1 & (p'^1)^T \\ \vdots & \vdots \\ 1 & (p'^{k+1})^T \end{bmatrix} \begin{bmatrix} \sigma \\ 2\rho \end{bmatrix} = \mathbf{0}. \tag{10.5}$$

But since (G', p') is generic, the points $p'^1, \ldots, p'^{k+1}$ are affinely independent. Hence, the matrix in (10.5) is nonsingular and thus $\sigma = 0$ and $\rho = \mathbf{0}$. Therefore, $\Phi = \mathbf{0}$ and the result follows.

$\square$

The second characterization of bar frameworks that admit nontrivial affine flexes, given in the following lemma, is in terms of $\bar{E}(\bar{G})$, the missing edges of the underlying graph, and Gale matrix Z of configuration p. This characterization allows us to replace the generic assumption in Theorem 10.2 by a general position assumption and a rank assumption on the stress matrix.

Lemma 10.4 (Alfakih [9]) *Let (G, p) be an r-dimensional bar framework with n nodes, $r \leq n - 2$. Let Z be a Gale matrix of (G, p). Then the following two statements are equivalent:*

(i) (G, p) admits a nontrivial affine flex.
(ii) There exists a nonzero $y \in \mathbb{R}^{\bar{m}}$ such that

$$V^T \mathscr{E}(y) Z = \mathbf{0}, \tag{10.6}$$

where $\mathscr{E}(y)$ is defined in (9.9) and V is defined in (3.11).

Proof. In light of Lemma 3.8, this is just a restatement of Lemma 10.2.

$\square$

By definition, if (G, p) admits a nontrivial affine flex, then (G, p) is infinitesimally flexible. Thus, in light of Theorem 9.8 and Lemma 10.4, it should come as no surprise that if $V^T \mathscr{E}(y) Z = \mathbf{0}$ for some nonzero y, then $Z^T \mathscr{E}(y) Z = \mathbf{0}$.

Example 10.1 *Consider the frameworks of Fig. 10.1. A Gale matrix of both frameworks is $Z = [1 \ -2 \ 1 \ 0]^T$. Note that $V^T \mathscr{E}(y) Z = \mathbf{0}$ is equivalent to $\mathscr{E}(y) Z = e\zeta$. Thus, for framework (a), the system of equation $\mathscr{E}(y) Z = e\zeta$, i.e.,*

$$\begin{bmatrix} 0 & 0 & 0 & y_{14} \\ 0 & 0 & 0 & 0 \\ 0 & 0 & 0 & y_{34} \\ y_{14} & 0 & y_{34} & 0 \end{bmatrix} \begin{bmatrix} 1 \\ -2 \\ 1 \\ 0 \end{bmatrix} = \begin{bmatrix} \zeta \\ \zeta \\ \zeta \\ \zeta \end{bmatrix}$$

has a solution $y_{14} = -y_{34} = 1$, $\zeta = 0$. Accordingly, framework (a) admits a nontrivial affine flex. On the other hand, for framework (b), the only solution of

$$\begin{bmatrix} 0 & 0 & y_{13} & 0 \\ 0 & 0 & 0 & 0 \\ y_{13} & 0 & 0 & 0 \\ 0 & 0 & 0 & 0 \end{bmatrix} \begin{bmatrix} 1 \\ -2 \\ 1 \\ 0 \end{bmatrix} = \begin{bmatrix} \zeta \\ \zeta \\ \zeta \\ \zeta \end{bmatrix}$$

is the trivial solution $y_{13} = \zeta = 0$. *Accordingly, framework* (b) *does not admit a nontrivial affine flex.*

It is worth pointing out that matrix Φ of Lemma 10.3 can be written explicitly in terms of vector y of Lemma 10.4 and vice versa. To this end, Eq. (10.3) is equivalent to $H \circ \mathcal{K}(P\Phi P^T) = \mathbf{0}$. Hence,

$$P\Phi P^T = \mathcal{E}(y) + ae^T + ea^T,$$

where $2a = \text{diag}(P\Phi P^T)$. Consequently, by multiplying this equation from the left by V^T and from the right by Z, we obtain Eq. (10.6). Moreover, assuming that $P^T e = \mathbf{0}$, we have

$$\Phi = (P^T P)^{-1} P^T \mathcal{E}(y) P (P^T P)^{-1},$$

$$\mathcal{E}(y) = -\frac{1}{2}\mathcal{K}(P\Phi P^T).$$

Lemma 10.4 is particularly useful when the framework (G, p) is in general position. In fact, Lemma 10.4 is used to identify several cases where a bar framework does not admit a nontrivial affine flex. But before we proceed, let us prove a stronger version of Lemma 8.4 by dropping the requirement that Ω is PSD.

Lemma 10.5 *Let* (G, p) *be an r-dimensional bar framework with n nodes and assume that* (G, p) *is in general position in* $\mathbb{R}^r$. *Let* Ω *be a stress matrix of* (G, p) *of rank* $n - r - 1$. *Then* $\deg(i) \geq r + 1$ *for every node i of G.*

Proof. Let $\Omega = Z\Psi Z^T$, then Ψ is nonsingular. By way of contradiction, assume that $\deg(v) \leq r$ for some node v. Then the vth column of Ω must have at least $\bar{r} = n - 1 - r$ zero entries. Wlog, assume that $v = n$ and that the first $\bar{r}$ entries of the nth column of Ω are all zeros; i.e., the first $\bar{r}$ entries of $Z\Psi z^n$ are all zeros. But by Lemma 3.1, the square submatrix of Z indexed by the first $\bar{r}$ rows and columns is nonsingular. Thus, we have a contradiction since Ψ is nonsingular and $z^n \neq \mathbf{0}$. □

Before we present another case of frameworks not admitting a nontrivial affine flex, we need the following crucial lemma, which we will use repeatedly in the sequel.

Lemma 10.6 ([16]) *Let* (G, p) *be an r-dimensional bar framework on n nodes, $r \leq n - 2$, and let* Ω *be a nonzero stress matrix of* (G, p). *Then the systems of equations* $V^T \mathcal{E}(y) \Omega = \mathbf{0}$ *and* $\mathcal{E}(y) \Omega = \mathbf{0}$ *are equivalent, where* $\mathcal{E}(y)$ *is as defined in (9.9).*

Proof. Obviously, if $\mathcal{E}(y)\Omega = \mathbf{0}$ for some y, then $V^T \mathcal{E}(y)\Omega = \mathbf{0}$. Now assume that $V^T \mathcal{E}(y)\Omega = \mathbf{0}$. Then $\mathcal{E}(y)\Omega = e\zeta^T$ for some ζ in $\mathbb{R}^n$. Hence, to complete the proof, it suffices to show that $\zeta = \mathbf{0}$. To this end, recall that $\Omega_{ij} = 0$ if $\{i, j\} \in \bar{E}(\bar{G})$ and $\mathcal{E}(y)_{ij} = 0$ if either $i = j$ or $\{i, j\} \in E(G)$. Therefore, for $i = 1, \ldots, n$, we have

$$(\mathscr{E}(y)\Omega)_{ii} = \sum_{j=1}^{n} \mathscr{E}(y)_{ij}\Omega_{ji}$$

$$= \mathscr{E}(y)_{ii}\Omega_{ii} + \sum_{j:\{i,j\}\in E(G)} \mathscr{E}(y)_{ij}\Omega_{ji} + \sum_{j:\{i,j\}\in \bar{E}(\bar{G})} \mathscr{E}(y)_{ij}\Omega_{ji}$$

$$= 0.$$

Thus, $\mathrm{diag}(\mathscr{E}(y)\Omega) = \zeta = 0$.

$\square$

Theorem 10.3 (Alfakih and Ye [20]) *Let (G,p) be an r-dimensional bar framework with n nodes, $r \leq n-2$. If (G,p) is in general position in $\mathbb{R}^r$ and if (G,p) admits a stress matrix Ω of rank $n-1-r$, then (G,p) does not admit a nontrivial affine flex.*

Proof. Let Ω be a stress matrix of (G,p) with rank $\bar{r} = n-r-1$. Let Z be the matrix consisting of the first $\bar{r}$ columns of Ω. Then by Lemma 8.3, Z is a Gale matrix of (G,p). Let $V^T \mathscr{E}(y)Z = 0$ for some y. Then, it follows from Lemma 10.6 that

$$\mathscr{E}(y)Z = 0 \tag{10.7}$$

But System (10.7) consists of n equations, one for each node of G, where the equation corresponding to node i is given by

$$\sum_{j:\{i,j\}\in \bar{E}(\bar{G})} y_{ij}z^j = 0. \tag{10.8}$$

Now, by Lemma 10.5, $\deg(i) \geq r+1$ for every node i of G and thus the number of nodes of G not adjacent to i is at most $\bar{r}-1$. Therefore, the LHS of (10.8) is a linear combination of at most $\bar{r}-1$ of the Gale transforms $z^1,\ldots,z^n$. But by Lemma 3.1, each $\bar{r}$ of the vectors $z^1,\ldots,z^n$ are linearly independent. Consequently, the only solution of Eq. (10.7) is $y = 0$ and the result follows.

$\square$

Theorem 10.3 was strengthened and generalized in [17]. The third case of frameworks not admitting a nontrivial affine flex is given in the following theorem.

Theorem 10.4 *Let (G,p) be an r-dimensional bar framework with n nodes, $r \leq n-2$. Assume that (G,p) is in general position in $\mathbb{R}^r$. If $\deg(i) \geq r$ for all nodes i of G and $\deg(v) = n-1$ for some node v, then (G,p) does not admit a nontrivial affine flex.*

Proof. Let $\mathscr{E}(y)Z = e\zeta^T$ for some $\zeta \in \mathbb{R}^{n-1-r}$. If $\deg(v) = n-1$, then the vth row of $\mathscr{E}(y)$ has all zeros. Therefore, $\zeta = 0$ and the rest of the proof proceeds as in the proof of Theorem 10.3.

$\square$

Finally, we present the fourth case of frameworks not admitting a nontrivial affine flex.

Theorem 10.5 *Let* (G,p) *be an* r-*dimensional bar framework with* n *nodes,* $r \leq$ $n-2$. *If* (G,p) *has a clique of size* $r+1$ *and if the nodes of this clique are affinely independent, then* (G,p) *does not admit a nontrivial affine flex.*

Proof.   Wlog assume that the nodes of this clique are $1,\dots,r+1$ and thus $p^1,\dots,p^{r+1}$ are affinely independent. Let Z be a Gale matrix of (G,p). Then by Lemma 3.7, we can assume that Z is of the form $Z = \begin{bmatrix} \bar{Z} \\ I_{\bar{r}} \end{bmatrix}$, where $\bar{r} = n-1-r$. As always, let e_k denote the vector of all 1s in $\mathbb{R}^k$. Then, clearly $e_{r+1}^T \bar{Z} = -e_{\bar{r}}^T$. Assume that $\mathscr{E}(y)$ is partitioned as

$$\mathscr{E}(y) = \begin{bmatrix} \mathbf{0} & \mathscr{E}_1(y_1) \\ \mathscr{E}_1^T(y_1) & \mathscr{E}_2(y_2) \end{bmatrix},$$

where $\mathscr{E}_2(y_2)$ is $\bar{r} \times \bar{r}$. Then, $\mathscr{E}(y)Z = e\zeta^T$ implies that

$$\mathscr{E}_1(y_1) = e_{r+1}\zeta^T \text{ and } \mathscr{E}_1^T(y_1)\bar{Z} + \mathscr{E}_2(y_2) = e_{\bar{r}}\zeta^T.$$

Hence, $\mathscr{E}_2(y_2) = e_{\bar{r}}\zeta^T + \zeta e_{\bar{r}}^T$. But $\text{diag}(\mathscr{E}_2(y_2)) = \mathbf{0}$. Therefore, $\zeta = \mathbf{0}$ and thus $\mathscr{E}_1(y_1) = \mathbf{0}$ and $\mathscr{E}_2(y_2) = \mathbf{0}$. Consequently, $y = \mathbf{0}$ and the result follows.
□

Consider the r-dimensional bar framework (G,p) representing an ad hoc wireless sensor network. If the number of anchors of (G,p) is $\geq r+1$ and if these anchors are affinely independent, then these anchors can be thought of as inducing a clique of G. Consequently, by Theorem 10.5, framework (G,p) does not admit a nontrivial affine flex.

Next, we turn to dimensional rigidity.

10.3 Dimensional Rigidity

The stress matrix Ω plays a crucial role in the dimensional rigidity problem. The following theorem presents a sufficient condition for dimensional rigidity in terms of Ω. As Example 10.2 below shows, this sufficient condition is not necessary in general. We will elaborate on this point later in this section.

Theorem 10.6 (Alfakih [6]) *Let* (G,p) *be an* r-*dimensional bar framework with* n *nodes,* $r \leq n-2$. *If* (G,p) *admits a PSD stress matrix* Ω *of rank* $n-1-r$, *then* (G,p) *is dimensionally rigid.*

Proof.   We prove the contrapositive statement. Hence, assume that (G,p) is not dimensionally rigid and let X be the projected Gram matrix of (G,p). Therefore, there exists $y \neq \mathbf{0}$ such that $\mathscr{X}(y) = X + \mathscr{M}(y) \succeq \mathbf{0}$ and $\text{rank}(\mathscr{X}(y)) \geq r+1$. Let W and U be the two matrices whose columns form orthonormal bases of $\text{col}(X)$ and $\text{null}(X)$, respectively, and thus $Q = [W\ U]$ is orthogonal. Hence,

$$Q^T \mathscr{X}(y)Q = \begin{bmatrix} W^T X W + W^T \mathscr{M}(y)W & W^T \mathscr{M}(y)U \\ U^T \mathscr{M}(y)W & U^T \mathscr{M}(y)U \end{bmatrix} \succeq 0.$$

Consequently, $U^T \mathscr{M}(y)U \succeq 0, \neq 0$ and $\mathrm{null}(U^T \mathscr{M}(y)U) \subseteq \mathrm{null}(W^T \mathscr{M}(y)U)$. Now by Lemma 3.8, $U^T \mathscr{M}(y)U \succeq 0, \neq 0$ if and only if

$$Z^T \mathscr{E}(y)Z = \sum_{\{i,j\} \in \bar{E}(\bar{G})} y_{ij}(z^i z^{j^T} + z^j z^{i^T}) \succeq 0, \neq 0. \tag{10.9}$$

But, by the homogeneous Farkas lemma (Corollary 2.4), (10.9) holds if and only if there does not exist $\Psi \succ 0$ such that $z^{i^T} \Psi z^j = 0$ for all $\{i, j\} \in \bar{E}(\bar{G})$. Consequently, in light of Theorem 8.4, (10.9) holds if and only if (G, p) admits no PSD stress matrix Ω of rank $n - 1 - r$.

$\square$

The reader is encouraged to find a simple geometric proof of Theorem 10.6. By combining Theorems 10.6 and 10.1 we obtain the following sufficient condition for universal rigidity.

Theorem 10.7 (Connelly [57, 60] and Alfakih [6]) *Let (G, p) be an r-dimensional bar framework with n nodes, $r \leq n - 2$. If the following two conditions hold:*

(i) (G, p) admits a PSD stress matrix Ω of rank $n - 1 - r$.
(ii) (G, p) does not admit a nontrivial affine flex,

then (G, p) is universally rigid.

Theorem 10.7 will be strengthened below (Theorem 10.13). Also, combining Theorems 10.6, 10.1 and 10.3, we have

Theorem 10.8 (Alfakih and Ye [20]) *Let (G, p) be an r-dimensional bar framework with n nodes, $r \leq n - 2$. If (G, p) is in general position in $\mathbb{R}^r$ and if (G, p) admits a PSD stress matrix Ω of rank $n - 1 - r$, then (G, p) is universally rigid.*

Recall that the maximum possible rank of a stress matrix is $n - 1 - r$. Now as we mentioned earlier and as the next example shows, the converse of Theorem 10.6 is not true in general [6]. That is, if (G, p) is dimensionally rigid, then it may or may not admit a PSD stress matrix of maximal rank. This issue will be investigated in detail in this and the next section. An important point to bear in mind is that if (G, p) is dimensionally rigid, then Theorem 8.5 guarantees that (G, p) admits a stress matrix Ω of rank ≥ 1. What is not guaranteed, however, is that $\mathrm{rank}(\Omega) = n - 1 - r$. In fact, by Theorem 5.10 and as will be discussed later in this section, Ω attains its maximum rank if and only if the singularity degree of $\mathscr{F}$, the Cayley configuration spectrahedron of (G, p), is 1.

Now suppose that (G, p) does not admit a PSD stress matrix Ω of rank $n - r - 1$. Then, by Corollary 2.4, there exists a nonzero y such that $U^T \mathscr{M}(y)U$ is a nonzero PSD matrix. However, it could happen, as illustrated in the following example, that $\mathrm{null}(U^T \mathscr{M}(y)U) \nsubseteq \mathrm{null}(W^T \mathscr{M}(y)U)$ and consequently we cannot conclude that $\mathscr{X}(y) = X + \mathscr{M}(y)$ is PSD, i.e., we cannot conclude that (G, p) is not dimensionally rigid.

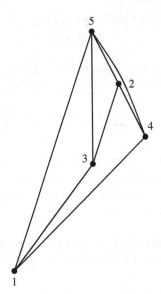

Fig. 10.2 The bar framework (G,p) of Example 10.2. Edge $\{4,5\}$ is drawn as an arc to make edges $\{2,4\}$ and $\{2,5\}$ visible. (G,p) is dimensionally rigid, however, it does not admit a PSD stress matrix of rank $n-1-r=2$

Example 10.2 ([6]) *Consider the framework (G,p) depicted in Fig. 10.2. A configuration matrix and a Gale matrix of (G,p) are given by*

$$P = \begin{bmatrix} -3 & -5 \\ 1 & 2 \\ 0 & -1 \\ 2 & 0 \\ 0 & 4 \end{bmatrix} \text{ and } Z = \begin{bmatrix} 2 & 0 \\ 0 & 2 \\ -6 & 0 \\ 3 & -1 \\ 1 & -1 \end{bmatrix}.$$

Obviously, (G,p) is dimensionally rigid, in fact it is also universally rigid. In order to find a stress matrix Ω, we have to find a 2×2 symmetric matrix Ψ such that $z^{1^T} \Psi z^2 = 0$ and $z^{3^T} \Psi z^4 = 0$. Thus, $\Psi = \begin{bmatrix} 0 & 0 \\ 0 & 1 \end{bmatrix}$ and hence (G,p) does not admit a PSD stress matrix of rank $n-1-r=2$.

Now let $y_{12} = 1$ and $y_{34} = -2/3$. Then

$$Z^T \mathscr{E}(y)Z = y_{12}(z^1 z^{2^T} + z^2 z^{1^T}) + y_{34}(z^3 z^{4^T} + z^4 z^{3^T}) = \begin{bmatrix} 24 & 0 \\ 0 & 0 \end{bmatrix}.$$

Moreover,

$$P^T \mathscr{E}(y)Z = y_{12}(p^1 z^{2^T} + p^2 z^{1^T}) + y_{34}(p^3 z^{4^T} + p^4 z^{3^T}) = \begin{bmatrix} 10 & -6 \\ 6 & -32/3 \end{bmatrix}.$$

Clearly, $null(Z^T \mathscr{E}(y)Z)/\subseteq null(P^T \mathscr{E}(y)Z)$, *i.e.,* $null(U^T \mathscr{M}(y)U)/\subseteq null$
$(W^T \mathscr{M}(y)U)$.

Observe that the framework (G,p) of Example 10.2, obviously, is not in general position since p^2, p^4, and p^5 are collinear. This raises the following question: Is the converse of Theorem 10.6 true under the general position assumption? As the following example shows [17], the answer is no.

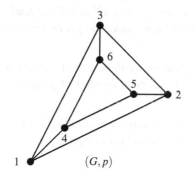
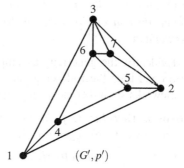

Fig. 10.3 The bar frameworks of Example 10.3

Example 10.3 ([17]) *Consider the framework* (G,p) *depicted in Fig. 10.3. A configuration matrix and a stress matrix of* (G,p) *are given by*

$$P = \begin{bmatrix} -2 & -2 \\ 2 & 0 \\ 0 & 2 \\ -1 & 1 \\ 1 & 0 \\ 0 & 1 \end{bmatrix} \quad and \quad \Omega = \begin{bmatrix} 3I_3 + E_3 & -6I_3 \\ -6I_3 & 12I_3 - 2E_3 \end{bmatrix},$$

where I_3 *and* E_3 *are, respectively, the identity matrix and the matrix of all 1's of order 3. Since* $\deg(i) = 3$ *for each node i and since* (G,p) *is in general position, it follows that, for each node i, the system of equations*

$$\sum_{j:\{i,j\} \in E(G)} \omega_{ij}(p^i - p^j) = 0$$

has a unique solution, up to multiplication by a scalar. Consequently, if we set $\Omega_{12} = -\omega_{12} = 1$, *then* Ω *is unique. Now* $(3I_3 + E_3)$ *is PD and*

$$(12I_3 - 2E_3) - 6I_3(3I_3 + E_3)^{-1} 6I_3 = (12I_3 - 2E_3) - 12(I_3 - \frac{1}{6}E_3) = 0.$$

Hence, by Schur complement, Ω *is PSD of rank 3. Therefore, by Theorem 10.8, framework* (G,p) *is universally rigid and thus dimensionally rigid.*

Now consider the framework (G', p') depicted also in Fig. 10.3. (G', p') is obtained from (G, p) by adding one node, namely 7, and connecting it to nodes 2,3, and 6 such that (G', p') is in general position. Hence, (G', p') is also universally rigid and thus dimensionally rigid. Let ω' be a stress on (G', p') and set $\Omega'_{12} = \Omega_{12} = 1$. Then, by the uniqueness of Ω, it follows that $\Omega'_{13} = \Omega_{13}$, $\Omega'_{14} = \Omega_{14}$. Hence, $\Omega'_{45} = \Omega_{45}$ and $\Omega'_{46} = \Omega_{46}$, which in turn implies that $\Omega'_{52} = \Omega_{52}$ and $\Omega'_{56} = \Omega_{56}$. Again by the uniqueness of Ω, this implies that $\Omega'_{63} = \Omega_{63}$ and $\Omega'_{67} = 0$. Consequently, $\Omega'_{73} = \Omega'_{72} = 0$ and thus $\Omega'_{23} = \Omega_{23}$. As a result, the row and the column of Ω' corresponding to node 7 have all zeros. Consequently, Ω' is PSD of rank 3. Hence, the converse of Theorem 10.6 does not hold true under the general position assumption.

Evidently, the genericity assumption is much stronger than that of general position and thus the following characterization of universal rigidity of generic bar frameworks should come as no surprise.

Theorem 10.9 *Let (G, p) be an r-dimensional bar framework on n nodes, $r \leq n - 2$. Assume that (G, p) is generic. Then the following statements are equivalent:*

(i) (G, p) is universally rigid.
(ii) (G, p) admits a PSD stress matrix Ω of rank $n - r - 1$.

The fact that Statement (ii) implies Statement (i) is an immediate consequence of Theorem 10.7, Theorem 10.2 and Lemma 10.5. Also, it trivially follows from Theorem 10.8. On the other hand, the fact that Statement (i) implies Statement (ii) was conjectured in [9] and proved by Gortler and Thurston in [90]. The following is a rough sketch of their proof. Assume that Statement (i) holds and let D_p be the EDM defined by (G, p). Let $d = \pi(D_p)$, where π is defined in (5.3). Then the embedding dimension of D_p is r and $\text{rank}(X) \leq r$ for all $X \succeq 0$ such that $\pi(\mathscr{K}_V(X)) = d$. Moreover, Theorem 5.9 implies that $\pi(\mathscr{K}_V(\mathscr{S}_+^{n-1})) = \pi(\mathscr{D}^n)$ is closed. Thus, by Theorem 5.10, it suffices to show that $\text{face}(d, \pi(\mathscr{K}_V(\mathscr{S}_+^{n-1})))$ is exposed. But d is generic since (G, p) is generic. Therefore, by Straszewicz Theorem (Theorem 1.43), $\text{face}(d, \pi(\mathscr{K}_V(\mathscr{S}_+^{n-1})))$ is exposed.

Connelly and Gortler obtained a characterization of dimensional rigidity without the genericity assumption [63]. Before we present a refined version of this characterization, we need the following definition. An $n \times n$ symmetric matrix Ω is said to be a *quasi-stress matrix* of (G, p) if it satisfies the following properties:

(a) $\Omega_{ij} = 0$ for all $\{i, j\} \in \bar{E}(\bar{G})$.

(b) $\Omega e = 0$.

(c) $P^T \Omega P = 0$, $\qquad\qquad\qquad\qquad\qquad\qquad\qquad$ (10.10)

where P is a configuration matrix of (G, p) such that $P^T e = 0$.

An immediate consequence of this definition is that if a quasi-stress matrix Ω is PSD, then Ω is a stress matrix since in this case $P^T \Omega P = 0$ implies that $\Omega P = 0$. The following theorem is a refined version of the Connelly–Gortler characterization of dimensional rigidity [63].

Theorem 10.10 ([14]) *Let (G,p) be an r-dimensional framework on n vertices in $\mathbb{R}^r$, $r \leq n-2$. Let P be a configuration matrix of (G,p) such that $P^T e = \mathbf{0}$. Then (G,p) is dimensionally rigid if and only if there exist nonzero quasi-stress matrices: $\Omega^0, \Omega^1, \ldots, \Omega^k$, for some $k \leq n-r-2$, such that:*

(i) $\Omega^0 \succeq \mathbf{0}$, $\mathscr{U}_1^T \Omega^1 \mathscr{U}_1 \succeq \mathbf{0}$, $\ldots$, $\mathscr{U}_k^T \Omega^k \mathscr{U}_k \succeq \mathbf{0}$,
(ii) $rank(\Omega^0) + rank(\mathscr{U}_1^T \Omega^1 \mathscr{U}_1) + \cdots + rank(\mathscr{U}_k^T \Omega^k \mathscr{U}_k) = n-r-1$,
(iii) $P^T \Omega^1 \rho_1 = \mathbf{0}$, $\ldots$, $P^T \Omega^k \rho_k = \mathbf{0}$,

where $\rho_1, \mathscr{U}_1, \ldots, \mathscr{U}_k$ and $\xi_1, \ldots, \xi_k$ are full column rank matrices defined as follows: $col(\rho_1) = null\left(\begin{bmatrix} \Omega^0 \\ P^T \\ e^T \end{bmatrix} \right)$, $col(\xi_i) = null(\rho_i^T \Omega^i \rho_i)$, $\mathscr{U}_i = [P \ \ \rho_i]$ for $i = 1, \ldots, k$ and $\rho_{i+1} = \rho_i \xi_i$ for all $i = 1, \ldots, k-1$.

Proof. Let $\mathscr{X}(\mathscr{F}) = \{X \in \mathscr{S}_+^{n-1} : \pi(\mathscr{K}_V(X)) = d\}$. Let F^{ij} be as defined in (8.17), i.e., $F^{ij} = (e^i - e^j)(e^i - e^j)^T$. Then $\mathscr{X}(\mathscr{F}) = \{X \in \mathscr{S}_+^{n-1} : \text{trace}(V^T F^{ij} V X) = d_{ij} \text{ for all } \{i,j\} \in E(G)\}$. Therefore, by setting $A^i = V^T F^{ij} V$, it follows from the semidefinite Farkas lemma (Theorem 2.22) that (G,p) is dimensionally rigid iff there exist nonzero matrices $\Omega^0, \ldots, \Omega^k$ such that:

(i) $\Omega^l = \sum_{\{i,j\} \in E(G)} \omega_{ij}^l F^{ij}$ for $l = 0, 1, \ldots, k$ for some scalars ω_{ij}^l,
(ii) $V^T \Omega^0 V \succeq \mathbf{0}$, $\mathscr{U}_1'^T V^T \Omega^1 V \mathscr{U}_1' \succeq \mathbf{0}$, $\ldots$, $\mathscr{U}_k'^T V^T \Omega^k V \mathscr{U}_k' \succeq \mathbf{0}$,
(iii) $rank(V^T \Omega^0 V) + rank(\mathscr{U}_1'^T V^T \Omega^1 V \mathscr{U}_1') + \cdots + rank(\mathscr{U}_k'^T V^T \Omega^k V \mathscr{U}_k') = n - r - 1$,
(iv) $\text{trace}(PP^T \Omega^l) = 0$ for $l = 0, 1, \ldots, k$,

where $\mathscr{U}_1', \ldots, \mathscr{U}_{k+1}'$ and $\mathscr{W}_0', \ldots, \mathscr{W}_k'$ are full column rank matrices such that: for $i = 0, 1, \ldots, k$, we have $col(\mathscr{W}_i') = null(\mathscr{U}_i'^T V^T \Omega^i V \mathscr{U}_i')$, $\mathscr{U}_{i+1}' = \mathscr{U}_i' \mathscr{W}_i'$ and $\mathscr{U}_0' = I_{n-1}$. Let us set $\mathscr{U}_i = V \mathscr{U}_i'$ for all $i = 1, \ldots, k$.

Since $\Omega^0 e = \mathbf{0}$, it follows that $V^T \Omega^0 V$ is PSD iff Ω^0 is PSD and $rank(V^T \Omega^0 V) = rank(\Omega^0)$. Moreover, $\text{trace}(P^T \Omega^0 P) = 0$ implies that $P^T \Omega^0 P = \mathbf{0}$, which in turn implies that $\Omega^0 P = \mathbf{0}$. Consequently, Ω^0 is a PSD stress matrix of (G,p).

Now if $rank(\Omega^0) = n - r - 1$, then we are done. Thus, assume that $rank(\Omega^0) = n - r - 1 - \delta_1$, where $\delta_1 \geq 1$. Let $null(\Omega^0) = col([P \ e \ \rho_1])$, where ρ_1 is $n \times \delta_1$, and assume that $e^T \rho_1 = \mathbf{0}$ and $P^T \rho_1 = \mathbf{0}$. Since $col(\mathscr{W}_0') = null(V^T \Omega^0 V) = null(\Omega^0 V)$ and since $P^T e = \mathbf{0}$, it follows that $col([V^T P \ V^T \rho_1])$ is in $null(\Omega^0 V)$. But $rank(V^T \Omega^0 V) = n - r - 1 - \delta_1$. Therefore, $dim(null(\Omega^0 V)) = r + \delta_1$ and thus $\mathscr{U}_1' = \mathscr{W}_0' = [V^T P \ V^T \rho_1]$. Moreover, $\mathscr{U}_1 = V \mathscr{U}_1' = [P \ \rho_1]$. Now since

$$\mathscr{U}_1^T \Omega^1 \mathscr{U}_1 = \begin{bmatrix} P^T \Omega^1 P & P^T \Omega^1 \rho_1 \\ \rho_1^T \Omega^1 P & \rho_1^T \Omega^1 \rho_1 \end{bmatrix} \succeq \mathbf{0},$$

it follows that $P^T \Omega^1 P \succeq \mathbf{0}$ and $\rho_1^T \Omega^1 \rho_1 \succeq \mathbf{0}$. But, $\text{trace}(PP^T \Omega^1) = 0$. Therefore, $P^T \Omega^1 P = \mathbf{0}$ and thus $P^T \Omega^1 \rho_1 = \mathbf{0}$. Accordingly, Ω^1 is a quasi-stress matrix of (G,p).

If $\rho_1^T \Omega^1 \rho_1$ is nonsingular, i.e., if it has rank δ_1, then we are done. Otherwise, let $\mathrm{col}(\xi_i) = \mathrm{null}(\rho_i^T \Omega^i \rho_i)$. Then

$$\mathrm{col}(\mathcal{W}_1') = \mathrm{null}(\mathcal{U}_1^T \Omega^1 \mathcal{U}_1) = \mathrm{null}(\begin{bmatrix} 0 & 0 \\ 0 & \rho_1^T \Omega^1 \rho_1 \end{bmatrix}) = \mathrm{col}(\begin{bmatrix} I & 0 \\ 0 & \xi_1 \end{bmatrix}).$$

Therefore, $\mathcal{U}_2' = \mathcal{U}_1' \mathcal{W}_1' = [V^T P \ V^T \rho_1 \xi_1]$ and thus $\mathcal{U}_2 = V \mathcal{U}_2' = [P \ \rho_1 \xi_1] = [P \ \rho_2]$. The rest of the proof for $\Omega^2, \ldots, \Omega^k$ proceeds along the same line.

$\square$

Example 10.4 *Let (G, p) be the dimensionally rigid framework depicted in Fig. 10.2 and considered in Example 10.2. Then*

$$\Omega^0 = [0 \ 2 \ 0 \ -1 \ -1]^T [0 \ 2 \ 0 \ -1 \ -1]$$

and thus $\rho_1 = [6 \ 4 \ -18 \ 7 \ 1]^T$. Therefore,

$$\mathcal{U}_1 = [P \ \rho_1] = \begin{bmatrix} -3 & -5 & 6 \\ 1 & 2 & 4 \\ 0 & -1 & -18 \\ 2 & 0 & 7 \\ 0 & 4 & 1 \end{bmatrix}.$$

To calculate $\Omega^1 = (\omega_{ij}^1)$, set $\omega_{25}^1 = \omega_{45}^1 = 0$. Hence, $P^T \Omega^1 P = 0$ and $P^T \Omega^1 \rho_1 = 0$ is a system of five equation in six variables whose solution yields

$$\Omega^1 = \begin{bmatrix} 2 & 0 & -6 & 3 & 1 \\ 0 & 30 & -18 & -12 & 0 \\ -6 & -18 & 18 & 0 & 6 \\ 3 & -12 & 0 & 9 & 0 \\ 1 & 0 & 6 & 0 & -7 \end{bmatrix}.$$

Therefore,

$$\mathcal{U}_1^T \Omega^1 \mathcal{U}_1 = \begin{bmatrix} P^T \Omega^1 P & P^T \Omega^1 \rho_1 \\ \rho_1^T \Omega^1 P & \rho_1 \Omega^1 \rho_1 \end{bmatrix} = \begin{bmatrix} 0 & 0 \\ 0 & 10082 \end{bmatrix}.$$

As a result, $\mathrm{rank}(\Omega^0) + \mathrm{rank}(\mathcal{U}_1^T \Omega^1 \mathcal{U}_1) = 2$. Note that Ω^1 is not PSD and thus, it is only a quasi-stress matrix.

We conclude this section by pointing out that unlike local rigidity, universal rigidity is not a generic property of bar frameworks. That is, as shown in the following example, for some graph G, there exist generic configurations p and p' such that (G, p) is universally rigid, while (G, p') is not universally rigid. It is worthy of note that framework (G, p) admits a PSD stress matrix of rank 2, but it does not admit a PSD stress matrix of rank 1.

Example 10.5 *Consider the two frameworks depicted in Fig. 10.4. A configuration matrix and a Gale matrix of (G, p) are*

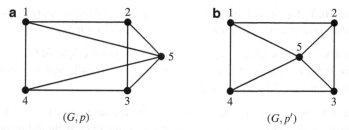

Fig. 10.4 An example of 2 two-dimensional bar frameworks. Framework (**a**) is universally rigid, while framework (**b**) is not universally rigid. Framework (**b**) has three- and four-dimensional equivalent frameworks. Both frameworks are locally rigid

$$P = \begin{bmatrix} -2 & 1 \\ 1 & 1 \\ 1 & -1 \\ -2 & -1 \\ 2 & 0 \end{bmatrix} \text{ and } Z = \begin{bmatrix} 3 & 0 \\ 0 & 3 \\ 8 & 5 \\ -5 & -2 \\ -6 & -6 \end{bmatrix}.$$

Thus, (G, p) has a unique, up to a multiplication by a scalar, stress matrix $\Omega = Z\Psi Z$, where $\Psi = \begin{bmatrix} 5 & -8 \\ -8 & 20 \end{bmatrix}$. As a result, Ω is PSD and with rank 2. In fact, as long as node 5 is not in the convex hull of the other four nodes, (G, p) is dimensionally rigid. On the other hand, a configuration matrix and a Gale matrix of (G, p') are

$$P' = \begin{bmatrix} -2 & 1 \\ 1 & 1 \\ 1 & -1 \\ -2 & -1 \\ 0 & 0 \end{bmatrix} \text{ and } Z' = \begin{bmatrix} 3 & 0 \\ 0 & 3 \\ 4 & 1 \\ -1 & 2 \\ -6 & -6 \end{bmatrix}.$$

Thus, (G, p') has a unique, up to a multiplication by a scalar, stress matrix $\Omega' = Z'\Psi'Z'$, where $\Psi' = \begin{bmatrix} 1 & -4 \\ -4 & -2 \end{bmatrix}$. Accordingly, (G, p) does not admit a nonzero PSD stress matrix and thus, by Theorem 8.5, there exists a four-dimensional framework that is equivalent to (G, p'). As a result, (G, p') is not dimensionally rigid and remains so, as long as node 5 lies in the convex hull of the other four nodes.

10.4 $(r+1)$-lateration Bar Frameworks

In this section, we investigate classes of graphs for which the converse of Theorem 10.8 is true. In other words, we investigate classes of graphs whose corresponding universally rigid bar frameworks in general position admit a PSD stress matrix of maximal rank. We start first with the generalization of trilateration graphs.

Definition 10.3 *A graph G on n nodes, $n \geq r+1$, is said to be an $(r+1)$-lateration graph if there exists a permutation π of the nodes of G such that:*

(i) *The first $(r+1)$ nodes, $\pi(1), \ldots, \pi(r+1)$, induce a clique in G.*
(ii) *Each remaining node $\pi(j)$, for $j = r+2, \ldots, n$, is adjacent to exactly $(r+1)$ nodes in the set $\{\pi(1), \ldots, \pi(j-1)\}$.*

As a result, an $(r+1)$-lateration graph on n nodes has $n(r+1) - (r+1)(r+2)/2$ edges and the nodes $\pi(1), \ldots, \pi(r+2)$ induce a clique in G. It should be pointed out that in Condition (ii) of Definition 10.3, if the neighbors of $\pi(j)$ in $\{\pi(1), \ldots, \pi(j-1)\}$ induce a clique, i.e., if $\pi(j)$ and its neighbors, for each $j = r+2, \ldots, n$, induce a clique of size $r+2$, then the $(r+1)$-lateration graph is called an $(r+1)$-*tree graph* [39]. Note that 1-tree graphs are the usual trees.

Evidently, an r-dimensional bar framework in general position in $\mathbb{R}^r$ whose underlying graph is an $(r+1)$-lateration graph is universally rigid [177, 201]. Moreover, the converse of Theorem 10.8 is true for such frameworks.

Theorem 10.11 (Alfakih et al. [22]) *Let (G, p) be an r-dimensional bar framework on n nodes in general position in $\mathbb{R}^r$, $r \leq n-2$, such that G is an $(r+1)$-lateration graph. Then (G, p) admits a positive semidefinite stress matrix Ω of rank $n-1-r$.*

The proof of Theorem 10.11 is constructive, i.e., an algorithm is presented to construct the desired stress matrix Ω. Our proof follows closely the one given in [13]. Recall from Theorem 8.4 that if there exists a symmetric matrix Ψ such that $(Z\Psi Z^T)_{ij} = 0$ for each $\{i, j\} \in \bar{E}(\bar{G})$, then $Z\Psi Z^T$ is a stress matrix of (G, p). Wlog, assume that G has a lateration order $1, \ldots, n$. Thus, the nodes $1, \ldots, r+1, r+2$ induce a clique and for $j = r+3, \ldots, n$, node j is adjacent to $r+1$ nodes in the set $\{1, \ldots, j-1\}$. As always, let $\bar{r} = n-1-r$.

Clearly, ZZ^T is a PSD matrix of rank $\bar{r}$. Hence, if it happens that $(ZZ^T)_{ij} = 0$ for each $\{i, j\} \in \bar{E}(\bar{G})$, then ZZ^T is the desired stress matrix and we are done. Otherwise, we generate a sequence of matrices $ZZ^T = \Omega^n, \Omega^{n-1}, \ldots, \Omega^k, \ldots, \Omega^{r+2}$, where each matrix Ω^k of this sequence satisfies: (i) $\Omega^k = Z\Psi^k Z^T$ for some symmetric matrix Ψ^k, (ii) Ω^k is PSD and of rank $\bar{r}$, and (iii) each entry in the last $n-k$ columns (rows) of Ω^k corresponding to a missing edge is zero. Consequently, Ω^{r+2} is the desired stress matrix. In other words, the above algorithm repeatedly modifies Ψ^k, starting from $\Psi^n = I_{\bar{r}}$, in order to "zero out" the entries of $Z\Psi^k Z^T$ which should be zero, but are not. This "zeroing out" is done one column (row) at a time, starting from the nth column. Obviously, the algorithm terminates when the $(r+2)$th column (row) is reached since the nodes $1, \ldots, r+2$ induce a clique. That is, no "zeroing out" is needed in the upper left $(r+2) \times (r+2)$ square submatrix of the desired stress matrix.

For $j \geq r+3$, let

$$\bar{N}(j) = \{i \in V(G) : i < j \text{ and } \{i, j\} \in \bar{E}(\bar{G})\}.$$

Thus, $|\bar{N}(j)| = j - r - 2$ since j is adjacent to $r+1$ nodes in the set $\{1, \ldots, j-1\}$.

We first show how to obtain Ω^{n-1} by "zeroing out" the entries, corresponding to missing edges, in the nth column (nth row) of Ω^n. Let $\bar{Z}^n$ be the submatrix of Z whose rows are indexed by the nodes in $\bar{N}(n) \cup \{n\}$. Then $\bar{Z}^n$ is a square matrix of order $\bar{r}$ and thus nonsingular (Corollary 3.1). Let b^n be the vector in $\mathbb{R}^{\bar{r}}$, where

$$b_i^n = \begin{cases} -\Omega_{in}^n & \text{if } i \in \bar{N}(n) \\ 1 & \text{if } i = n, \end{cases}$$

and let ξ_n be the unique solution of

$$\bar{Z}^n \xi_n = b^n.$$

Then we have the following lemma.

Lemma 10.7 *Let* $\Omega^{n-1} = \Omega^n + Z\xi_n\xi_n^T Z^T = Z(I_{\bar{r}} + \xi_n\xi_n^T)Z^T$. *Then*

(i) Ω^{n-1} *is PSD and of rank* $n-1-r$.
(ii) $\Omega_{in}^{n-1} = 0$ *for each* $i < n$ *such that* $\{i,n\} \in \bar{E}(\bar{G})$.

Proof. Part (i) is obvious since $I_{\bar{r}} + \xi_n\xi_n^T$ is PD. Also, part (ii) is immediate since for $i < n$ such that $\{i,n\} \in \bar{E}(\bar{G})$, we have $\Omega_{in}^{n-1} = \Omega_{in}^n + b_i^n b_n^n = \Omega_{in}^n + b_i^n = 0$.

$\square$

Similarly, we continue constructing $\Omega^{n-2}, \ldots, \Omega^k, \ldots, \Omega^{r+2}$ by, respectively, "zeroing out" the entries of columns $n-1, \ldots, r+3$ corresponding to missing edges. More precisely, suppose that $\Omega^k = Z\Psi^k Z^T$ is PSD and of rank $\bar{r}$ and that $\Omega_{ij}^k = 0$ for all $j = k+1, \ldots, n$ such that $\{i,j\} \in \bar{E}(\bar{G})$. Let $\bar{Z}^k$ be the submatrix of Z whose rows are indexed by the nodes in $\bar{N}(k) \cup \{k, k+1, \ldots, n\}$. Then $\bar{Z}^k$ is a nonsingular square matrix of order $\bar{r}$. Let b^k be the vector in $\mathbb{R}^{\bar{r}}$, where

$$b_i^k = \begin{cases} -\Omega_{ik}^k & \text{if } i \in \bar{N}(k) \\ 1 & \text{if } i = k \\ 0 & \text{if } i = k+1, \ldots, n. \end{cases}$$

Now let ξ_k be the unique solution of

$$\bar{Z}^k \xi_k = b^k.$$

Lemma 10.8 *Let* $\Omega^{k-1} = \Omega^k + Z\xi_k\xi_k^T Z^T$. *Then*

(i) Ω^{k-1} *is PSD and of rank* $n-1-r$.
(ii) $\Omega_{ij}^{k-1} = 0$ *for each* $i < j$ *and for all* $j = k, \ldots, n$ *such that* $\{i,j\} \in \bar{E}(\bar{G})$.

Proof. Part (i) is obvious. Now for each $i < k$ such that $\{i,k\} \in \bar{E}(\bar{G})$ we have $\Omega_{ik}^{k-1} = \Omega_{ik}^k + b_i^k b_k^k = 0$. Moreover, for each $i < j$ and $j = k+1, \ldots, n$ such that $\{i,j\} \in \bar{E}(\bar{G})$ we have $\Omega_{ij}^{k-1} = \Omega_{ij}^k + b_i^k b_j^k = \Omega_{ij}^k = 0$, i.e., the entries in columns $k+1, \ldots, n$ of Ω^{k-1} are unchanged from Ω^k.

$\square$

Proof of Theorem 10.11. Clearly, the matrix

$$\Omega^{r+2} = \Omega^{r+3} + Z\xi_{r+3}\xi_{r+3}^T Z^T = Z(I_{\bar{r}} + \xi_n\xi_n^T + \cdots + \xi_{r+3}\xi_{r+3})Z^T$$

is PSD and of rank $\bar{r}$. Moreover, $\Omega_{ij}^{r+2} = 0$ for all $\{i,j\} \in \bar{E}(\bar{G})$. Hence, Ω^{r+2} is the desired stress matrix of (G, p).

□

As we show next, if the $(r+1)$-lateration graph is an $(r+1)$-tree graph, then a PSD stress matrix of rank $n - r - 1$ can be obtained directly without the need for the "zeroing out" steps in the above algorithm [22].

Let

$$N(k) = \{i \in V(G) : i < k \text{ and } \{i,k\} \in E(G)\}$$

and for $j = 1, \ldots, n - r - 1$, let $x = (x_{ij}) \in \mathbb{R}^{r+1}$ be the solution of the system of equations

$$\sum_{i: i \in N(j+r+1)} x_{ij} \begin{bmatrix} p^i \\ 1 \end{bmatrix} = - \begin{bmatrix} p^{j+r+1} \\ 1 \end{bmatrix}. \tag{10.11}$$

Note that the set $\{i : i \in N(j+r+1)\}$ has cardinality $r+1$ and thus under the general position assumption, System of Eq. (10.11) has a unique solution. Now let $\hat{Z} = (\hat{z}_{ij})$ be the Gale matrix defined as follows:

$$\hat{z}_{ij} = \begin{cases} 1 & \text{if } i = j+r+1 \\ x_{ij} & \text{if } i \in N(j+r+1) \\ 0 & \text{otherwise.} \end{cases}$$

We claim that if $\{i,j\} \in \bar{E}(\bar{G})$, then $\hat{z}_{ik}\hat{z}_{jk} = 0$ for every k. To see this, assume, to the contrary, that $\hat{z}_{ik} \neq 0$ and $\hat{z}_{jk} \neq 0$ for some k. Then we have either one of the following four cases:

Case 1: $i = k+r+1$ and $j = k+r+1$. Thus $i = j$, a contradiction.

Case 2: $i = k+r+1$ and $j \in N(k+r+1)$. Thus $j \in N(i)$, a contradiction.

Case 3: $j = k+r+1$ and $i \in N(k+r+1)$. Thus $i \in N(j)$, a contradiction.

Case 4: $i \in N(k+r+1)$ and $j \in N(k+r+1)$. Hence, it follows from the definition of an $(r+1)$-tree graph that the nodes i and j belong to a clique in G. Thus, $\{i,j\} \in E(G)$, a contradiction.

Accordingly, let $\Omega = \hat{Z}\hat{Z}^T$. Then, for each $\{i,j\} \in \bar{E}(\bar{G})$, we have

$$\Omega_{ij} = \sum_{k=1}^{n-r-1} \hat{z}_{ik}\hat{z}_{jk} = 0.$$

As a result, Ω is the desired stress matrix.

Recall that chordal graphs have a perfect elimination ordering. Hence, with a slight modification, the above procedure for $(r+1)$-tree graphs can be used to construct PSD stress matrices of maximal rank for universally rigid bar frameworks whose underlying graphs are chordal [11]. Finally, we point out that the universal rigidity problem for complete bipartite bar frameworks is investigated in [118, 64].

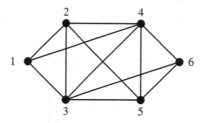

Fig. 10.5 The framework of Example 10.6 whose underlying graph is a trilateration graph. Observe that in this case, G is a 3-tree graph

Example 10.6 *Consider the bar framework* (G,p) *depicted in Fig. 10.5, where G is a 3-lateration graph with missing edges* $\bar{E}(\bar{G}) = \{\{1,5\},\{1,6\},\{2,6\}\}$. *A configuration matrix and a Gale matrix of* (G,p) *are*

$$P = \begin{bmatrix} -2 & 0 \\ -1 & 1 \\ -1 & -1 \\ 1 & 1 \\ 1 & -1 \\ 2 & 0 \end{bmatrix} \quad and\ Z = \begin{bmatrix} 1 & 0 & 0 \\ 0 & 1 & 0 \\ 0 & 0 & 1 \\ -2 & -2 & -1 \\ -2 & -1 & -2 \\ 3 & 2 & 2 \end{bmatrix}.$$

$\Omega^6 = ZZ^T$. *Thus,* $\bar{Z}^6$ *is the square submatrix of Z whose rows are indexed by 1,2, and 6. Therefore,*

$$\bar{Z}^6 = \begin{bmatrix} 1 & 0 & 0 \\ 0 & 1 & 0 \\ 3 & 2 & 2 \end{bmatrix}, b^6 = \begin{bmatrix} -3 \\ -2 \\ 1 \end{bmatrix}, \text{ and hence } \xi^6 = (\bar{Z}^6)^{-1} b^6 = \begin{bmatrix} -3 \\ -2 \\ 7 \end{bmatrix}.$$

Consequently,

$$\Omega^5 = \begin{bmatrix} 10 & 6 & -21 & -11 & 16 & 0 \\ 5 & -14 & -8 & 11 & 0 \\ & 50 & 20 & -44 & 9 \\ & & 18 & -10 & -9 \\ & & & 45 & -18 \\ & & & & 18 \end{bmatrix}.$$

Hence, $\bar{Z}^5$ *is the square submatrix of Z whose rows are indexed by 1,5, and 6; and* $b^5 = \begin{bmatrix} -16 \\ 1 \\ 0 \end{bmatrix}$. *Therefore,* $\xi^5 = \begin{bmatrix} -16 \\ 17 \\ 7 \end{bmatrix}$ *and thus*

$$\Omega^4 = \begin{bmatrix} 266 & -266 & -133 & 133 & 0 & 0 \\ & 294 & 105 & -161 & 28 & 0 \\ & & 99 & -43 & -37 & 9 \\ & & & 99 & -19 & -9 \\ & & & & 46 & -18 \\ & & & & & 18 \end{bmatrix}$$

is a PSD stress matrix of (G,p) of rank 3. On the other hand, since G is a 3-tree graph, we have

$$\hat{Z} = \begin{bmatrix} 2 & 0 & 0 \\ -2 & 1 & 0 \\ -1 & -1 & 1/2 \\ 1 & -1 & -1/2 \\ 0 & 1 & -1 \\ 0 & 0 & 1 \end{bmatrix}.$$

Hence, $\hat{Z}\hat{Z}^T$ is a PSD stress matrix of (G,p) of rank 3.

10.5 Universally Linked Nodes

The notion of universal rigidity applies to the bar framework as a whole. In this section, we discuss the equivalent notion for a pair of nonadjacent nodes.

Definition 10.4 Let $\{k,l\}$ be a missing edge of bar framework (G,p). Nodes k and l are said to be universally linked if $||p^k - p^l|| = ||q^k - q^l||$ in every bar framework (G,q) that is equivalent to (G,p).

In other words, even though nodes k and l are nonadjacent, the distance between them remains the same as if they are joined by an edge. Consequently, (G,p) is universally rigid if and only if every pair of nonadjacent nodes of G is universally linked.

The following lemma is an immediate consequence of the definition and Theorem 8.2.

Lemma 10.9 (Alfakih [16]) Let $\mathscr{F}$ be the Cayley configuration spectrahedron of bar framework (G,p) and let $\{k,l\} \in \bar{E}(\bar{G})$. Then k and l are universally linked if and only if $\mathscr{F}$ is contained in the subspace $\{y \in \mathbb{R}^{\bar{m}} : y_{kl} = 0\}$ of $\mathbb{R}^{\bar{m}}$.

The fact that (G,p) is universally rigid iff $\mathscr{F} = \{0\}$ follows as an immediate corollary of Lemma 10.9. The following lemmas are needed to establish a sufficient condition for universal linkedness.

Lemma 10.10 Let $\mathscr{F}$ be the Cayley configuration spectrahedron of bar framework (G,p) and let Ω be a nonzero positive semidefinite stress matrix of (G,p). Then

$$\Omega V \mathscr{X}(y) V^T = \mathbf{0} \text{ for all } y \in \mathscr{F},$$

where $\mathscr{X}(y)$ is as defined in (8.9).

Proof.   Let $y \in \mathscr{F}$. Then

$$\text{trace}(\Omega V \mathscr{X}(y) V^T) = \text{trace}(\Omega V X V^T) - \frac{1}{2} \sum_{\{i,j\} \in \bar{E}(\bar{G})} y_{ij} \text{trace}(\Omega E^{ij}) = 0.$$

Therefore, $\Omega V \mathscr{X}(y) V^T = \mathbf{0}$ since both Ω and $V \mathscr{X}(y) V^T$ are PSD.

$\square$

Lemma 10.11 *Let $\mathscr{F}$ be the Cayley configuration spectrahedron of bar framework (G,p) and let Ω be a nonzero positive semidefinite stress matrix of (G,p). Then $\mathscr{F}$ is contained in the subspace*

$$\{y \in \mathbb{R}^{\tilde{m}} : \Omega \mathscr{E}(y) = \mathbf{0}\},$$

where $\mathscr{E}(y)$ is as defined in (9.9).

Proof.   Lemma 10.10 implies that $\Omega V \mathscr{X}(y) V^T = -\Omega V V^T \mathscr{E}(y) V V^T / 2 = \mathbf{0}$ for all $y \in \mathscr{F}$. Thus, $\Omega \mathscr{E}(y) V = \mathbf{0}$ since $\Omega(I - ee^T/n) = \Omega$ and since V^T has full row rank. Hence, by Lemma 10.6, $\Omega \mathscr{E}(y) = \mathbf{0}$.

$\square$

The following theorem establishes a sufficient condition for universal linkedness.

Theorem 10.12 (Alfakih [16]) *Let (G,p) be an r-dimensional bar framework on n nodes, $r \leq n-2$, and let $\{k,l\} \in \bar{E}(\bar{G})$. Further, let Ω be a nonzero positive semidefinite stress matrix of (G,p). If the following condition holds*

$$\text{there does not exist } y_{kl} \neq 0 \text{ such that } \Omega \mathscr{E}(y) = \mathbf{0}, \tag{10.12}$$

then nodes k and l are universally linked.

Proof.   Suppose that Condition 10.12 holds. Then, by Lemma 10.11, $\mathscr{F}$ is contained in the subspace $\{y \in \mathbb{R}^{\tilde{m}} : y_{kl} = 0\}$. Therefore, the result follows from Lemma 10.9.

$\square$

An important point to bear in mind is that Condition 10.12 is equivalent to the following condition

$$\text{there does not exist } y_{kl} \neq 0 \text{ such that } \mathscr{E}(y) W = \mathbf{0}, \tag{10.13}$$

where W is any matrix whose columns form a basis of $\text{col}(\Omega)$.

Example 10.7 *Consider the framework (G,p) depicted in Fig. 10.6. Then a basis of a stress matrix of (G,p) is given by $W = [-2 \ 1 \ 0 \ 1 \ 0]^T$. Thus $\mathscr{E}(y) W = \mathbf{0}$, i.e.,*

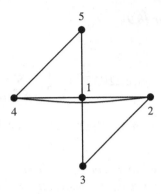

Fig. 10.6 The bar framework of Example 10.7. The edge $\{2,4\}$ is drawn as an arc to make edges $\{1,2\}$ and $\{1,4\}$ visible. Nodes 2 and 5; and nodes 3 and 4 are universally linked while nodes 3 and 5 are not universally linked

$$\begin{bmatrix} 0 & 0 & 0 & 0 & 0 \\ 0 & 0 & 0 & 0 & y_{25} \\ 0 & 0 & 0 & y_{34} & y_{35} \\ 0 & 0 & y_{34} & 0 & 0 \\ 0 & y_{25} & y_{35} & 0 & 0 \end{bmatrix} \begin{bmatrix} -2 \\ 1 \\ 0 \\ 1 \\ 0 \end{bmatrix} = \begin{bmatrix} 0 \\ 0 \\ 0 \\ 0 \\ 0 \end{bmatrix}$$

has a solution $y_{34} = y_{25} = 0$ and y_{35} is free. Consequently, nodes 3 and 4; and nodes 2 and 5 are universally linked, while nodes 3 and 5 are not universally linked. Obviously, (G, p) is not universally rigid since it can fold across the edge $\{2,4\}$.

As a corollary of Theorem 10.12, we can strengthen Theorem 10.7 as follows.

Theorem 10.13 (Alfakih [16]) *Let (G, p) be an r-dimensional bar framework on n nodes, $r \leq n-2$, and let Ω be a nonzero positive semidefinite stress matrix of (G, p). If the following condition holds:*

$$\text{there does not exist } y \neq \mathbf{0} \text{ such that } \Omega \mathscr{E}(y) = \mathbf{0}, \qquad (10.14)$$

then (G, p) is universally rigid.

Theorem 10.13 is stronger than Theorem 10.7 since Ω does not have to be of maximal rank, i.e., $\text{rank}(\Omega)$ need not be $n-1-r$. Moreover, if $\text{rank}(\Omega) = n-1-r$, then Condition 10.14 is equivalent to the assertion that the framework does not admit a nontrivial affine flex (Lemmas 10.4 and 10.6) since any matrix whose columns form a basis of $\text{col}(\Omega)$ is a Gale matrix of (G, p).

Example 10.8 *Consider the framework (G, p) depicted in Fig. 10.2. As discussed in Example 10.2, (G, p) is universally rigid even though it does not admit a PSD stress matrix of rank 2. Thus, the universal rigidity of (G, p) cannot be inferred from Theorem 10.7. However, in this case $\mathscr{E}(y)W = 0$, i.e.,*

$$\begin{bmatrix} 0 & y_{12} & 0 & 0 & 0 \\ y_{12} & 0 & 0 & 0 & 0 \\ 0 & 0 & 0 & y_{34} & 0 \\ 0 & 0 & y_{34} & 0 & 0 \\ 0 & 0 & 0 & 0 & 0 \end{bmatrix} \begin{bmatrix} 0 \\ 2 \\ 0 \\ -1 \\ -1 \end{bmatrix} = \begin{bmatrix} 0 \\ 0 \\ 0 \\ 0 \\ 0 \end{bmatrix}$$

has a unique solution $y_{12} = y_{34} = 0$. Accordingly, the universal rigidity of (G, p) can be inferred from Theorem 10.13.

Condition 10.14 of Theorem 10.13 has interpretations in terms of the Strong Arnold Property of matrices and in terms of the notion of nondegeneracy of semidefinite programming [26, 132].

Let G be a given graph and let A be a nonzero matrix in $\mathscr{S}^n$ such that $A_{ij} = 0$ for all $\{i, j\} \in \bar{E}(\bar{G})$. A is said to satisfy the *Strong Arnold Property (SAP)* [56] if $Y = \mathbf{0}$ is the only matrix in $\mathscr{S}^n$ that satisfies: (i) $Y_{ij} = 0$ if $i = j$ or if $\{i, j\} \in E(G)$, i.e., $Y = \mathscr{E}(y)$ for some y, and (ii) $AY = \mathbf{0}$. Therefore, Condition 10.14 is equivalent to the assertion that the stress matrix Ω satisfies the SAP.

Let Ω be a nonzero PSD stress matrix of (G, p). Let $\mathscr{F}$ be the Cayley configuration spectrahedron of framework (G, p) and consider the following SDP problem

$$(P) \qquad \min \qquad 0$$
$$\text{subject to } X + \mathscr{M}(y) \succeq 0.$$

Thus, every y in $\mathscr{F}$ is an optimal solution of (P) since the objective function is identically 0. Assume that (G, p) admits a nonzero PSD stress matrix Ω. Then, clearly, $V^T \Omega V$ is an optimal solution of the dual SDP problem

$$(D) \qquad \max \qquad -\text{trace}(XY)$$
$$\text{subject to } \text{trace}(M^{ij}Y) = 0 \ \text{ for all } \{i, j\} \in \bar{E}(\bar{G}),$$
$$Y \succeq \mathbf{0}.$$

Let U' be the matrix whose columns form an orthonormal basis of $\text{null}(V^T \Omega V)$. Then, by Eq. (2.5), Ω is nondegenerate iff

$$\{\mathscr{M}(y) : y \in \mathbb{R}^{\bar{m}}\} \cap \{C : C = U' \Phi U'^T\} = \{\mathbf{0}\}$$

iff the only solution of the system

$$V^T \Omega V \mathscr{M}(y) = -\frac{1}{2} V^T \Omega \mathscr{E}(y) V = \mathbf{0} \qquad (10.15)$$

is $y = \mathbf{0}$. Hence, by multiplying (10.15) from the left by V and using Lemma 10.6, it follows that Ω is nondegenerate iff Condition 10.14 holds. As a result, Theorem 10.13 follows from Theorem 2.15.

Epilogue

The primary focus of this monograph is on Euclidean distance matrices (EDMs). Among the various aspects of EDMs discussed are: characterizations, classes, eigenvalues, geometry, and completions. As is well known, EDMs of order n are in one-to-one correspondence with positive semidefinite (PSD) matrices of order $n-1$. As a result, a substantial portion of this monograph is dedicated to various aspects of PSD matrices.

EDM completion problems are best understood in the context of rigidity theory, which has a long history going at least as far back as Cauchy. Hence, the secondary focus of this monograph is on rigidity theory. The use of Cartesian coordinates has been the traditional and dominant approach in the study of rigidity theory. In particular, a configuration of n points in r dimensions is usually represented as a vector in rn dimensions. By exploiting the correspondence between EDMs and PSD matrices, and by representing point configurations by their projected Gram matrices, this monograph discusses a new approach to rigidity theory. This new approach gives rise to alternative tools such as the dual rigidity matrix and Gale matrices. Also, this approach puts the semidefinite programming (SDP) machinery at our disposal by reducing the universal rigidity problem to an SDP problem.

Two topics in rigidity theory, namely global rigidity and tensegrity frameworks, are notably absent from this monograph. An r-dimensional bar framework (G, p) is *globally rigid* if every r-dimensional bar framework (G, p') that is equivalent to (G, p) is actually congruent to (G, p). In other words, (G, p) is globally rigid if there does not exist a configuration p' satisfying: (i) $H \circ D_p = H \circ D_{p'}$, (ii) the embedding dimension of $D_{p'} = r$, and (iii) $D_p \neq D_{p'}$. Here, H is the adjacency matrix of graph G, $(\circ)$ is the Hadamard product, and D'_p is the EDM defined by p'. Hence, a globally rigid r-dimensional bar framework can have equivalent bar frameworks of dimensions $\geq r+1$. Consequently, the notion of global rigidity, while stronger than that of local rigidity, is weaker than the notion of universal rigidity.

Restricting the embedding dimension of $D_{p'}$ is equivalent to restricting the rank of the corresponding projected Gram matrix. This renders the global rigidity problem nonconvex and thus not amenable to the SDP machinery. As a result, tackling

© Springer Nature Switzerland AG 2018

A.Y. Alfakih, *Euclidean Distance Matrices and Their Applications in Rigidity Theory*, https://doi.org/10.1007/978-3-319-97846-8

the global rigidity problem requires tools different from those used in this monograph. For a discussion of the global rigidity problem, the reader is referred to [58, 61, 114, 91, 174, 183] and the references therein.

A *tensegrity graph* is a simple graph where the nodes are labelled $1, \ldots, n$, and where each edge is labelled as either a bar, a cable or a strut. Thus, a tensegrity graph is denoted by $G = (V, B \cup C \cup S)$, where B is the set of bars, C is the set of cables, and S is the set of struts. A *tensegrity framework* is a tensegrity graph whose nodes are mapped onto points $p^1, \ldots, p^n$ in $\mathbb{R}^r$. In any flexing of a tensegrity framework, the distance between the end points of a bar must stay the same, while the distance between the end points of a cable (strut) can decrease (increase) or stay the same, but not increase (decrease). Consequently, a bar is equivalent to a cable plus a strut. Furthermore, a bar framework is a tensegrity framework that consists only of bars. The various notions of rigidity for bar frameworks extend easily to tensegrity frameworks. However, special care must be taken in the definition of a stress matrix. In particular, whereas the stress on a bar can be either positive or negative, the stress on a cable (strut) must be positive (negative). Rigidity theory for tensegrity frameworks is not discussed in this monograph due to space limitations. Instead, the reader is referred to [162, 158, 62] and the references therein.

References

1. A.Y. Alfakih, Graph rigidity via Euclidean distance matrices. Linear Algebra Appl. **310**, 149–165 (2000)
2. A.Y. Alfakih, On rigidity and realizability of weighted graphs. Linear Algebra Appl. **325**, 57–70 (2001)
3. A.Y. Alfakih, On the uniqueness of Euclidean distance matrix completions: the case of points in general position. Linear Algebra Appl. **397**, 265–277 (2005)
4. A.Y. Alfakih, On the nullspace, the rangespace and the characteristic polynomial of Euclidean distance matrices. Linear Algebra Appl. **416**, 348–354 (2006)
5. A.Y. Alfakih, A remark on the faces of the cone of Euclidean distance matrices. Linear Algebra Appl. **414**, 266–270 (2006)
6. A.Y. Alfakih, On dimensional rigidity of bar-and-joint frameworks. Discret. Appl. Math. **155**, 1244–1253 (2007)
7. A.Y. Alfakih, On the dual rigidity matrix. Linear Algebra Appl. **428**, 962–972 (2008)
8. A.Y. Alfakih, On the eigenvalues of Euclidean distance matrices. Comput. Appl. Math. **27**, 237–250 (2008)
9. A.Y. Alfakih, On the universal rigidity of generic bar frameworks. Contrib. Discret. Math. **5**, 7–17 (2010)
10. A.Y. Alfakih, On bar frameworks, stress matrices and semidefinite programming. Math. Program. Ser. B **129**, 113–128 (2011)
11. A.Y. Alfakih, On stress matrices of chordal bar frameworks in general positions, 2012. arXiv/1205.3990
12. A.Y. Alfakih, A remark on the Manhattan distance matrix of a rectangular grid, 2012. arXiv/1208.5150
13. A.Y. Alfakih, Universal rigidity of bar frameworks in general position: a Euclidean distance matrix approach, in *Distance Geometry: Theory, Methods, and Applications*, ed. by A. Mucherino, C. Lavor, L. Liberti, N. Maculan (Springer, Berlin, 2013), pp. 3–22

© Springer Nature Switzerland AG 2018
A.Y. Alfakih, *Euclidean Distance Matrices and Their Applications in Rigidity Theory*, https://doi.org/10.1007/978-3-319-97846-8

14. A.Y. Alfakih, On Farkas lemma and dimensional rigidity of bar frameworks. Linear Algebra Appl. **486**, 504–522 (2015)

15. A.Y. Alfakih, On yielding and jointly yielding entries of Euclidean distance matrices. Linear Algebra Appl. **556**, 144–161 (2018)

16. A.Y. Alfakih, Universal rigidity of bar frameworks via the geometry of spectrahedra. J. Glob. Optim. **67**, 909–924 (2017)

17. A.Y. Alfakih, V.-H. Nyugen, On affine motions and universal rigidity of tensegrity frameworks. Linear Algebra Appl. **439**, 3134–3147 (2013)

18. A.Y. Alfakih, H. Wolkowicz, Two theorems on Euclidean distance matrices and Gale transform. Linear Algebra Appl. **340**, 149–154 (2002)

19. A.Y. Alfakih, H. Wolkowicz, Some necessary and some sufficient trace inequalities for Euclidean distance matrices. Linear Multilinear Algebra **55**, 499–506 (2007)

20. A.Y. Alfakih, Y. Ye, On affine motions and bar frameworks in general positions. Linear Algebra Appl. **438**, 31–36 (2013)

21. A.Y. Alfakih, A. Khandani, H. Wolkowicz, Solving Euclidean distance matrix completion problems via semidefinite programming. Comput. Optim. Appl. **12**, 13–30 (1999)

22. A.Y. Alfakih, N. Taheri, Y. Ye, On stress matrices of $(d+1)$-lateration frameworks in general position. Math. Program. **137**, 1–17 (2013)

23. S. AlHomidan, *Hybrid Methods for Optimization Problems with Positive Semidefinite Matrix Constraints*. PhD thesis, University of Dundee, 1993

24. S. AlHomidan, R. Fletcher, Hybrid methods for finding the nearest Euclidean distance matrix, in *Recent Advances in Nonsmooth Optimization* (World Scientific Publishing, River Edge, 1995), pp. 1–17

25. S. AlHomidan, H. Wolkowicz, Approximate and exact completion problems for Euclidean distance matrices using semidefinite programming. Linear Algebra Appl. **406**, 109–141 (2005)

26. F. Alizadeh, J.A. Haeberly, M.L. Overton, Complementarity and nondegeneracy in semidefinite programming. Math. Program. Ser. B **77**, 111–128 (1997)

27. F. Alizadeh, J.A. Haeberly, M.L. Overton, Primal-dual interior-point methods for semidefinite programming: convergence rates, stability and numerical results. SIAM J. Optim. **8**, 746–768 (1998)

28. L. Asimow, B. Roth, The rigidity of graphs. Trans. Am. Math. Soc. **245**, 279–289 (1978)

29. L. Asimow, B. Roth, The rigidity of graphs II. J. Math. Anal. Appl. **68**, 171–190 (1979)

30. L. Auslander, R.E. MacKenzie, *Introduction to Differentiable Manifolds* (McGraw-Hill, New York, 1963)

31. M. Bakonyi, C.R. Johnson, The Euclidean distance matrix completion problem. SIAM J. Matrix Anal. Appl. **16**, 646–654 (1995)

32. R. Balaji, R.B. Bapat, On Euclidean distance matrices. Linear Algebra Appl. **424**, 108–117 (2007)

33. R.B. Bapat, *Graphs and Matrices* (Springer, London, 2010)

34. G. Barker, Theory of cones. Linear Algebra Appl. **39**, 263–291 (1981)

35. G.P. Barker, D. Carlson, Cones of diagonally dominant matrices. Pac. J. Math. **57**, 15–31 (1975)

36. G. Barker, R. Hill, R. Haertel, On the completely positive and positive semidefinite preserving cones. Linear Algebra Appl. **56**, 221–229 (1984)

37. F.L. Bauer, C.T. Fike, Norms and exclusion theorems. Numer. Math. **2**, 137–141 (1960)

38. F. Bavaud, On the Schoenberg transformations in data analysis: theory and illustrations. J. Classif. **28**, 297–314 (2011)

39. L.W. Beineke, R.E. Pippert, The number of labeled k-dimensional trees. J. Combin. Theory **6**, 200–205 (1969)

40. J. Bénasséni, A variance inequality ensuring that a pre-distance matrix is Euclidean. Linear Algebra Appl. **416**, 365–372 (2006)

41. D.S. Bernstein, *Matrix Mathematics: Theory, Facts and Formulas* (Princeton University Press, Princeton, 2009)

42. N. Biggs, *Algebraic Graph Theory* (Cambridge University Press, London, 1993)

43. P. Biswas, Y. Ye, Semidefinite programming for ad hoc wireless sensor network localization, in *Proceedings 3rd IPSN*, 2004, pp. 46–54

44. J.R.S. Blair, B. Peyton, An introduction to chordal graphs and clique trees, in *Graph Theory and Sparse Matrix Computation*, ed. by J.A. George, J.R. Gilbert, J.W.-H. Liu. IMA Volumes in Mathematics and Its Applications, vol. 56 (Springer, New York, 1993), pp. 1–29

45. L.M. Blumenthal, *Theory and Applications of Distance Geometry* (Clarendon Press, Oxford, 1953)

46. L.M. Blumenthal, B.E. Gillam, Distribution of points in n-space. Am. Math. Mon. **50**, 181–185 (1943)

47. J.A. Bondy, U.S.R. Murty, *Graph Theory* (Springer, New York, 2008)

48. I. Borg, P.J.F. Groenen, *Modern Multidimensional Scaling* (Springer, New York, 2005)

49. J.M. Borwein, H. Wolkowicz, Facial reduction for a non-convex programming problem. J. Austral. Math. Soc. Ser. A **30**, 369–380 (1981)

50. J.M. Borwein, H. Wolkowicz, Regularizing the abstract convex program. J. Math. Anal. Appl. **83**, 495–530 (1981)

51. A.L. Cauchy, sur les polygons et les polyèdres, second mémoire. J. Ecole Polytech. **9**, 87–98 (1813)

52. A. Cayley, On a theorem in the geometry of position. Camb. Math. J. **2**, 267–271 (1841)

53. Y. Chabrillac, J.-P. Crouzeix, Definiteness and semidefiniteness of quadratic forms revisited. Linear Algebra Appl. **63**, 283–292 (1984)

54. Y.-L. Cheung, *Preprocessing and Reduction for Semidefinite Programming via Facial Reduction: Theory and Practice*. PhD thesis, University of Waterloo, 2013

55. M.T. Chu, Inverse eigenvalue problems. SIAM Rev. **40**, 1–39 (1998)

56. Y. Colin De Verdière, Sur un nouvel invariant des graphes et un critère de planarité. J. Combin. Theory Ser B **50**, 11–21 (1990)

57. R. Connelly, Rigidity and energy. Invent. Math. **66**, 11–33 (1982)
58. R. Connelly, On generic global rigidity, in *Applied Geometry and Discrete Mathematics*. DIMACS: Series in Discrete Mathematics and Theoretical Computer Science (American Mathematical Society, Providence, 1991), pp. 147–155
59. R. Connelly, Rigidity, in *Handbook of Convex Geometry*, ed. by P.M. Gruber, J.M. Wills (North-Holand, Amsterdam, 1993), pp. 223–271
60. R. Connelly, Tensegrity structures: why are they stable? in *Rigidity Theory and Applications*, ed. by M.F. Thorpe, P.M. Duxbury (Kluwer Academic/Plenum Publishers, New York, 1999), pp. 47–54
61. R. Connelly, Generic global rigidity. Discret. Comput. Geom. **33**, 549–563 (2005)
62. R. Connelly, What is a tensegrity. Not. Am. Math. Soc. **60**, 78–80 (2013)
63. R. Connelly, S. Gortler, Iterative universal rigidity. Discret. Comput. Geom. **53**, 847–877 (2015)
64. R. Connelly, S. Gortler, Universal rigidity of complete bipartite graphs. Discret. Comput. Geom. **57**, 281–304 (2017)
65. H. Crapo, W. Whiteley, Plane self stresses and projected polyhedra I: the basic pattern. Struct. Topol. **20**, 55–77 (1993)
66. G.M. Crippen, T.F. Havel, *Distance Geometry and Molecular Conformation* (Wiley, New York, 1988)
67. F. Critchley, On certain linear mappings between inner-product and squared distance matrices. Linear Algebra Appl. **105**, 91–107 (1988)
68. J.-P. Crouzeix, J.A. Ferland, Criteria for quasi-convexity and pseudo-convexity: relations and comparisons. Math. Program. **23**, 193–205 (1982)
69. J. Dattorro, *Convex Optimization and Euclidean Distance Geometry* (Meboo Publishing, Palo Alto, 2005)
70. M. Deza, M. Laurent, *Geometry of Cuts and Metrics, Algorithms and Combinatorics*, vol. 15 (Springer, Berlin, 1997)
71. Y. Ding, N. Krislock, J. Qian, H. Wolkowicz, Sensor network localization, Euclidean distance matrix completions, and graph realization. Optim. Eng. **11**, 45–66 (2010)
72. G.A. Dirac, On rigid circuit graphs. Abh. Math. Sem. Univ. Hamburg **25**, 71–76 (1961)
73. P.G. Doyle, J.L. Snell, *Random Walks and Electric Networks* (Mathematical Association of America, Washington, 1984)
74. D. Drusvyatskiy, H. Wolkowicz, The many faces of degeneracy in conic optimization. Found. Trends Optim. **3**, 77–170 (2017)
75. D. Drusvyatskiy, G. Pataki, H. Wolkowicz, Coordinate shadows of semidefinite and Euclidean distance matrices. SIAM J. Optim. **25**, 1160–1178 (2015)
76. M. Dur, B. Jargalsaikhan, G. Still, Genericity results in linear conic programming, a tour d'horizon. Math. Oper. Res. **42**, 77–94 (2017)
77. H. Fang, D.P. O'leary, Euclidean distance matrix completion problems. Optim. Methods Softw. **27**, 695–717 (2012)

78. I. Fáry, On straight line representation of planar graphs. Acta Sci. Math. Szeged **11**, 229–233 (1948)

79. M. Fiedler, A geometric approach to the Laplacian matrix of a graph, in *Combinatorial and Graph-Theoretical Problems in Linear Algebra*, ed. by R.A. Brualdi, S. Friedland, V. Klee (Springer, New York, 1993), pp. 73–98

80. M. Fiedler, Elliptic matrices with zero diagonal. Linear Algebra Appl. **197/198**, 337–347 (1994)

81. M. Fiedler, Moore-Penrose involutions in the classes of Laplacians and simplices. Linear Multilinear Algebra **39**, 171–178 (1995)

82. D.R. Fulkerson, O.A. Gross, Incidence matrices and interval graphs. Pac. J. Math. **15**, 835–855 (1965)

83. H.N. Gabow, H.H. Westermann, Forests, frames and games, in *Proceedings of the 20th Annual ACM Symposium on the Theory of Computing*, 1988, pp. 407–421

84. D. Gale, Neighboring vertices on a convex polyhedron, in *Linear Inequalities and Related System* (Princeton University Press, Princeton, 1956), pp. 255–263

85. M.R. Garey, D.S. Johnson, *Computers and Intractability, A Guide to the Theory of NP-Completeness* (W. H. Freeman and Company, New York, 1979)

86. H. Gluck, Almost all simply connected surfaces are rigid, in *Geometric Topology*. Lecture Notes in Mathematics, vol. 438 (Springer, New York, 1975), pp. 225–239

87. C.D. Godsil, Compact graphs and equitable partitions. Linear Algebra Appl. **255**, 259–266 (1997)

88. C.D. Godsil, B.D. McKay, Feasibility conditions for the existence of walk-regular graphs. Linear Algebra Appl. **30**, 51–61 (1980)

89. M.C. Golumbic, *Algorithmic Graph Theory and Perfect Graphs*. Annals of Discrete Mathematics, vol. 57 (Elsevier, Amsterdam, 2004)

90. S.J. Gortler, D.P. Thurston, Characterizing the universal rigidity of generic frameworks. Discret. Comput. Geom. **51**, 1017–1036 (2014)

91. S.J. Gortler, D.P. Thurston, A.D. Healy, Characterizing generic global rigidity. Am. J. Math. **132**, 897–939 (2010)

92. J.C. Gower, Euclidean distance geometry. Math. Sci. **7**, 1–14 (1982)

93. J.C. Gower, Properties of Euclidean and non-Euclidean distance matrices. Linear Algebra Appl. **67**, 81–97 (1985)

94. R.L. Graham, L. Lovász, Distance matrix polynomials of trees. Adv. Math. **29**, 60–88 (1978)

95. R.L. Graham, H.O. Pollak, On the addressing problem for loop switching. Bell Syst. Tech. J. **50**, 2495–2519 (1971)

96. R.L. Graham, P.M. Winkler, On isometric embeddings of graphs. Trans. Am. Math. Soc. **288**, 527–536 (1985)

97. J. Graver, B. Servatius, H. Servatius, *Combinatorial Rigidity* (American Mathematical Society, Providence, 1993)

98. R. Grone, C.R. Johnson, E.M. Sá, H. Wolkowicz, Positive definite completions of partial Hermitian matrices. Linear Algebra Appl. **58**, 109–124 (1984)

99. B. Grünbaum, *Convex Polytopes* (Wiley, New York, 1967)

100. T.F. Havel, I.D. Kuntz, G.M. Crippen, Theory and practice of distance geometry. Bull. Math. Biol. **45**, 665–720 (1983)

101. T.L. Hayden, P. Tarazaga, Distance matrices and regular figures. Linear Algebra Appl. **195**, 9–16 (1993)

102. T.L. Hayden, J. Wells, Approximation by matrices positive semidefinite on a subspace. Linear Algebra Appl. **109**, 115–130 (1988)

103. T.L. Hayden, J. Wells, W.-M. Liu, P. Tarazaga, The cone of distance matrices. Linear Algebra Appl. **144**, 153–169 (1991)

104. T.L. Hayden, J. Lee, J. Wells, P. Tarazaga, Block matrices and multispherical structure of distance matrices. Linear Algebra Appl. **247**, 203–216 (1996)

105. T.L. Hayden, R. Reams, J. Wells, Methods for constructing distance matrices and the inverse eigenvalue problem. Linear Algebra Appl. **295**, 97–112 (1999)

106. B. Hendrickson, The molecule problem: exploiting the structure in global optimization. SIAM J. Optim. **5**, 835–857 (1995)

107. L. Henneberg, *Die graphische statik der starren systeme* (B. G. Teubner, Leipzig, 1911)

108. R. Hill, S. Waters, On the cone of positive semidefinite matrices. Linear Algebra Appl. **90**, 81–88 (1987)

109. J.-B. Hiriart-Urruty, C. Lemaréchal, *Fundamentals of Convex Analysis* (Springer, Berlin, 2001)

110. A.J. Hoffman, On the polynomial of a graph. Am. Math. Mon. **70**, 30–36 (1963)

111. J.E. Hopcroft, P.J. Kahn, A paradigm for robust geometric algorithms. Algorithmica **7**, 339–380 (1992)

112. R.A. Horn, C.R. Johnson, *Matrix Analysis* (Cambridge University Press, Cambridge, 1985)

113. R.A. Horn, C.R. Johnson, *Topics in Matrix Analysis* (Cambridge University Press, Cambridge, 1991)

114. B. Jackson, T. Jordán, Connected rigidity matroids and unique realizations of graphs. J. Combin. Theory Ser. B **94**, 1–29 (2005)

115. D.J. Jacobs, B. Hendrickson, An algorithm for the two-dimensional rigidity percolation: the pebble game. J. Comput. Phys. **137**, 346–365 (1997)

116. G. Jaklič, J. Modic, On properties of cell matrices. Appl. Math. Comput. **216**, 2016–2023 (2010)

117. G. Jaklič, J. Modic, A note on methods for constructing distance matrices and the inverse eigenvalue problem. Linear Algebra Appl. **437**, 2781–2792 (2012)

118. T. Jordán, V.-H. Nguyen, On universal rigid frameworks on the line. Contrib. Discret. Math. **10**, 10–21 (2015)

119. H.W.E. Jung, Ueber die kleinste kugel, die eine raumliche figur einschliesst. J. Reine Angew. Math. **123**, 241–257 (1901)

120. H. Kharaghani, B. Tayfeh-Rezaie, A Hadamard matrix of order 428. J. Combin. Des. **13**, 435–440 (2005)

121. D.J. Klein, M. Randić, Resistance distance. J. Math. Chem. **12**, 81–95 (1993)

122. N. Krislock, H. Wolkowicz, Explicit sensor network localization using semidefinite representations and facial reductions. SIAM J. Optim. **20**, 2679–2708 (2010)

123. H. Kurata, S. Matsuura, Characterization of multispherical and block structures of Euclidean distance matrices. Linear Algebra Appl. **439**, 3177–3183 (2013)

124. H. Kurata, T. Sakuma, A group majorization ordering for Euclidean distance matrices. Linear Algebra Appl. **420**, 586–595 (2007)

125. H. Kurata, P. Tarazaga, Multispherical Euclidean distance matrices. Linear Algebra Appl. **433**, 534–546 (2010)

126. H. Kurata, P. Tarazaga, The cell matrix closest to a given Euclidean distance matrices. Linear Algebra Appl. **485**, 194–207 (2015)

127. G. Laman, On graphs and rigidity of plane skeletal structures. J. Eng. Math. **4**, 331–340 (1970)

128. M. Laurent, Cuts, matrix completion and graph rigidity. Math. Program. **79**, 255–283 (1997)

129. M. Laurent, A connection between positive semidefinite and Euclidean distance matrix completion problems. Linear Algebra Appl. **273**, 9–22 (1998)

130. M. Laurent, Polynomial instances of the positive semidefinite and Euclidean distance matrix completion problems. SIAM J. Matrix Anal. Appl. **22**, 874–894 (2000)

131. M. Laurent, S. Poljak, On a positive semidefinite relaxation of the cut polytope. Linear Algebra Appl. **223/224**, 439–461 (1995)

132. M. Laurent, A. Varvitsiotis, Positive semidefinite matrix completion, universal rigidity and the strong Arnold property. Linear Algebra Appl. **452**, 292–317 (2014)

133. A. Lee, I. Streinu, Pebble game algorithms and sparse graphs. Discret. Math. **308**, 1425–1437 (2008)

134. C.-K. Li, T. Milligan, Uniqueness of the solutions of some completion problems. Technical report, Dept. of Mathematics, The College of William and Mary, 2003

135. L. Liberti, C. Lavor, *Euclidean Distance Geometry, An Introduction* (Springer, Cham, 2017)

136. M. Liu, G. Pataki, Exact duality in semidefinite programming based on elementary reformations. SIAM J. Optim. **25**, 1441–1454 (2015)

137. L. Lovász, Semidefinite programs and combinatorial optimization, in *Recent Advances in Algorithms and Combinatorics* (Springer, New York, 2003), pp. 137–194

138. L. Lovász, Geometric representations of graphs. Unpublished lecture notes, 2016

139. L. Lovász, Y. Yemini, On generic rigidity in the plane. SIAM J. Algebraic Discret. Methods **3**, 91–98 (1982)

140. G. Marsaglia, G.P.H. Styan, When does rank (a+b)= rank a + rank b? Can. Math. Bull. **15**, 451–452 (1972)

141. J.C. Maxwell, On reciprocal figures and diagrams of forces. Philos. Mag. **4**, 250–261 (1864)
142. J.C. Maxwell, On reciprocal figures, frameworks and diagrams of forces. Trans. R. Soc. Edinb. **26**, 1–40 (1870)
143. K. Menger, Untersuchungen uber allegemeine Metrik. Math. Ann. **100**, 75–163 (1928)
144. K. Menger, New foundation of Euclidean geometry. Am. J. Math. **53**, 721–745 (1931)
145. J. Milnor, *Singular Points of Complex Hypersurfaces*. Annals of Mathematics Studies, vol. 61 (Princeton University Press, Princeton, 1968)
146. H. Mittelmann, J. Peng, Estimating bounds for quadratic assignment problems associated with Hamming and Manhattan distance matrices based on semidefinite programming. SIAM J. Optim. **20**, 3408–3426 (2010)
147. J.J. Moré, Z. Wu, Global for distance geometry problems. SIAM J. Optim. **7**, 814–836 (1997)
148. J.J. Moré, Z. Wu, Distance geometry optimization for protein structures. J. Glob. Optim. **15**, 219–234 (1999)
149. J.J. Moreau, Décomposition orthogonale d'un espace hilbertien selon deux cônes mutuellement polaires. C. R. Acad. Sci Paris **255**, 238–240 (1962)
150. A. Mowshowitz, The adjacency matrix and the group of a graph, in *New Directions in the Theory of Graphs*, ed. by F. Harary (Academic Press, New York, 1973), pp. 129–148
151. A.M. Nazari, F. Mahdinasab, Inverse eigenvalue problem of distance matrix via orthogonal matrix. Linear Algebra Appl. **450**, 202–216 (2014)
152. A. Neumaier, Distances, graphs and designs. Eur. J. Combin. **1**, 163–174 (1980)
153. A. Neumaier, Distance matrices, dimension and conference graphs. Nederl. Akad. Wetensch. Indag. Math. **43**, 385–391 (1981)
154. G. Pataki, The geometry of semidefinite programing, in *Handbook of Semidefinite Programming: Theory, Algorithms and Applications*, ed. by H. Wolkowicz, R. Saigal, L. Vandenberghe (Kluwer Academic Publishers, Boston, 2000), pp. 29–65
155. J.E. Prussing, The principal minor test for semidefinite matrices. J. Guid. Control Dyn. **9**, 121–122 (1986)
156. M. Ramana, A.J. Goldman, Some geometric results in semi-definite programming. J. Glob. Optim. **7**, 33–50 (1995)
157. A. Recski, A network theory approach to the rigidity of skeletal structures 1: modelling and interconnection. Discret. Appl. Math. **7**, 313–324 (1984)
158. A. Recski, Combinatorial conditions for the rigidity of tensegrity frameworks, in *Horizons of Combinatorics*. Bolyai Society Mathematical Studies, vol. 17 (Springer, Berlin, 2008), pp. 163–177
159. J. Richter-Gebert, *Realization Spaces of Polytopes*. Lecture Notes in Mathematics, vol. 1643 (Springer, Berlin, 1996)
160. R.T. Rockafellar, *Convex Analysis* (Princeton University Press, Princeton, 1970)

161. D.J. Rose, R.E. Tarjan, G.S. Leuker, Algorithmic aspects of vertex elimination on graphs. SIAM J. Comput. **5**, 266–283 (1976)
162. B. Roth, W. Whiteley, Tensegrity frameworks. Trans. Am. Math. Soc. **265**, 419–446 (1981)
163. H. Sachs, Über teiler, faktoren und charakteristische polynome von graphen. Teil I. Wiss. Z. TH Ilmenau **12**, 7–12 (1966)
164. J.B. Saxe, Embeddability of weighted graphs in k-space is strongly NP-hard. *Proceedings of the 17th Allerton Conference in Communications, Control, and Computing*, 1979, pp. 480–489
165. H. Schneider, Positive operators and an inertia theorem. Numer. Math. **7**, 11–17 (1965)
166. R. Schneider, *Convex Bodies: The Brunn-Minkowski Theory* (Cambridge University Press, Cambridge, 1993)
167. I.J. Schoenberg, Remarks to Maurice Fréchet's article: Sur la définition axiomatique d'une classe d'espaces vectoriels distanciés applicables vectoriellement sur l'espace de Hilbert. Ann. Math. **36**, 724–732 (1935)
168. I.J. Schoenberg, On certain metric spaces arising from Euclidean spaces by a change of metric and their imbedding in Hilbert space. Ann. Math. **38**, 787–793 (1937)
169. I.J. Schoenberg, Metric spaces and completely monotone functions. Ann. Math. **39**, 811–841 (1938)
170. I.J. Schoenberg, Metric spaces and positive definite functions. Trans. Am. Math. Soc. **44**, 522–536 (1938)
171. I.J. Schoenberg, Linkages and distance geometry. Nedrel. Akad. Wetensch. proc. Ser A. 72, Indag. Math. **31**, 43–52 (1969)
172. A.J. Schwenk, Computing the characteristic polynomial of a graph, in *Graphs and Combinatorics*. Lecture Notes in Mathematics, Vol. 406 (Springer, Berlin, 1974), pp. 153–162
173. B. Servatius, H. Servatius, Generic and abstract rigidity, in *Rigidity Theory and Applications*, ed. by M.F. Thorpe, P.M. Duxbury (Kluwer, New York, 2002)
174. B. Servatius, H. Servatius, Rigidity, global rigidity, and graph decomposition. Eur. J. Combin. **31**, 1121–1135 (2010)
175. S. Seshu, M.B. Reed, *Linear Graphs and Electrical Networks* (Addison-Wesley, Reading, 1961)
176. M. Sitharam, H. Gao, Characterizing graphs with convex and connected Cayley configuration spaces. Discret. Comput. Geom. **43**, 594–625 (2010)
177. A.M.-C. So, *Semidefinite Programming Approach to the Graph Realization Problem: Theory, Applications and Extensions*. PhD thesis, Stanford University, 2007
178. A.M.-C. So, Y. Ye, Semidefinite programming approach to tensegrity theory and realizability of graphs, in *Proceedings of the 17th Annual ACM-SIAM Symposium on Discrete Algorithms (SODA 2006)*, 2006, pp. 766–775
179. A.M.-C. So, Y. Ye, Theory of semidefinite programming for sensor network localization. Math. Prog. Ser. B **109**, 367–384 (2007)
180. S. Straszewicz, Uber exponierte punkte abgeschlossener punktmengen. Fundam. Math. **24**, 139–143 (1935)

181. J.F. Sturm, Error bounds for linear matrix inequalities. SIAM J. Optim. **10**, 1228–1248 (2000)

182. G.P.H. Styan, G.E. Subak-Sharpe, Inequalities and equalities associated with the Campbell-Youla generalized inverse of the indefinite admittance matrix of resistive networks. Linear Algebra Appl. **250**, 349–370 (1997)

183. S.-I. Tanigawa, Sufficient conditions for the global rigidity of graphs. J. Combin. Theory Ser. B **113**, 123–140 (2015)

184. P. Tarazaga, Faces of the cone of Euclidean distance matrices: characterizations, structure and induced geometry. Linear Algebra Appl. **408**, 1–13 (2005)

185. P. Tarazaga, J.E. Gallardo, Some properties of Euclidean distance matrices and elliptic matrices. Tech. Report CRPC-TR97678, 1997

186. P. Tarazaga, T.L. Hayden, J. Wells, Circum-Euclidean distance matrices and faces. Linear Algebra Appl. **232**, 77–96 (1996)

187. P. Tarazaga, H. Kurata, On cell matrices: a class of Euclidean distance matrices. Appl. Math. Comput. **238**, 468–474 (2014)

188. P. Tarazaga, B. Sterba-Boatwright, K. Wijewardena, Euclidean distance matrices: special subsets, systems of coordinates and multibalanced matrices. Comput. Appl. Math. **26**, 415–438 (2007)

189. R. Thomas, Lecture notes on topology of Euclidean spaces. Georgia Tech., 1993

190. M.W. Trosset, Distance matrix completion by numerical optimization. Comput. Optim. Appl. **17**, 11–22 (2000)

191. A. Wallace, Algebraic approximation of curves. Can. J. Math. **10**, 242–278 (1958)

192. H.S. White, Cremona's work. Bull. Am. Math. Soc. **24**, 238–243 (1918)

193. W. Whiteley, Motions and stresses of projected polyhedra. Struct. Topol. **7**, 13–38 (1982)

194. W. Whiteley, Matroids and rigid structures, in *Matroid Applications, Encyclopedia of Mathematics and Its Applications*, ed. by N. White, vol. 40 (Cambridge University Press, Cambridge, 1992), pp. 1–53

195. W. Whiteley, B. Roth, Tensegrity frameworks. Trans. Am. Math. Soc. **265**, 419–446 (1981)

196. H. Wolkowicz, G.P.H. Styan, Bounds for eigenvalues using traces. Linear Algebra Appl. **29**, 471–506 (1980)

197. H. Wolkowicz, G.P.H. Styan, More bounds for eigenvalues using traces. Linear Algebra Appl. **31**, 1–17 (1980)

198. H. Wolkowicz, R. Saigal, L. Vandenberghe (eds.), *Handbook of Semidefinite Programming. Theory, Algorithms and Applications* (Kluwer Academic Publishers, Boston, 2000)

199. Y. Yemini, Some theoretical aspects of location-location problems, in *Proceedings of the IEEE Symposium on Foundations of Computer Science*, 1979, pp. 1–8

200. G. Young, A.S. Householder, Discussion of a set of points in terms of their mutual distances. Psychometrika **3**, 19–22 (1938)

201. Z. Zhu, A.M.-C. So, Y. Ye, Universal rigidity: towards accurate and efficient localization of wireless networks, in *Proceedings IEEE INFOCOM*, 2010

Index

A
Affine flex, 212
Affine motion, 212
Algebraic independence, 203
Analytic flex, 186
Antipodal, 92

B
Bar framework, 170
  congruent, 171
  equivalent, 171

C
Cayley configuration space, 171
Cayley configuration spectrahedron, 172
Cauchy interlacing theorem, 6
Cauchy Schwarz inequality, 2
Cayley–Menger determinant, 66
Clique, 164, 181
Clique sum, 14
Column space, 9
Combinatorial designs, 139
Cone, 15
  dual, 35
  of feasible directions, 26
  normal, 25, 117
  pointed, 15
  polar, 23, 35, 116
  polyhedral, 116
  tangent, 26, 116
Conic at infinity, 216
Continuous flex, 186
Coordinate shadow, 119

E
EDM, 51
  cell matrix, 99, 116

centrally symmetric, 125
completion problem, 163
r-completion problem, 163
configuration matrix, 51
cospectral, 133
degree of, 136
embedding dimension, 51
isomorphic, 133
multispherical, 111
nonspherical, 89
regular, 97
spherical, 89
strength of, 136
EDM completion, 163
EDM entry
  unyielding, 152
  yielding, 153
  yielding interval of, 152
Effective resistance, 106
Eigenpair, 4
Elliptope, 116
Equilibrium load, 192
Equitable partition, 127

F
Face path, 182
Facial reduction, 44, 48
Farkas lemma, 35
Feasible
  region, 38
  solution, 38

G
Gale matrix, 59–61, 63, 90, 91, 94, 100, 109,
  111, 119, 122, 149, 153–156, 159, 160,
  176–178, 180, 185, 195–198, 201, 217,
  219, 220, 222, 226, 227, 230, 231, 234

© Springer Nature Switzerland AG 2018
A.Y. Alfakih, *Euclidean Distance Matrices and Their Applications in Rigidity Theory*, https://doi.org/10.1007/978-3-319-97846-8

Printed in the United States
By Bookmasters